AF389418

Integrated Human Exposure to Air Pollution

Integrated Human Exposure to Air Pollution

Editors

Nuno Canha
Marta Almeida
Evangelia Diapouli

MDPI • Basel • Beijing • Wuhan • Barcelona • Belgrade • Manchester • Tokyo • Cluj • Tianjin

Editors
Nuno Canha
Universidade de Lisboa
Portugal
University of Aveiro
Portugal

Marta Almeida
Universidade de Lisboa
Portugal

Evangelia Diapouli
National Centre for Scientific
Research "Demokritos"
Greece

Editorial Office
MDPI
St. Alban-Anlage 66
4052 Basel, Switzerland

This is a reprint of articles from the Special Issue published online in the open access journal *International Journal of Environmental Research and Public Health* (ISSN 1660-4601) (available at: https://www.mdpi.com/journal/ijerph/special_issues/IHETAP).

For citation purposes, cite each article independently as indicated on the article page online and as indicated below:

LastName, A.A.; LastName, B.B.; LastName, C.C. Article Title. *Journal Name* **Year**, *Volume Number*, Page Range.

ISBN 978-3-0365-1082-8 (Hbk)
ISBN 978-3-0365-1083-5 (PDF)

Contents

About the Editors

Nuno Canha is a junior researcher at the Center for Nuclear Sciences and Technologies (C2TN) of Instituto Superior Técnico from Universidade de Lisboa (IST/UL), Portugal. He holds a MSc in Chemistry from the IST/UL, Portugal (2008) and a Ph.D. in Environmental Sciences from the Delft University of Technology, The Netherlands (2014). His Ph.D. focused on the indoor air quality (IAQ) of primary schools from urban and rural environments, targeting new sampling methods for particulate matter (PM), its chemical characterisation and the identification of the contribution of different pollution sources to the PM levels, by source apportionment techniques. His post-doc research focused on PM toxicity potential and the assessment of air pollutants exposure in different micro-environments, such as the sleeping environment. Today, his research focus is on indoor and outdoor air pollution sources of PM, their characterisation and assessment of their contribution for the integrated human exposure to air pollutants, along with the understanding of their health impact (by identifying their toxic potential).

Nuno Canha has participated in several European projects from different funding programmes/agencies related to air quality. Overall, he has participated in 16 national and international research projects. He published 56 scientific articles in peer review journals and his works were presented on the most rated international conferences on atmospheric pollution and IAQ. He participated in 20 international training courses, 4 scientific and 3 organizing committees of international conferences. He has always actively promoted and participated in several science communication activities to different types of public.

Marta Almeida graduated in Environmental Engineering in 1998 from the Universidade Nova de Lisboa (FCT/UNL) and completed her Ph.D. in Environmental Sciences at Universidade de Aveiro in 2004. She worked at the Instituto de Soldadura e Qualidade, where, in addition to her research activities, she was responsible for consulting projects related to air quality management for more than 40 national and international industries. Since 2008, she has been working at Instituto Superior Técnico from Universidade de Lisboa (IST/UL), where she is a Principal Researcher. She focuses her research on the integration of air quality, climate changes and health in urban systems. As part of her research work, she has coordinated seven European and national projects and participated in 24 other projects on air quality management, indoor air quality, exposure to pollutants, source apportionment, and more recently, on low carbon economy. Marta Almeida published 130 articles in internationally circulated magazines with scientific arbitration. She has been acting as an expert evaluator in international and national programmes. Marta Almeida participated in the organization and the Scientific Committee of several international conferences and courses and coordinates and teaches the course "Air Quality Management" of the Master in Environmental Engineering at IST.

Evangelia Diapouli is Researcher (Level C) at the Institute of Nuclear & Radiological Sciences and Technology, Energy & Safety, of the National Centre for Scientific Research "Demokritos", in Athens, Greece. She holds a Chemical Engineering Diploma from the National Technical University of Athens, Greece (2000), an MSc in Environmental Engineering from Johns Hopkins University, USA (2002), and a PhD in Air Pollution from the National Technical University of Athens, Greece (2008). She has more than 15 years of experience in the study of atmospheric aerosols. Her research activities include the physico-chemical characterisation of ambient and indoor particulate pollution, aerosol source apportionment, and population and occupational risk assessment. She has been actively involved in European initiatives for the development of standard methods for aerosol characterisation, including the quantification of carbonaceous species in PM and the adoption of a European harmonized methodology in the application of receptor models for PM source apportionment. She has participated in several European and national projects, and has acted as project coordinator or coordinator for the NCSR "Demokritos" team. She is the co-author of more than 60 papers in peer review journals and more than 90 papers in national and international conferences.

International Journal of
*Environmental Research
and Public Health*

MDPI

Editorial

Integrated Human Exposure to Air Pollution

Nuno Canha [1,2,*], Evangelia Diapouli [3] and Susana Marta Almeida [1]

1   Centro de Ciências e Tecnologias Nucleares (C²TN), Instituto Superior Técnico, Universidade de Lisboa, Estrada Nacional 10, Km 139.7, 2695-066 Bobadela LRS, Portugal; smarta@ctn.tecnico.ulisboa.pt
2   Centre for Environmental and Marine Studies (CESAM), Department of Environment and Planning, University of Aveiro, 3810-193 Aveiro, Portugal
3   National Centre for Scientific Research "Demokritos", Agia Paraskevi, 15341 Athens, Greece; ldiapouli@ipta.demokritos.gr
*   Correspondence: nunocanha@ctn.tecnico.ulisboa.pt

Citation: Canha, N.; Diapouli, E.; Almeida, S.M. Integrated Human Exposure to Air Pollution. *Int. J. Environ. Res. Public Health* **2021**, *18*, 2233. https://doi.org/10.3390/jerph18052233

Received: 8 February 2021
Accepted: 20 February 2021
Published: 24 February 2021

Publisher's Note: MDPI stays neutral with regard to jurisdictional claims in published maps and institutional affiliations.

Air pollution is one of the major environmental health problems that people face nowadays, affecting everyone in the world. The World Health Organization has estimated that, in 2016, ambient air pollution caused 4.2 million premature deaths worldwide, in both cities and rural areas, while household air pollution caused 3.8 million deaths, mainly in low and middle-income countries [1].

Usually, citizen exposure to air pollutants is calculated based only on concentrations of pollutants monitored using air quality monitoring stations from environment national agencies. These monitoring stations focus on outdoor air quality and, most of them, are located in urban centres. However, this approach fails to account for all components of exposure since:

- There is a high variability of air pollutant concentrations within a city, depending on the city topography and the existence (or not) of specific pollution sources (such high traffic areas);
- The time–activity patterns and the use of spaces are very heterogeneous within the population;
- People spend around 90% of their time in indoor environments;

Therefore, human exposure during a full day cannot be reflected only by outdoor exposure and should consider all micro-environments where individuals spend their time (e.g., home, workplace, transports, leisure, and others) and the time spent in them.

Moreover, the characterization of air pollutants in indoor and outdoor environments is essential to understand the integrated human exposure to them and, thus, to improve those exposure assessments.

The Special Issue "Integrated human exposure to air pollution" aimed to increase the knowledge about human exposure in different micro-environments or when citizens are performing specific tasks, to demonstrate methodologies for the understanding of pollution sources and their impact on indoor and ambient air quality, and, ultimately, to identify the most effective mitigation measures to decrease human exposure and protect public health. Taking advantage on the latest available tools, such as internet of things (IoT), low-cost sensors and a wide access to online platforms and apps by the citizens, new methodologies and approaches can be implemented to understand which factors can influence human exposure to air pollution. This knowledge made available to the citizens, along with the awareness of the impact of air pollution on human life and earth systems, can empower them to act, individually or collectively, to promote behavior changes aiming to reduce pollutants' emissions.

Overall, this Special Issue gathers fourteen peer-reviewed and open access articles that provide new insights regarding these important topics within the scope of human exposure to air pollution. A total of five main areas were discussed and explored within this special issue, which are described below, and, hopefully, can contribute to the advance of knowledge in this field.

1. New Methodologies to Assess Human Exposure

In recent years, the use of low-cost sensors for assessing environmental parameters was a landmark regarding personal exposure assessments due to their portability, low price and easy handling, allowing to add multi sensors to monitor different parameters, along with global positioning system (GPS) sensor and transmission of acquired data to online platforms. Such combination allows to gather different types of information/data in cloud environments, usually in a great amount, which contributed to the concept of big data in the field of human exposure assessments regarding air pollution. However, a critical issue on the use of such type of monitoring devices is their validation towards reference methods to assure the reliability of results.

A study conducted in Texas (USA) [2] focused on this issue where low cost particle sensors (targeting suspended particulate matter) were developed, calibrated and tested in real life conditions. The developed monitoring unit used an Arduino interface, a commercially available dust sensor (namely, Sharp GP2Y1010AU0F), a Global Positioning System (GPS) sensor (to provide the location of the measurements) and a communications module to transmit data to an online platform. The authors also provided a historical and critical overview of the evolution of these types of low-cost sensors and their performance, based on a literature review. The dust sensors employed in the monitoring units were calibrated regarding a reference aerosol monitor, using a mixing chamber and generation of sodium chloride particles, where the sensors showed standard deviations for replicate measurements of 3-6 $\mu g/m^3$. To show the applicability of the developed set-up, measurements in real life conditions were conducted both at indoor and outdoor environments in Lubbcok, Texas. Analysis of the results allowed to identify, for instance, increased exposure levels to particulate matter nearby restaurants and when cooking at home. The gathered data also allowed to build maps with spatial location of the monitored PM levels.

Using a case study in Madrid (Spain), a new methodology, based on the Internet of Things (IoT) approach and using low cost sensors of air pollutants ($PM_{2.5}$, CO_2 and VOC) along with available outdoor air quality data, was proposed for estimating personal air pollution exposure (PAPE) [3]. Individual's PAPE was estimated based on combined outdoor and indoor data with spatio-temporal activity patterns. The understanding of instantaneous PAPE, by means of a forecasting approach, may be useful for individual awareness, as for example, to decide travel routes in order to minimize their exposure as it was shown in this case study. Overall, this study provided a new framework for an Air Quality Decision Support System (AQDSS), taking advantage of the nowadays technologies and showing its applicability.

2. Citizens Empowerment Regarding Air Pollution

The importance of public awareness on air pollution issues and its impacts was highlighted in a study conducted in Shanghai (China), where the relationship between air pollution levels and the expression of concern of the city's residents was evaluated [4]. Using information of a period of two thousand days (with a daily frequency) (from 2013 to 2019) regarding the air quality index provided by the local environmental monitoring stations and an index about the internet searches of the residents based in selected keywords, an empirical analysis was conducted using a vector autoregression model.

Using this strategy, three main findings were reported about the awareness levels of citizens regarding this topic: (1) degradation of air quality was perceived by citizens within one day; (2) information regarding air quality issues in another major city (Beijing) conducted to a concern on the inhabitants of Shanghai about local air quality; and (3) higher concern levels of the citizens had a beneficial impact on local air quality, which can be explained by their environmental awareness that promoted pro-environmental behaviors along with public pressure toward governmental actions to tackle air pollution. This study highlights the importance of knowledge transfer about air pollution and their impacts to the citizens, which is reflected by a higher awareness level that may lead to actions to improve air quality and, consequently, to minimize citizens' exposure to air pollutants and its negative impacts on human health.

3. Outdoor Pollution Sources

Identifying pollution sources is crucial to define mitigation measures to improve air quality at specific locations. In urban areas, one of the main pollution sources is traffic and several cities worldwide are implementing traffic restrictions taking into account the vehicles category and their pollution potential. One example is the multi-year progressive traffic policy implemented in an extended limited traffic zone of the municipality of Milan (Italy) that started in the beginning of 2019, with the restriction of circulation of some classes of highly polluting vehicles (Euro 0 petrol vehicles and Euro 0 to 3 diesel vehicles). Another important step in these processes is a continuous evaluation of those measures in order to improve them or change them according to the desired results. For this reason, monitoring data is essential along with statistical tools which can provide information regarding trends and forecast outcomes. Such type of analysis was conducted for the Milan case [5], where the early-stage impact of this policy on two specific vehicle-generated pollutants (namely, total nitrogen oxides (NO_x) and nitrogen dioxide (NO_2)) was evaluated.

For this, daily environmental data gathered by Lombardy Regional Agency for Environmental Protection (ARPA Lombardia) was divided in pre-policy data (from 2014 to 2018) and in the early-stage policy assessment (focusing on eight months of data after the policy implementation). Several factors were considered in the evaluation of models, such as weather conditions, socio-economic factors and pollutants' concentration in neighboring cities. Despite the fact that the results showed an average reduction of concentrations after the policy implementation, this study highlighted that these changes could be due to other factors than the policy implementation, taking into account that the evaluated short time window may not reflect the citizens' adaptation to the new rules. Moreover, a negative impact on air quality of some specific areas was found which may be explained by the congestion of public transport buses (promoted by the restrictions) and by the change of the traffic temporal dynamics (since the aforementioned restriction is limited to business hours). Overall, this study characterized the initial phase of the policy implementation, where future assessments should be done to assess its impact when fully implemented, and, additionally, hourly data should also be considered to understand the hourly traffic dynamics and their impact on the local air quality.

A study conducted in Setúbal (Portugal) provided a ten year evaluation (2003–2012) of local air quality trends (focusing on PM_{10}, $PM_{2.5}$, O_3, NO, NO_2 and NO_x) and identified potential pollution sources of particulate matter, using a source apportionment technique, namely Positive Matrix Factorization [6]. Overall, air quality improved over the years with a decreasing trend of air pollutant concentrations, with the exception of O_3. However, despite this improvement, levels of PM_{10}, O_3 and nitrogen oxides still do not fully comply with the requirements of European legislation and the guideline values of World Health Organization (WHO). This study identified the main anthropogenic sources contributing to local PM levels as traffic, industry and wood burning, highlighting the need to address the sources by specific mitigation measures in order to minimize their impact on the local air quality.

Taking into account that road dust resuspension is a current air quality management challenge in Europe, a study conducted in Viana do Castelo (Portugal) focused on this issue and characterized road dust samples collected in suburban and urban streets, defined PM_{10} emissions factors for different road types and identified the main anthropogenic contributions to them [7]. Estimated PM_{10} emission factors ranged from 49 mg·veh^{-1}·km^{-1} (asphalt) to 330 mg·veh^{-1}·km^{-1} (cobble stone). Road dusts in urban streets revealed the contribution from traffic emissions (regarding the levels of Cu, Zn and Pb), while in suburban streets, the contribution from agriculture practices was highlighted by As levels in the finest road dust fraction. Hazard quotients were also assessed and a probability of induction of non-carcinogenic adverse health effects in children was found due to ingestion of Zr, while As in the suburban street was found to represent a human health risk of 1.58×10^{-4}, which reveals the need for adoption of local mitigation measures.

Besides chemical pollutants, air pollution also includes atmospheric bioaerosols, such as bacteria and yeasts. To understand how human hazard to these microorganisms changed over time, their temporal evaluation (focusing on a temporal gap of 10 years) was assessed in the South Western Siberia [8]. Bioaerosol samples obtained in 2006–2008 and 2012–2016 were characterized (by means of growth, morphological and biochemical properties) and the indices of hazards for both bacteria and yeasts were calculated. Overall, hazard to humans of culturable microorganisms in the atmospheric aerosol has not changed significantly over these 10 years (trends are undistinguishable from zero with a confidence level of more than 95%) despite a noticeable decrease in the average annual number of culturable microorganisms per cubic meter (6–10 times for the 10 year period).

Particulate matter (PM) pollution is a common environmental problem in urban areas and, in particular, in urban areas with fast development. In order to understand how the economic growth, urbanization and industrialization affect PM levels, a study targeted the specific case of Liaoning Province (China) where it evaluated PM levels and socio-economic indicators during a period of 16 years (from 2000 to 2015) [9]. Applying statistical tools, such as Granger causality test, it was shown that, in terms of the causal interactions economic activities, industrialization and urbanization processes all showed positive long term impacts on changes of $PM_{2.5}$ levels. In terms of contributions, industrialization contributed the most to the variability of $PM_{2.5}$ levels in studied sixteen years, followed by economic growth. These outcomes highlight the need for policy makers to explore more targeted policies to address the regional air pollution issue considering the local development characteristics.

4. Indoor Pollution Sources

Indoor micro-environments are the environments where humans spend most of their daily time. Therefore, it is important to understand the different mechanisms that can increase indoor concentration levels (from outdoor infiltration and human activities to specific emissions from products—such as candles—or materials used in within buildings in order to identify the best measures to decrease the exposure associated to them.

Smoking is a worldwide concern, both regarding direct or indirect health issues and also regarding its contribution to the deterioration of indoor air quality. In the last years, the use of electronic cigarettes (e-cigarettes) and heat-not-burn tobacco (HNBT) has been widely adopted as popular nicotine delivery systems (NDS), mainly among adult demographics. A study aimed to understand the effect on indoor air quality by traditional tobacco cigarettes (TCs) and new smoking alternatives, focusing in a multi-pollutant approach (particulate matter, black carbon—BC, carbon monoxide—CO—and carbon dioxide—CO_2) in two real life scenarios: home and car [10]. Results showed that particle emissions between the different NDS were significantly different, with TCs corresponding to higher PM10 and ultrafine particle concentrations and HNBT corresponding to the lowest levels registered. Considering that black carbon and CO are released by combustion processes, significantly lower levels were found for e-cigarettes and HNBT since no combustion occurs during their use. However, despite the fact that the new generation of cigarettes results to substantially lower levels of air pollutants compared to TCs, their use still constitutes a source of indoor air pollutants, which should be tackled.

Several actions can be taken in order to reduce indoor air pollutants. For instance, in some countries, such as South Korea, filter-type air purifiers are used to eliminate fine particulate matter from indoor environments. A study assessed the performance of this type of solution to reduce particulate matter levels in a person's breathing zone, according to two approaches: a movable air purifier responding to the movement of persons in the room or a fixed air purifier [11]. The results showed a decrease of 51 $\mu g/m^3$ and 68 $\mu g/m^3$ from the implementation of these two strategies, respectively.

5. Health Impacts of Human Exposure to Air Pollution

Understanding the health impacts of air pollution by means of statistical models is an important step to provide to the scientific community, policy makers and citizens in general the knowledge about the real burden of human exposure to air pollutants. This valuable information is, certainly, the foundation for promoting behavior change and public acceptance to the implementation of measures to tackle the human exposure. Therefore, improving such type of assessment models is very important in order to provide the most accurate and reliable information. Considering only outdoor concentration of air pollutants as a surrogate for total population exposure and neglecting micro-environments where people spend most of their time, may lead to biased and under-evaluated outcomes.

A study focused on this issue on the Greater London Area (UK) [12], including different micro-environments where people spend time at (based on time-activity and housing stock data), identified a misclassification of 1174–1541 mean predicted mortalities attributable to $PM_{2.5}$ exposure in the studied area when only outdoor concentration was considered. Indoor exposure to $PM_{2.5}$ was found to be the largest contributor to total population exposure concentrations accounting for 83%, followed by the exposure in the London Underground with 15%, despite the time spent on it being only 0.4%. Overall, this study confirmed that increasing the models' complexity and incorporating relevant micro-environments regarding the potential human exposure can significantly reduce the misclassification of health burden assessments.

The use of statistical models can also be employed to explore potential associations between exposure to air pollutants and health outcomes to highlight potential causal relationships. A study conducted in Wuhan city (China) evaluated the time-series of air pollutants ($PM_{2.5}$, PM_{10}, SO_2, NO_2, CO and O_3) and focused on the tuberculosis' incidence in order to identify its potential association with short-term exposure to air pollution, based on kriged data and single and multi-pollutant models [13]. Single pollutant models showed that an increase of 10 $\mu g/m^3$ in concentrations of $PM_{2.5}$, PM_{10} and O_3 promoted an increase of the associated tuberculosis risk, while in the multi-pollutant model, only $PM_{2.5}$ showed a statistically significant effect on tuberculosis incidence. Despite the complexity of the mechanism linking air pollution and tuberculosis incidence, this study provides insights on potential associations, which should be explored in future research.

A cross-sectional study conducted in Selangor (Malaysia) focused on assessing the allergy and lung inflammation promoted by exposure to different indoor pollutants ($PM_{2.5}$, PM_{10}, NO_2 and CO_2) in scholar environments, based on marker expression on eosinophils and neutrophils [14]. A total of 96 students from eight suburban and urban schools answered to standardized questionnaires and their fractional exhaled nitric oxide (FeNO) was evaluated, along with allergic skin prick test and sputum induction (which was later analysed for the expression of CD11b, CD35, CD63 and CD66b on eosinophils and neutrophils). By using chemometric and regression analysis, it was found that exposure to $PM_{2.5}$ and NO_2 was likely associated with the degranulation of eosinophils and neutrophils, following the activation mechanisms that led to the inflammatory reactions.

Adverse birth outcomes (stillbirth, preterm births and low birth weight) due to exposure to household air pollution from unclean cooking fuel in Nigeria were also assessed in a study using Bayesian models, taking into account geographic variability [15]. Using data obtained in a national cross-sectional survey, unclean fuel was the primary source of cooking for 89.3% of the women and the risk of stillbirth was significantly higher for mothers using this type of cooking fuel. It was also found that mothers who attained at least primary education had reduced risk of stillbirth, while for the increasing age of the mother the risk of stillbirth increased. Geographical differences were also found with mothers living in the Northern states having a significantly higher risk of adverse births outcomes.

Funding: The discussion about Integrated Human Exposure to Air Pollution contemplated in this special issue was promoted and partially funded by the LIFE Index-Air project (LIFE15 ENV/PT/000674). This work reflects only the authors' view and EASME is not responsible for any use that may be made of the information it contains. N. Canha acknowledges the national funding through FCT—Fundação para a Ciência e a Tecnologia, I.P. (Portugal) for his Postdoc grant (SFRH/BPD/102944/2014) and his contract IST-ID (IST-ID/098/2018). The FCT support is also acknowledged by C^{2}TN/IST (UIDB/04349/2020+UIDP/04349/2020) and CESAM (UIDB/50017/2020+UIDP/50017/2020).

Acknowledgments: The guest editors of this special issue of International Journal of Environmental Research and Public Health are grateful to all of the authors, reviewers, and MDPI staff.

Conflicts of Interest: The authors declare no conflict of interest.

References

1. Schraufnagel, D.E.; Balmes, J.R.; Cowl, C.T.; De Matteis, S.; Jung, S.H.; Mortimer, K.; Perez-Padilla, R.; Rice, M.B.; Riojas-Rodriguez, H.; Sood, A.; et al. Air Pollution and Noncommunicable Diseases: A Review by the Forum of International Respiratory Societies' Environmental Committee, Part 1: The Damaging Effects of Air Pollution. *Chest* **2019**, *155*, 409–416 [CrossRef] [PubMed]

2. Agrawaal, H.; Jones, C.; Thompson, J.E. Personal Exposure Estimates via Portable and Wireless Sensing and Reporting of Particulate Pollution. *Int. J. Environ. Res. Public Health* **2020**, *17*, 843. [CrossRef] [PubMed]

3. Arano, K.A.G.; Sun, S.; Ordieres-Mere, J.; Gong, B. The Use of the Internet of Things for Estimating Personal Pollution Exposure. *Int. J. Environ. Res. Public Health* **2019**, *16*, 3130. [CrossRef] [PubMed]

4. Dong, D.; Xu, X.; Xu, W.; Xie, J. The Relationship Between the Actual Level of Air Pollution and Residents' Concern about Air Pollution: Evidence from Shanghai, China. *Int. J. Environ. Res. Public Health* **2019**, *16*, 4784. [CrossRef] [PubMed]

5. Maranzano, P.; Fassò, A.; Pelagatti, M.; Mudelsee, M. Statistical Modeling of the Early-Stage Impact of a New Traffic Policy in Milan, Italy. *Int. J. Environ. Res. Public Health* **2020**, *17*, 1088. [CrossRef] [PubMed]

6. Silva, A.V.; Oliveira, C.M.; Canha, N.; Miranda, A.I.; Almeida, S.M. Long-Term Assessment of Air Quality and Identification of Aerosol Sources at Setúbal, Portugal. *Int. J. Environ. Res. Public Health* **2020**, *17*, 5447. [CrossRef] [PubMed]

7. Candeias, C.; Vicente, E.; Tomé, M.; Rocha, F.; Ávila, P.; Célia, A. Geochemical, Mineralogical and Morphological Characterisation of Road Dust and Associated Health Risks. *Int. J. Environ. Res. Public Health* **2020**, *17*, 1563. [CrossRef] [PubMed]

8. Safatov, A.; Andreeva, I.; Buryak, G.; Ohlopkova, O.; Olkin, S.; Puchkova, L.; Reznikova, I.; Solovyanova, N.; Belan, B.; Panchenko, M.; et al. How Has the Hazard to Humans of Microorganisms Found in Atmospheric Aerosol in the South of Western Siberia Changed over 10 Years? *Int. J. Environ. Res. Public Health* **2020**, *17*, 1651. [CrossRef] [PubMed]

9. Shi, T.; Hu, Y.; Liu, M.; Li, C.; Zhang, C.; Liu, C. How Do Economic Growth, Urbanization, and Industrialization Affect Fine Particulate Matter Concentrations? An Assessment in Liaoning Province, China. *Int. J. Environ. Res. Public Health* **2020**, *17*, 5441. [CrossRef] [PubMed]

10. Savdie, J.; Canha, N.; Buitrago, N.; Almeida, S.M. Passive Exposure to Pollutants from a New Generation of Cigarettes in Real Life Scenarios. *Int. J. Environ. Res. Public Health* **2020**, *17*, 3455. [CrossRef]

11. Park, H.; Park, S.; Seo, J. Evaluation on Air Purifier's Performance in Reducing the Concentration of Fine Particulate Matter for Occupants according to its Operation Methods. *Int. J. Environ. Res. Public Health* **2020**, *17*, 5561. [CrossRef] [PubMed]

12. Kazakos, V.; Luo, Z.; Ewart, I. Quantifying the Health Burden Misclassification from the Use of Different PM2.5 Exposure Tier Models: A Case Study of London. *Int. J. Environ. Res. Public Health* **2020**, *17*, 1099. [CrossRef] [PubMed]

13. Huang, S.; Xiang, H.; Yang, W.; Zhu, Z.; Tian, L.; Deng, S.; Zhang, T.; Lu, Y.; Liu, F.; Li, X.; et al. Short-Term Effect of Air Pollution on Tuberculosis Based on Kriged Data: A Time-Series Analysis. *Int. J. Environ. Res. Public Health* **2020**, *17*, 1522. [CrossRef] [PubMed]

14. Mohd Isa, K.N.; Hashim, Z.; Jalaludin, J.; Lung Than, L.T.; Hashim, J.H. The Effects of Indoor Pollutants Exposure on Allergy and Lung Inflammation: An Activation State of Neutrophils and Eosinophils in Sputum. *Int. J. Environ. Res. Public Health* **2020**, *17*, 5413. [CrossRef] [PubMed]

15. Roberman, J.; Emeto, T.I.; Adegboye, O.A. Adverse Birth Outcomes Due to Exposure to Household Air Pollution from Unclear Cooking Fuel among Women of Reproductive Age in Nigeria. *Int. J. Environ. Res. Public Health* **2021**, *18*, 634. [CrossRef]

International Journal of
*Environmental Research
and Public Health*

Article

Personal Exposure Estimates via Portable and Wireless Sensing and Reporting of Particulate Pollution

Harsshit Agrawaal, Courtney Jones and J.E. Thompson *

Department of Chemistry & Biochemistry, Texas Tech University, Box 41061, Lubbock, TX 79409-1061, USA; Harsshit.Agrawaal@ttu.edu (H.A.); Courtney.K.Jones@ttu.edu (C.J.)
* Correspondence: jon.thompson@ttu.edu

Received: 28 November 2019; Accepted: 27 January 2020; Published: 29 January 2020

Abstract: Low-cost, portable particle sensors (n = 3) were designed, constructed, and used to monitor human exposure to particle pollution at various locations and times in Lubbock, TX. The air sensors consisted of a Sharp GP2Y1010AU0F dust sensor interfaced to an Arduino Uno R3, and a FONA808 3G communications module. The Arduino Uno was used to receive the signal from calibrated dust sensors to provide a concentration ($\mu g/m^3$) of suspended particulate matter and coordinate wireless transmission of data via the 3G cellular network. Prior to use for monitoring, dust sensors were calibrated against a reference aerosol monitor (RAM-1) operating independently. Sodium chloride particles were generated inside of a 3.6 m^3 mixing chamber while the RAM-1 and each dust sensor recorded signals and calibration was achieved for each dust sensor independently of others by direct comparison with the RAM-1 reading. In an effort to improve the quality of the data stream, the effect of averaging replicate individual pulses of the Sharp sensor when analyzing zero air has been studied. Averaging data points exponentially reduces standard deviation for all sensors with n < 2000 averages but averaging produced diminishing returns after approx. 2000 averages. The sensors exhibited standard deviations for replicate measurements of 3–6 $\mu g/m^3$ and corresponding 3σ detection limits of 9–18 $\mu g/m^3$ when 2000 pulses of the dust sensor LED were averaged over an approx. 2 min data collection/transmission cycle. To demonstrate portable monitoring, concentration values from the dust sensors were sent wirelessly in real time to a *ThingSpeak* channel, while tracking the sensor's latitude and longitude using an on-board Global Positioning System (GPS) sensor. Outdoor and indoor air quality measurements were made at different places and times while human volunteers carried sensors. The measurements indicated walking by restaurants and cooking at home increased the exposure to particulate matter. The construction of the dust sensors and data collected from this research enhance the current research by describing an open-source concept and providing initial measurements. In principle, sensors can be massively multiplexed and used to generate real-time maps of particulate matter around a given location.

Keywords: air quality; crowd-sensing; crowd-sourced sensing; environmental analysis; pollution; particulate matter; dust sensor; human exposure; Arduino; wireless networks; IoT

---

1. Introduction

1.1. Summary of Problem

High concentrations of air pollution are due to fine solids, gases, or liquid aerosols locally releasing into the atmosphere or being produced at a faster rate than the environment can dilute, absorb, or dissipate the material [1]. If the rate of production is sufficiently high, substances can build up and reach a high concentration in the air that can contribute to a host of adverse health effects for humans such as cardiovascular mortality and respiratory distress [2–6]. Miller et al. found an increase of 10 μg/m^3 in particulate matter mass concentration was associated with a 14% (95% confidence interval, 3%–26%) increase in nonfatal cardiovascular events and with a 32% (95% CI, 1%–73%) increase in fatal cardiovascular events [7]. Exposure to air pollution whether long or short can lead to asthma, decreased lung function, and infections of the respiratory system. Goss et al. found PM$_{2.5}$ was associated with statistically significant declines in lung function and an increase in the odds of two or more pulmonary exacerbations in patients >6 years of age with cystic fibrosis [8]. Woodruff et al. found that each 10 μg/m^3 increase in PM$_{2.5}$ was associated with a near doubling of the risk of post neonatal death because of respiratory causes [9]. Hoek et al. found the risk of cardiopulmonary mortality nearly doubled for individuals who lived within 100 m of a freeway or within 50 m of a major urban road [10]. Clearly, particulate matter presents a clear and present danger to human health.

Aerosols are found everywhere including in the air over oceans, deserts, mountains, forests and ice, with the most abundant being from natural sources such as sea-spray and wind-blown soil dust [11]. For instance, dry areas of North Africa emit 800 Tgy^{-1} of soil dust each year and summertime winds can transport the Saharan dust across the Atlantic Ocean, the Caribbean, and to southern North America [12–15]. Urban areas add to the airborne particulate mass through direct emissions into the atmosphere and through emitting gases that can react in the atmosphere to form secondary aerosol. A significant body of research has gone into understanding and controlling the chemical and physical transformations that lead to secondary aerosol formation [16–18]. Despite the great success of air quality improvement programs implemented in Western cities over the preceding 50 years, urban environments typically have higher levels of particulate pollution compared to rural locations as cities remain global hubs of fossil fuel combustion and direct emissions [19–23]. An emerging threat is found in some of the world's poorest cities as the World Health Organization (W.H.O.) has recently reported 98% of cities in low and middle income countries with more than 100,000 inhabitants do not meet WHO air quality guidelines [24]. Aerosols are also present indoors and homemakers in economically disadvantaged environments can often be exposed to very high levels of particulate matter when using home cook-stoves to prepare meals [25,26]. Monitoring and controlling human exposure to PM$_{2.5}$ remains a crucial scientific challenge and this focus requires the continued development and application of portable and low–cost sensing platforms for widespread application.

1.2. Current Work in the Field

Historically, PM$_{2.5}$ concentrations have been measured using either inertial impactors, filter based sampling, or sophisticated laboratory-based measurement devices [27–31]. However, such approaches are either too expensive, too labor intense, too slow, too limited in scope, or all of the above to address the needs of the problem. A new paradigm is needed in which many end users can access accurate, real-time data at very low cost to society [32,33].

Several brands (Shinyei, Sharp, Nova, Plantower, Wuhan, Alphasense, Air Beam, etc.) of low-cost electronic sensors that use light scattering can be used to monitor particulate matter disbursed in air. These sensors come at different price points and sizes but are designed to use low power and can operate at high measurement frequency. The common limitation in the sensors is the accuracy and the reproducibility of the data, and very limited ability to measure particles below 200–300 nm in diameter due to low scattering cross section. In addition, low-cost sensors often have limited linear dynamic range, and often plateau or 'max out' at a few hundred microgram per cubic meter concentration. This feature of devices may underestimate the impact of high exposure events.

Currently, one of the most successful integrated approaches to particulate matter monitoring is called Airbeam (habitatmap.org). In this project, a palm-sized device costing $250 measures particulate pollution that a user is exposed to by a light scattering method and transmits data to a co-located Android device through a Bluetooth connection. The wireless connection on the secondary device and an Android OS App is used to 'AirCast' collected data to a central data server and maps are created by combining various end users' data streams. The integrated approach demonstrates the potential of the crowd-sourced architecture.

The excitement generated by the initial emergence of low-cost sensing platforms such as 'Airbeam' has led the Environmental Protection agency (EPA) of a major North American country to begin evaluating the performance of low-cost sensors co-located with sensors their government considers "reference" and "equivalent" methods (see https://www.epa.gov/air-sensor-toolbox). In an exclusive project, in 2019 this EPA began a study aimed to tackle questions about long-term performance of low-cost air sensors, an area that is rarely (if ever) explored in academic settings. The exclusive group of EPA funded staff are currently evaluating six different models of low-cost air sensors, placed at seven diverse locations throughout states within North America. The locations have diverse climates and air quality conditions, assuring that the project produces a dataset that investigators may use to assess how weather conditions affect sensor signals and long-term performance.

In Gunawan et al., a suite of portable sensors ($PM_{2.5}$, PM_{10}, CO, O_3) were used in Malaysia to derive the local air pollution index (API) at an end user's location [34]. The API was reported to the user in nearly real-time thru an LCD display. For carbon monoxide a model MQ-9 sensor was employed, while ozone was monitored with the MQ-131 gas sensor. The Sharp dust sensor was used to measure PM_{10} while a Shinyei PPD42NS was used to derive $PM_{1-2.5}$. The set-up was controlled by the Arduino microcontroller, and the authors reported a cellular modem and Global Positioning System (GPS) card could be added if desired. However, wireless communication was not a focus of the work. Unfortunately, the authors were unable to perform proper calibration of each sensor and had to rely on suggested calibration equations from the manufacturer. In addition, laboratory validation measurements were not completed and data quality unassessed.

In Reilly et al., an affordable monitoring device costing ≈$280 was constructed [35]. The device can send and collect data to a real-time mapping program wirelessly using the Global System for Mobile Communications (GSM) to a server, or as a text message to people nearby. The device consisted of the Redboard Arduino clone, a GSM board, a carbon monoxide sensor, an ozone sensor, and Sharp dust sensor for particulate matter (PM), a fan to promote airflow, and instrument case. While the authors report successful implementation of their device, environmental sensors were not calibrated nor rigorously tested in a laboratory. Instead, the authors used the recommended calibration equations from the product manufacturers to convert electrical signal to environmental measurement. Suspiciously high and constant PM levels were reported (approx. 150 $\mu g/m^3$) and ozone levels did not exhibit the characteristic diurnal pattern that is observed nearly everywhere in the troposphere. Consequently, despite the approach being successfully implemented, the quality of environmental sensor results is questionable.

In Zamora et al., Plantower AMS A003 sensors were tested in the laboratory to measure particulate matter of various composition in a series of well-defined and controlled experiments [36]. The sensors were precise; however, when the sensor results were compared with a reference method, accuracy varied as a function of aerosol composition and humidity. The sensors were highly precise with a R^2 values greater than 0.86 for all sources. However, the accuracy had a large range from 13% to greater than 90% compared with reference instruments, depending upon aerosol type. The sensors were more accurate when the particulate matter was spherical and smaller than 1 μm in size. The sensors' accuracy was greatly affected by the humidity. The work of Zamora et al. contributes significantly to the field because it draws attention to the limits of single angle nephelometry to estimate PM mass concentration vis-à-vis the aerosol composition and shape. While the sensors were able to provide usable data when in motion or in high or low temperatures, the focus of this manuscript was refreshingly on data quality. Results suggest that portable sensors should be calibrated versus an accepted reference method using authentic aerosol as a sample.

Liu et al. takes this approach by examining performance of Nova particulate matter sensors (SDS011) [37]. The sensors were tested on ambient samples in Oslo, using a co-located, official air quality monitor as a reference method. All of the sensors had results that were similar including inter-sensory correlations with R^2 values greater than 0.97. Again, high humidity greatly affected the sensors. There was a linear relationship between the sensors and the reference monitor. However, the R^2 varied over the range of 0.55 to 0.71. When a data correction using relative humidity and temperature was used, the R^2 value for each sensor increased from 0.71 to 0.80, 0.68 to 0.79 and 0.55 to 0.76, respectively. These sensors were also limited by the environmental conditions and can have high error if used outside of the manufacturer specifications. This work draws attention to the need to calibrate versus an accepted gravimetric reference method with authentic aerosol and the need to correct measurements (or at least flag them) for conditions of high humidity.

Liu et al. have compared Shinyei, Sharp, and Oneair optical sensors in laboratory experiments with polydisperse particles of a variety of compositions, concentration, and mean size [38]. These authors found the mass concentration normalized response of all optical sensors clearly changed with mean particle size even when considering the narrow range of very small particles between 70–95 nm. The composition of particles was even more crucial—particles of methylene blue, sodium fluorescein, and sodium chloride all produced sensitivities differing by at least 50% (with NaCl being lowest). These results clearly demonstrate that choice of calibrant aerosol is crucial for obtaining accurate sensor mass calibrations. The current manuscript is not exempt from such considerations or potential errors.

1.3. Contribution of the Current Work

The current study aims to build upon previous work from our laboratory in which a field-portable device for logging $PM_{2.5}$ mass concentration data was developed [39]. The previous device also used the Arduino, and a Sharp sensor with nephelometric detection, but logged all collected data to an SD card. This prevented real-time feedback to the user and complicates integrating/multiplexing many sensors together to form a network of streaming data. This limitation has recently been addressed thru our laboratory's development of an open-source data acquisition platform we have called *logIT* that wirelessly transmits data over the 3G cellular network [40]. Source code to implement *logIT* is available online for the user community [40]. The *logIT* platform has allowed multiplexing multiple sensors (here n = 3) to demonstrate multi-user simultaneous measurements of PM exposure throughout a mid-sized city. In addition, we build upon the previous literature work using the Sharp GP2Y1010AU dust sensors [34,35] by performing laboratory measurements to characterize and constrain device performance. This was not reported in [34,35] and results presented within this manuscript will allow the community to make more informed decisions regarding the performance characteristics of Sharp GP2Y1010AU dust sensors. The current manuscript also serves as a companion article to [38], as the current work highlights signal to noise considerations and intercomparisons of individual Sharp dust sensors themselves, while Liu et al. [38] provide much more through insight into calibration of low-cost sensors against a gravimetric equivalent method and the crucial effect of calibrant composition and size.

2. Methodology

Three Arduino Uno R3 ($14.99 USD each), FONA808 modules (Adafruit, $49.95 USD each) and dust sensors (GP2Y1010AU0F $15.23 USD each) were purchased and used as received without modification. The dust sensor is a nephelometric sensor illuminating sample with a near infrared LED with a wavelength of 860 nm and collecting light at 120 degrees from incident illumination as depicted in Figure 1A. The dust sensors were affixed to circuit boards with cyanoacrylate and wired into Arduino 'stackable shields' in house by laboratory staff. Stackable shields plug into the Arduino board directly by using header pins for both electrical and mechanical connection. The wiring connections used are illustrated in the circuit diagram of Figure 1B. The FONA808, Arduino Uno, and dust sensors were mounted into plastic project boxes as depicted in Figure 1C. For user comfort, a lanyard was attached to each box. The Arduino Uno was used to control FONA808 modules and dust sensors. FONA808 communication modules were used for reporting the particulate matter concentration to a Thingspeak.com channel. Technical details of this process [40]. The data was collected in real time and logged to Thingspeak.com until downloaded for analysis. All the calibrations were performed with wireless communication with 9V power supply powering the dust sensor modules.

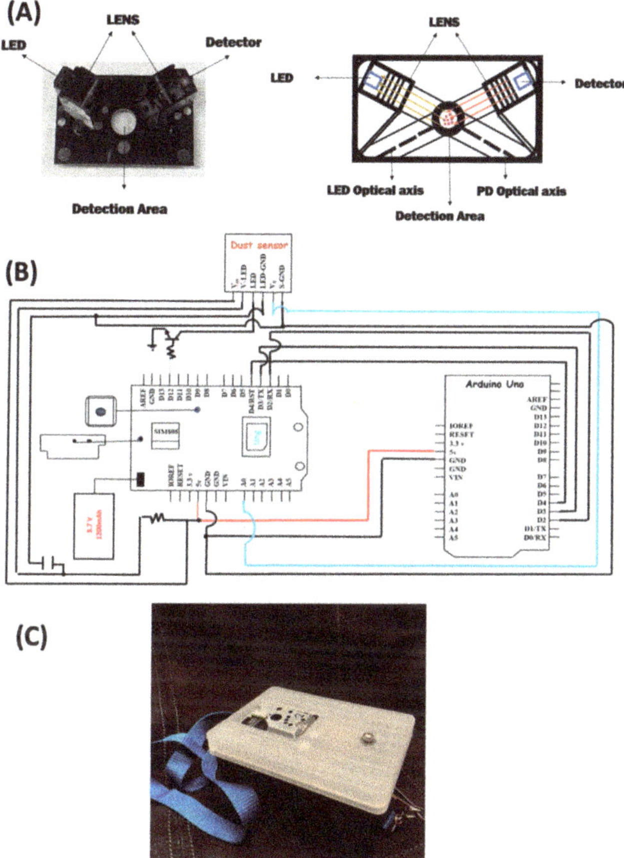

Figure 1. (**A**) Photo and drawing of Sharp GP2Y1010AU0F dust sensor. (**B**) Wiring schematic for the dust sensor and FONA module to the Arduino (**C**) Photograph of the portable project box containing the dust sensor, FONA module and Arduino. The project box is 12 cm × 9 cm × 6.5 cm and has a mass of 250 g.

The Sharp dust sensors produce an analog voltage as output signal. To convert the analog signal to a meaningful PM concentration, the dust sensor must be calibrated against references before making meaningful measurements. A commercial device, the reference aerosol monitor (RAM-1) from Monitoring Instruments for the Environment (MIE, Inc., Billerica, MA, USA) was chosen as a reference method to calibrate each sensor. The RAM-1 actively samples aerosol using a pump and estimates concentration using the light scattering principle. The RAM-1 analysis is not an EPA equivalent method for $PM_{2.5}$. We have performed four trials in which the RAM-1 sensor was used to compare the indicated concentration of a sodium chloride test aerosol within a chamber with gravimetric determination of aerosol mass concentration. A 37 mm quartz filter was used to collect a sodium chloride test aerosol at 15 SLPM using a mass flow controller and vacuum pump. Gravimetric measurements of (average) concentration over the sampling period could be obtained and compared directly with the RAM-1 values. For these 4 trials we find that the RAM reading is on average 107% (Std Dev. = 67%) of the gravimetric result with n = 4.

For calibration, all Sharp dust sensors were placed inside a 3.6 m^3 volume chamber lined with a fluoropolymer (FEP). Each dust sensor was individually calibrated in the 3.6 m^3 chamber. A single jet atomizer (TSI 9302) generated sodium chloride particles into the chamber with an atomizer pressure of 20 psi. The flow of particles entering the chamber was 5.7 L min^{-1}. The jet generated polydisperse sodium chloride particles into the chamber. After approximately 2 min, sufficient aerosol was produced and the atomizer jet was shut off for the remainder of the experiment. The relative humidity inside the chamber under these conditions was measured to be < 20%, indicating presence of a dry aerosol. Experimental runs at high relative humidity were not attempted, and performance under these conditions not explored in this study. The calibration data was then collected for 4–5 h as the initially high particle concentration dissipated due to impaction on the chamber walls. For each point in time, the individual sensors were compared to the indicated concentration of the reference aerosol instrument (RAM) placed inside the chamber. Each dust sensor was calibrated individually, and the process was repeated for each sensor. All of the data from the dust sensors were uploaded to the ThingSpeak channel in real time using the FONA808 module.

3. Results and Discussion

3.1. Optimizing Delay Time

The Sharp dust sensors operate by turning on a near-IR LED, waiting a user-specified delay period, and then sampling the scattered light signal prior to cycling the LED off again. The entire measurement cycle can be completed in under a millisecond and this measurement cycle can repeat itself many times. Indeed, signals can be averaged to improve limits of detection (see section below). The LED is turned on by a HI/LO (5 - 0 Volt) transition at a digital pin on the microcontroller, and a delay prior to signal acquisition is then initiated. While Sharp recommends a delay time of 280 μsec prior to collection of data in product literature, we have systematically studied the effect of delay time on sensitivity (slope of calibration line). Figure 2 reports the results of this study. As observed, measurements suggest that a 220 microsecond delay after the LED trigger allowed optimal sensitivity to be achieved for the individual sensors tested in this study. The cause of the shift in optimum delay time from the manufacturer's recommendation is not known. However, differences in stray capacitance between our apparatus and the manufacturer's test bed may be the cause. The result suggests end users may wish to optimize delay time for their own application. Regardless of the cause, the delay time has been optimized to provide maximum sensitivity for this study.

3.2. Sensor Calibration Results

Figure 3 illustrates plots of dust sensor signals vs. indicated PM mass concentration for the three dust sensors used in this study. In these experiments, signal from a commercial PM mass concentration monitor (RAM) was used as the reference/accepted concentration. Our analysis is limited in we assume no error, uncertainty or imprecision is present in the RAM data stream. For all sensors, there was a linear and positive correlation (R^2 > 0.92) observed between the concentration (μg m^{-3}) and dust sensor signal to 500 μg m^{-3}. A linear-least squares best-fit line was added to each dataset and used to determine the slope and intercept for each sensor. The slopes were 0.491, 0.446, and 0.506 digital counts per μg m^{-3} for sensors 1, 2 and 3. Since the Arduino analog acquisition is a 10 bit device operating over a full span of 1024 steps from 0 to 5 V, a digital step corresponds to 4.88 mV/count. Therefore, our sensors were observed to produce between 0.217 and 0.247 V per 100 μg/m^3 or roughly half of the signal specified by the product data sheet for the Sharp dust sensor which specifies 0.5 V signal per 100 μg/m^3 of PM mass concentration [41]. The exact cause of this discrepancy in sensitivity remains unclear. However, we note that Liu et al. [38] previously observed sodium chloride aerosol produced a much lower signal compared to other compositions. Varying the size distribution or particle shape of the test aerosol will lead to differences in differential scattering cross-section at 120 deg. and consequently, the observed slope of calibration lines. Emerging research on low-cost dust sensors [36,38] has provided clear evidence that the light scattering method can be subject to large biases when projecting mass concentrations when calibrations were conducted against aerosol with different properties (refractive index, shape, size, composition) than the ambient type encountered at a particular location or for a particular application. As such, this could be responsible for the low sensor responses we report.

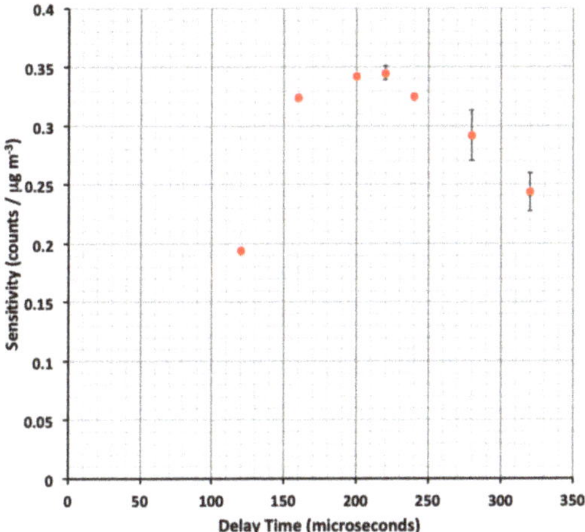

Figure 2. Plot of Sharp GP2Y1010AU0 dust sensor sensitivity vs. delay time prior to signal acquisition. The sensitivity is the slope of the best-fit calibration line when signal was plotted vs. PM mass concentration (μg/m^3). The optimal delay of 220 microseconds was employed for subsequent measurements. Error bars represent $\pm 1\sigma$ of replicate trials.

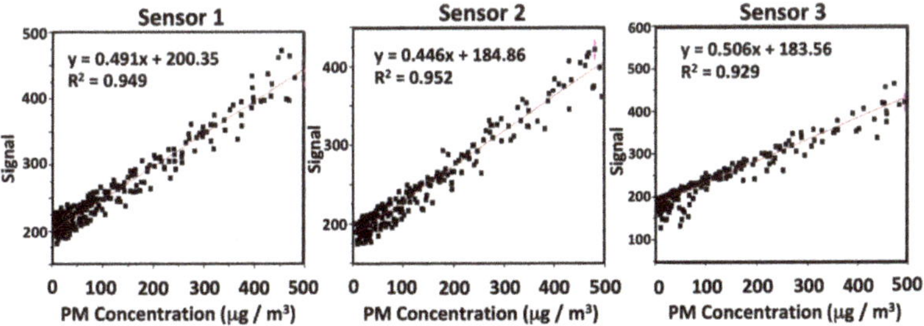

Figure 3. Calibration plots for sensors 1–3. Observed signal from the Sharp GP2Y1010AU0 dust sensors vs. PM concentration indicated by the RAM reference method for laboratory generated polydisperse NaCl aerosol. Observed signal is the digital count assigned by the Arduino board during analog-to-digital conversion. Differences were found in slope and intercept between the sensors tested, and $R^2 > 0.92$ observed. Relative humidity within the 3.6 m^3 chamber during calibration was $< 20\%$.

3.3. Effect of Signal Averaging

Since the data acquisition cycle for a single pulse of the LED is very short (ms time), averaging many pulses from the LED is an attractive approach to enhancing detection limits for the Sharp dust sensors. Code can easily be added to the Arduino sketch to accomplish signal averaging in the onboard memory of the microprocessor on the fly. Only the averaged result is then reported in the data stream. A study was carried out in which an air blank was analyzed, and the number of LED pulses averaged together were 100, 200, 500, 1000, 1500, 2000, 5000, and 10,000 individual pulses. Data points were reported for each average and the standard deviation (σ) of replicate averages computed. Then, the limit of detection (L.O.D.) for each sensor was computed by using the 3σ standard. As observed in the black trace of Figure 4, as the number of LED pulses averaged increased to roughly n = 2000, the measurement L.O.D. decreased an order of magnitude for all the sensors tested. Interestingly, averaging approx. 2000 pulses only marginally affects the total time required to report a single data point (shown in blue data series) because the wireless communication protocol alone requires nearly a minute on average to complete its reporting cycle. Figure 4 also demonstrates that if additional LED pulses > 2000 were averaged together, there were only marginal additional gains in limit of detection achieved. However, the time required for data acquisition when n > 2000 began to linearly increase with the number of averages since wireless communication was no longer the rate limiting step under these conditions. Consequently, 2000 averages were used in subsequent monitoring activities. Under these conditions, the standard deviation and limit of detection for the three sensors were between σ = 3–6 $\mu g/m^3$ and L.O.D. = 9–18 $\mu g/m^3$, respectively. These results indicate that Sharp dust sensors and microprocessor mediated signal averaging can be used to track PM pollution within environments where substantial particle pollution is expected. This could include routine workplace monitoring or personal exposure monitoring for citizens living in urban centers where levels regularly exceed 25–30 $\mu g/m^3$. However, it should be noted that the method we describe herein is not sensitive enough for monitoring in all environments (see below). Further improvements in precision, limits of detection, and sensor accuracy are still required to bring ambient portable sensing to its full potential.

3.4. Precision of Dust Sensors

Figure 5A reports a histogram of observed percent difference between individual Sharp dust sensor measurements and the reference method (RAM) measurements for test sodium chloride aerosol within a laboratory chamber. Figure 5 includes data points from all 3 Sharp dust sensors. Here, the percent difference was computed as (Sharp sensor measurement–reference measurement)/reference measurement and the result expressed as parts-per-hundred relative difference. As observed in Figure 5A, results were observed to follow a normal distribution. A Gaussian fit to the data indicated a standard deviation of σ = 30.8 for the entire dataset. This result suggests, considering all data, Sharp sensors report concentrations within 30.8% of the reference value about 68% of the time. The standard deviation also allows end users to define confidence intervals for their data points. In Figure 5B, a plot of percent difference between Sharp dust sensor and the reference measurement is plotted vs. the indicated PM mass concentration for the reference method. This plot indicates that the largest relative discrepancies between dust sensor indicated mass and reference monitor mass occurred when PM mass concentration was < 40 $\mu g\ m^{-3}$. Above this mass loading, relative difference between the dust sensor and reference monitor was very frequently < 20%. Considering only data collected when PM > 40 $\mu g\ m^{-3}$, the average percent difference between a Sharp dust sensor and reference measurement for single point comparison was 5.8% and the median absolute percent difference was 15.2%. This result indicates that unmodified Sharp dust sensors offer best precision when PM mass loadings are relatively high (e.g., PM > 40 $\mu g\ m^{-3}$).

3.5. Monitoring Experiments

The ultimate goal of this work is to improve understanding of human exposure to PM pollution through creating and implementation of a network of multiplexed, portable PM sensors. In this manuscript, we report a limited scope proof-of-concept application by having laboratory personnel carry the three dust sensors around Lubbock, TX. The staff engaged in normal life activities during sampling in an effort to demonstrate that data from multiple sensors can simultaneously be streamed to a Thingspeak channel, while the end user can access information about his/her own exposure in nearly real-time via a public Thingspeak web link. Figure 6 reports example time-series monitoring data that end users encountered during the effort. As observed, spikes in PM mass loadings were encountered by the users during random life events such as walking near an individual smoking cigarettes, cleaning an apartment space, burning incense at home, or cooking. It is often difficult to accurately parameterize exposure during such casual life experiences in models of human exposure to PM pollution, and portable PM sensors have great promise to improve quantification of PM for such circumstances. In an effort to improve understanding of human exposure in different environments we have performed measurements of PM mass concentration at a variety of locations and report a summary of these measurements in Table 1 of this manuscript. While a wide range of PM levels were encountered across locations (and even at single locations through time), the reported levels begin to improve understanding of typical levels of human exposure in various environments. For instance, results have suggested that human exposure to PM is often quite high near any cooking operation, or at some restaurants. The use of low-cost, portable sensors to uncover and document such knowledge can be used to develop exposure control strategies for workers in these environments.

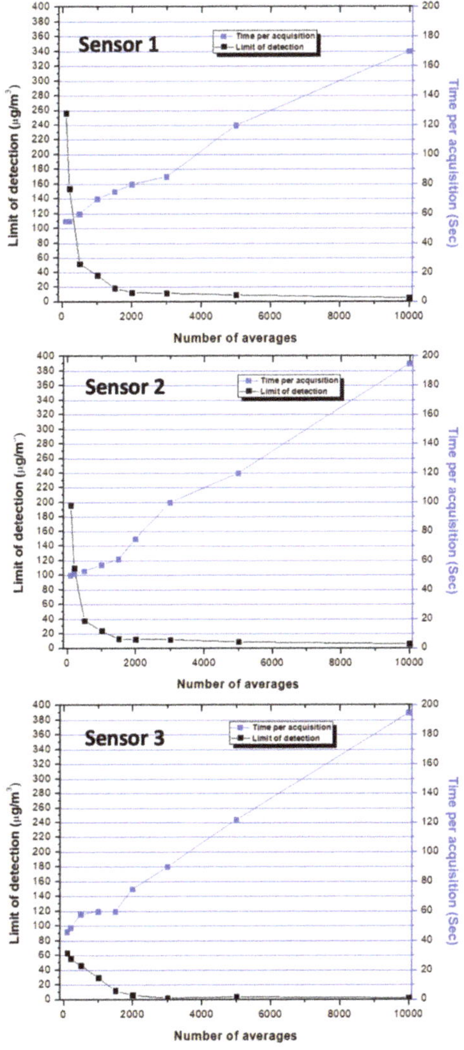

Figure 4. The effect of number of LED pulses averaged (N) on the limit of detection (black squares) for sensors 1–3. The second y-axis (blue) presents the seconds required to acquire and report data for the specified number of averages.

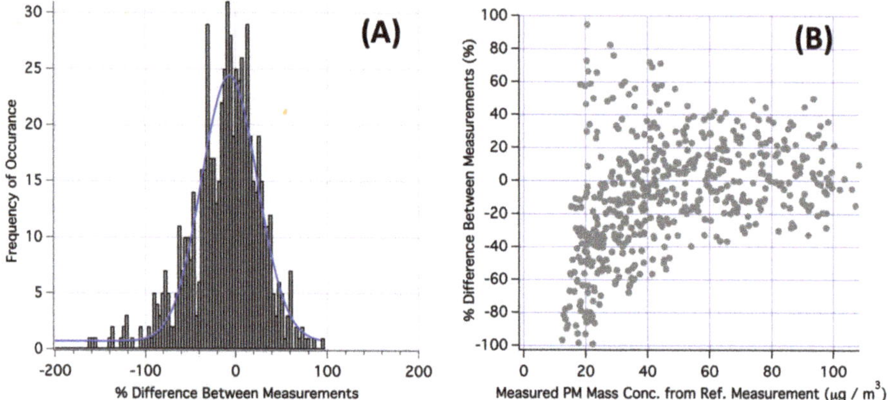

Figure 5. (**A**) Histogram of percent difference between individual Sharp GP2Y1010AU0 dust sensor measurements and the reference method measurements (RAM-1) for test sodium chloride aerosol within a chamber. A Gaussian fit to the data indicated σ = 30.8 for the entire comparison dataset. (**B**) Plot of percent difference between Sharp dust sensor and the reference measurement plotted vs. the indicated PM mass concentration for the reference method.

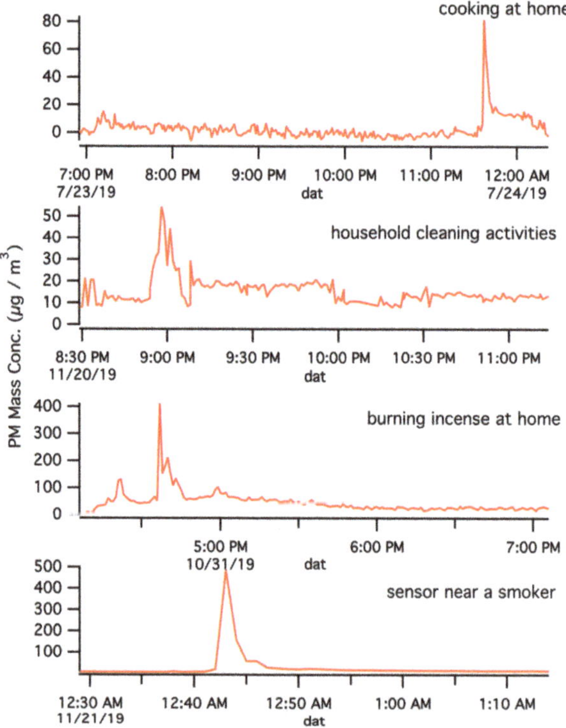

Figure 6. Sample monitoring results for end users during the trial implementation. Real-time personal exposure monitoring allows improved understanding of the times, locations, and extent of high exposure events. The sensors were in the same room adjacent to the end user during the events.

Another goal of future research is to begin documenting the temporal and spatial behavior of aerosol levels at specific locations. Figure 7 presents a map of PM levels encountered by end users as they carried the portable sensors with them through their daily routines. Since the platform developed transmits GPS coordinates along with PM levels, data from multiple sensors can be combined and plotted. Such plots can ultimately be used to better understand the spatial dynamics of particle pollution within a city, and comprehensive datasets can better educate the public in regard to differences in pollution exposure between regions of a city or neighborhoods. However, the results presented in this manuscript do not represent such an achievement. Understanding temporal and spatial averages for PM concentrations at specific locations is a complex task that can only be accomplished through many more measurements than what we report within this paper.

Table 1. Summary of results for locations tested.

Place/Activity		Range of Concentration ($\mu g/m^3$)		Statistical Indices ($\mu g/m^3$)	
		Maximum	Minimum	Mean Conc.	Median Conc.
Driving		159	Below L.O.D.	15	Below L.O.D.
Living Space	Location 1	84	Below L.O.D.	Below L.O.D.	Below L.O.D.
	Location 2	186	Below L.O.D.	Below L.O.D.	Below L.O.D.
	Location 3	113	Below L.O.D.	27	18
Parking Lot	Location 1	28	Below L.O.D.	Below L.O.D.	Below L.O.D.
	Location 2	48	29	34	33
	Location 3	25	Below L.O.D.	18	19
	Location 4	16	Below L.O.D.	Below L.O.D.	Below L.O.D.
Church	Location 1	Below L.O.D.	Below L.O.D.	Below L.O.D.	Below L.O.D.
Room Renovation	Location 1	85	14	28	25
Workshop	Location 1	70	Below L.O.D.	Below L.O.D.	Below L.O.D.
Restaurant	Location 1	346	20	46	39
	Location 2	44	Below L.O.D.	24	24
	Location 3	24	Below L.O.D.	Below L.O.D.	Below L.O.D.
	Location 4	81	Below L.O.D.	46	47
Grocery Stores	Location 1	25	Below L.O.D.	Below L.O.D.	Below L.O.D.
	Location 2	22	Below L.O.D.	17	17
	Location 3	41	Below L.O.D.	26	26
	Location 4	Below L.O.D.	Below L.O.D.	Below L.O.D.	Below L.O.D.
University building	inside	39	Below L.O.D.	Below L.O.D.	Below L.O.D.
	outside	18	Below L.O.D.	Below L.O.D.	Below L.O.D.
Coffee shop	Location 1	25	Below L.O.D.	16	16
	Location 2	17	Below L.O.D.	Below L.O.D.	Below L.O.D.
	Location 3	Below L.O.D.	Below L.O.D.	Below L.O.D.	Below L.O.D.

Note: Table 1 data is not representative of exposure in all specified microenvironments. Human exposure to PM is highly circumstantial based upon an individual's specific experience.

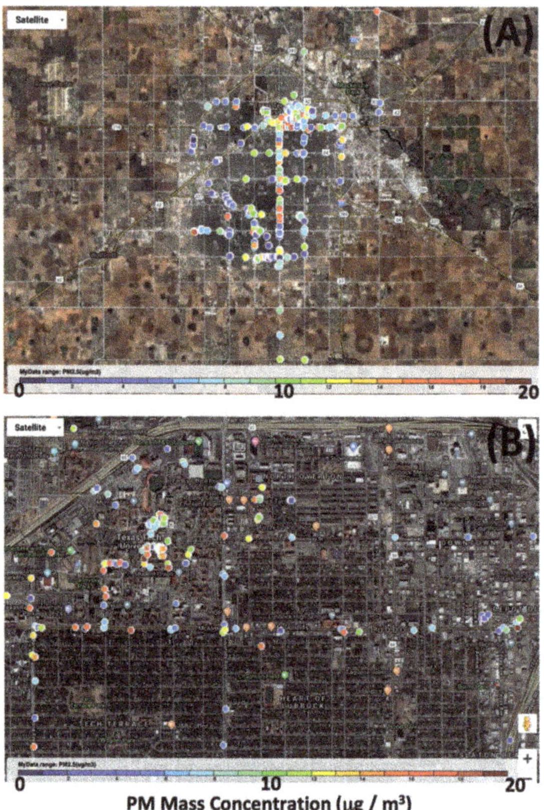

Figure 7. Maps of Lubbock, TX with user sampled PM single-point data superimposed. Mobile sensors transmitted GPS coordinates and dust sensor PM measurements to a webserver, and maps were created from data. Map in (**A**) is approx. 40 km wide by 28 km high. Map in (**B**) is zoomed to illustrate specific locations sampled by users during their experiences. Data point colors describe an individual observation PM mass concentration in $\mu g\ m^{-3}$ and are not meant to represent trends or averages.

4. Conclusions

Low-cost dust sensor modules (n = 3) were constructed using an Arduino Uno R3, a FONA808 module and a Sharp dust sensor. The modules employed the *LogIt* platform our laboratory has previously developed to wirelessly transmit GPS and particle mass concentration data to a web server [40]. The dust sensor was calibrated in the lab against a commercial optical PM mass concentration monitor and subsequently used to monitor the air quality in Lubbock, TX during daily activities, such as walking, cooking, sleeping, sweeping, traveling or working. The concentration of particulate matter was collected in real time with GPS coordinates. The optimal limits of detection for the unmodified Sharp dust sensors tested ranged from $8\ \mu g\ m^{-3}$–$17\ \mu g\ m^{-3}$. The particulate matter concentration remained relatively low, and constant during outside measurements, but data had moderate spikes inside dwellings during certain activities. High concentrations were often observed near restaurants. Multiplexing the sensors (using many simultaneously) can better constrain human exposure to PM, and subsequently better determine health effects of particle pollution.

In the future, our sensors should be calibrated and challenged against additional aerosol compositions and size distributions to improve performance and reliability. Because particle size distribution and composition may affect sensor response we advocate for direct calibration of the Sharp sensors alongside real-time EPA federal equivalent methods such as the TEOM or beta-attenuation monitor using authentic aerosol as the sample. Such direct comparison/calibration on authentic aerosol typical of a specific location will provide best performance. Additionally, the state of the science can be advanced by developing in-field calibration protocols for the low-cost sensors to routinely assess and check sensor calibrations.

Author Contributions: H.A. and C.J. contributed by collecting and reporting data and authoring manuscript text. J.E.T. contributed by editing manuscript, providing guidance, and project concept. All authors have read and agreed to the published version of the manuscript.

Funding: This research received no external funding.

Acknowledgments: The authors would like to thank Scott Hiemstra for design and construction of the project box used to contain the sensors.

Conflicts of Interest: The authors declare no conflict of interest.

References

1. Nathanson, J. Air Pollution. Encyclopedia Britannica [Online]. Posted 31 October 2018. Available online: https://www.britannica.com/science/air-pollution (accessed on 18 July 2019).
2. Brook, R.D.; Rajagopalan, S.; Pope, C.A. Particulate Matter Air Pollution and Cardiovascular Disease: An Update to the Scientific Statement from the American Heart Association. *Circulation* **2010**, *121*, 2331–2378. [CrossRef]
3. Thompson, J.E. Airborne Particulate Matter: Human Exposure & Health Effects. *J. Occup. Environ. Med.* **2018**, *60*, 392–423.
4. Pope, C.A.; Tuner, M.C.; Burnett, R.; Jerrett, M.; Gapstur, S.M.; Diver, W.R.; Krewski, D.; Brook, R.D. Relationships Between Fine Particulate Air Pollution, Cardiometabolic Disorders and Cardiovascular Mortality. *Circ. Res.* **2015**, *116*, 108–115. [CrossRef] [PubMed]
5. Chen, H.; Burnett, R.T.; Kwong, J.C.; Villeneuve, P.J.; Goldberg, M.S.; Brook, R.D.; van Dokelaar, A.; Jerrett, M.; Martin, R.V.; Kopp, A.; et al. Spatial association between ambient fine particulate matter and incident hypertension. *Circulation* **2014**, *129*, 562–569. [CrossRef] [PubMed]
6. World Health Organization. Ambient Air Pollution: Health Impacts. Available online: https://www.who.int/airpollution/ambient/health-impacts/en/ (accessed on 18 July 2019).
7. Miller, K.A.; Siscovick, D.S.; Sheppard, L. Air Pollution and Cardiovascular Disease Events in the Women's Health Initiative Observational (WHI-OS) Study. *Circulation* **2004**, *109*, 71. [CrossRef]
8. Goss, C.H.; Newsom, S.A.; Schildcrout, J.S. Effect of Ambient Air Pollution on Pulmonary Exacerbations and Lung Function in Cystic Fibrosis. *Am. J. Repir. Crit. Care Med.* **2004**, *169*, 816–821. [CrossRef] [PubMed]
9. Woodruff, T.J.; Parker, J.D.; Schoendorf, K.C. Fine Particulate Matter ($PM_{2.5}$) Air Pollution and Selected Causes of Postneonatal Infant Mortality in California. *Environ. Health Perspect.* **2006**, *114*, 786–790. [CrossRef] [PubMed]
10. Hoek, G.; Brunekreef, B.; Goldbohm, S.; Fischer, P.; van den Brandt, P.A. Association between mortality and indicators of traffic-related air pollution in The Netherlands: A cohort study. *Lancet* **2002**, *360*, 1203–1209. [CrossRef]
11. National Oceanic & Atmospheric Administration. Aerosols: Climate and Air Quality. Available online: https://www.esrl.noaa.gov/research/themes/aerosols/ (accessed on 18 July 2019).
12. Zhang, Q.; Thompson, J.E. A Model for Absorption of Solar Radiation by Mineral Dust within Liquid Cloud Drops. *J. Atmos. Solar-Terrestr. Phys.* **2015**, *133*, 121–128. [CrossRef]
13. Bozlaker, A.; Prospero, J.M.; Fraser, M.P.; Chellam, S. Quantifying the Contribution of Long-Range Saharan Dust Transport on Particulate Matter Concentrations in Houston, Texas, Using Detailed Elemental Analysis. *Environ. Sci. Technol.* **2013**, *47*, 10179–10187. [CrossRef]
14. Thompson, J.E. Photochemical and Climate Implications of Airborne Dust. In *Dust: Sources, Environmental Concerns and Control*; Nova Publishers: New York, NY, USA, 2011; ISBN 978-1-61942-566-8.
15. Redmond, H.; Dial, K.; Thompson, J.E. Light Scattering & Absorption by Wind Blown Dust: Theory, Measurement and Recent Data. *Aeolian Res.* **2010**, *2*, 5–26.

16. Koo, B.; Knipping, E.; Yarwood, G. 1.5-Dimensional volatility basis set approach for modeling organic aerosol in CAMx and CMAQ. *Atmos. Environ.* **2014**, *95*, 158–164. [CrossRef]

17. Jiang, J.; Aksoyoglu, S.; El-Haddad, I.; Ciarelli, G.; Denier van der Gon, H.A.C.; Canonaco, F.; Gilardoni, S.; Paglione, M.; Minguillón, M.C.; Favez, O.; et al. Sources of organic aerosols in Europe: A modelling study using CAMx with modified volatility basis set scheme. *Atmos. Chem. Phys. Discuss.* **2019**, *19*, 15247–15270. [CrossRef]

18. Wei, Y.; Cao, T.; Thompson, J.E. The Chemical Evolution & Physical Properties of Atmospheric Organic Aerosol: A Molecular Structure Based Approach. *Atmos. Environ.* **2012**, *62*, 199–207.

19. Gallardo, L.; Barraza, F.; Ceballos, A.; Galleguillos, M.; Huneeus, N.; Lambert, F.; Ibarra, C.; Munizaga, M.; O'Ryan, R.; Osses, M.; et al. Evolution of air quality in Santiago: The role of mobility and lessons from the science-policy interface. *Elem. Sci. Anth.* **2018**, *6*, 38. [CrossRef]

20. Power, A.L.; Tennant, R.K.; Jones, R.T.; Tang, Y.; Du, J.; Worsley, A.T.; Love, J. Monitoring Impacts of Urbanisation and Industrialisation on Air Quality in the Anthropocene Using Urban Pond Sediments. *Front. Earth Sci.* **2018**, *6*, 131. [CrossRef]

21. Eggleston, M.P.; Hackman, C.A.; Heyes, J.G.; Irwin, R.J.; Timmis, M.L. Trends in urban air pollution in the United Kingdom during recent decades. *Atmos. Environ. Part B Urban Atmos.* **1992**, *26*, 227–239. [CrossRef]

22. Thompson, J.E.; Hayes, P.L.; Jimenez, J.L.; Adachi, K.; Zhang, X.; Liu, J.; Weber, R.J.; Buseck, P.R. Aerosol Optical Properties at Pasadena, CA During CALNEX 2010. *Atmos. Environ.* **2012**, *55*, 190–200. [CrossRef]

23. Fathy El-Sharkawy, M.; Dahlawi, S.M. Study the effectiveness of different actions and policies in improving urban air quality: Dammam City as a case study. *J. Taibah Univ. Sci.* **2019**, *13*, 514–521. [CrossRef]

24. Air Pollution Levels Rising in Many of the World's Poorest Cities. Available online: https://www.who.int/news-room/detail/12-05-2016-air-pollution-levels-rising-in-many-of-the-world-s-poorest-cities (accessed on 24 November 2019).

25. Sanford, L.; Burney, J. Cookstoves illustrate the need for a comprehensive carbon market. *Environ. Res. Lett.* **2015**, *10*, 084026. [CrossRef]

26. Wathore, R.; Mortimer, K.; Grieshop, A.P. In-Use Emissions and Estimated Impacts of Traditional, Natural-and Forced-Draft Cookstoves in Rural Malawi. *Environ. Sci. Technol.* **2017**, *51*, 1929–1938. [CrossRef] [PubMed]

27. Lonati, G.; Giugliano, M.; Butelli, P.; Romele Ruggero Tardivo, L. Major chemical components of PM2.5 in Milan (Italy). *Atmos. Environ.* **2005**, *39*, 1925–1934. [CrossRef]

28. Wei, Y.; Ma, L.; Cao, T.; Zhang, Q.; Wu, J.; Buseck, P.R.; Thompson, J.E. Light Scattering and Extinction Measurements Combined with Laser-Induced Incandescence for the Real-Time Determination of Soot Mass Absorption Cross Section. *Anal. Chem.* **2013**, *85*, 9181–9188. [CrossRef] [PubMed]

29. Hueglina, C.; Gehriga, R.; Baltensperger, U.; Gysel, M.; Monn, C.; Vonmont, H. Chemical characterisation of PM2.5, PM10 and coarse particles at urban, near-city and rural sites in Switzerland. *Atmos. Environ.* **2005**, *39*, 637–651. [CrossRef]

30. Thompson, J.E.; Myers, K. Cavity Ring-Down Lossmeter using a Pulsed Light Emitting Diode Source and Photon Counting. *Meas. Sci. Technol.* **2007**, *18*, 147–154. [CrossRef]

31. Kumar, A.; Gupta, T. Development and Field Evaluation of a Multiple Slit Nozzle-Based High Volume PM2.5 Inertial Impactor Assembly (HVIA). *Aerosol Air Qual. Res.* **2015**, *15*, 1188–1200. [CrossRef]

32. Li, J.; Biswas, P. Optical Characterization Studies of a Low-Cost Particle Sensor. *Aerosol Air Qual. Res.* **2017**, *17*, 1691–1704. [CrossRef]

33. Sousan, S.; Koehler, K.; Thomas, G.; Park, J.H.; Hillman, M.; Halterman, A.; Peters, T.M. Inter-comparison of Low-cost Sensors for Measuring the Mass Concentration of Occupational Aerosols. *Aerosol Sci. Technol.* **2016**, *50*, 462–473. [CrossRef]

34. Gunawan, T.S.; Munir, Y.M.; Kartiwi, M.; Mansor, H. Design and Implementation of Portable Outdoor Air Quality Measurement using Arduino. *Int. J. Electr. Comput. Eng.* **2018**, *8*, 280–290. [CrossRef]

35. Reilly, M.K.; Birner, T.M.; Johnson, G.N. Measuring Air Quality using Wireless Self-Powered Devices. *Inst. Electr. Electr. Eng.* **2015**, 267–272. [CrossRef]

36. Zamora, L.M.; Xiong, F.; Gentner, D.; Kerkez, B. Field and Laboratory Evaluations of the Low-Cost Plantower Particulate Matter Sensor. *Environ. Sci. Technol.* **2018**, *53*, 838–849. [CrossRef] [PubMed]

37. Liu, H.; Schneider, P.; Haugen, R.; Vogt, M. Performance Assessment of a Low-Cost PM2.5 Sensor for a near Four-Month Period in Oslo, Norway. *Atmosphere* **2019**, *10*, 41. [CrossRef]

38. Liu, D.; Zhang, Q.; Jiang, J.; Chen, D.-R. Performance calibration of low-cost and portable particular matter (PM) sensors. *J. Aerosol. Sci.* **2017**, *112*, 1–10. [CrossRef]
39. Cao, T.; Thompson, J.E. Portable, Ambient PM2.5 Sensor for Human and/or Animal Exposure Studies. *Anal. Lett.* **2016**, *50*, 712–723. [CrossRef]
40. Agrawaal, H.; Thompson, J.E. Wireless Transmission and Logging of Measurement Data through Cellular Networks. *NCSLI Meas.* **2019**. [CrossRef]
41. Sharp GP2Y1010AU0F Datasheet. Available online: https://www.sparkfun.com/datasheets/Sensors/gp2y1010au_e.pdf (accessed on 24 November 2019).

 International Journal of
*Environmental Research
and Public Health*

Article

The Use of the Internet of Things for Estimating Personal Pollution Exposure

Keith April G. Arano [1], Shengjing Sun [2], Joaquin Ordieres-Mere [2] and and Bing Gong [3,*]

[1] Department of Management, Economics and Industrial Engineering, Politecnico di Milano, Via Lambruschini 4/B - (Building 26/B), 20156 Milan, Italy
[2] Department of Industrial Engineering, Business Administration and Statistics, E.T.S de Ingenieros Industriales, Universidad Politécnica de Madrid, Calle de Jose Gutierrez Abascal 2, 28006 Madrid, Spain
[3] Jülich Supercomputing Center Forschungszentrum Jülich GmbH, Wilhelm-Wohnen-Str, 52425 Jülich, Germany
* Correspondence: b.gong@fz-juelich.de

Received: 1 June 2019; Accepted: 29 July 2019; Published: 28 August 2019

Abstract: This paper proposes a framework for an Air Quality Decision Support System (AQDSS), and as a proof of concept, develops an Internet of Things (IoT) application based on this framework. This application was assessed by means of a case study in the City of Madrid. We employed different sensors and combined outdoor and indoor data with spatiotemporal activity patterns to estimate the Personal Air Pollution Exposure (PAPE) of an individual. This pilot case study presents evidence that PAPE can be estimated by employing indoor air quality monitors and e-beacon technology that have not previously been used in similar studies and have the advantages of being low-cost and unobtrusive to the individual. In future work, our IoT application can be extended to include prediction models, enabling dynamic feedback about PAPE risks. Furthermore, PAPE data from this type of application could be useful for air quality policy development as well as in epidemiological studies that explore the effects of air pollution on certain diseases.

Keywords: Personal Air Pollution Exposure (PAPE); air pollution monitoring; IoT; Air Quality Decision Support System; health impact

---

1. Introduction

Pollution and various forms of ecosystem contamination continue to be pressing issues across the globe [1]. China's rapid increase in urbanization in the last three decades, for example, has resulted into environmental challenges where air pollution is the leading problem [2]. Protecting the environment, therefore, is a serious undertaking that faces businesses and governments today. In recent years, there has been increasing pressure on institutions to measure and report environment-related parameters [3]. For this reason, there has been a significant increase in the number of reporting instruments used globally, of which sustainability reporting instruments account for the largest share owing to government regulations [4].

Environmental sustainability now underpins the policy-building initiatives of government institutions and businesses alike. In developed countries such as those of the European Union (EU), air pollution damage, which brings about a direct threat to public health, is expected to rise in the next decade. This has compelled the EU governments to give priority to air pollution level reduction above any other climate change policy plans [5]. In developing countries, however, there are still inadequate air quality policies and environmental monitoring plans. This is a major concern primarily because these are the regions that are more susceptible to increasing levels of air pollution [6]. There is therefore a challenge in finding economical solutions to monitor pollution levels and other relevant health parameters.

Air pollution from both outdoor and indoor sources constitutes the greatest environmental risk to human health around the globe [7]. About seven million premature deaths were attributed to air pollution in 2014, based on an estimate by the World Health Organization (WHO). It is projected that, by 2050, outdoor air pollution will be the number one cause of environment-related deaths worldwide [8]. For these reasons, governments across the globe have started to monitor the levels of major air pollutants, especially in metropolitan and urban areas.

In addition, people spend around 90% of their time indoors, and human exposure to indoor air pollutants may occasionally be more than 100 times higher than outdoor pollutant levels, according to the United States (US) Environmental Protection Agency (EPA). Indoor air pollution is equally detrimental, as statistics show that 4.3 million people per year die from exposure to household air pollution [9]. Exposure to poor indoor air is a significant cause of productivity loss in the US, as productivity decreases by 0.5 to 5% per workplace, generating a loss of 20 to 200 billion US dollars annually [10].

Monitoring Personal Air Pollution Exposure (PAPE), which refers to the amount (µg) of pollution being inhaled by an individual, has been a topic of growing interest worldwide, not only as a result of global health policies, but more importantly, due to the interest in understanding its effects on various cardiovascular and respiratory diseases. These diseases have been documented widely in existing epidemiological studies [11,12]. Nevertheless, this traditional evaluation of PAPE has not been directly undertaken for individuals, but rather for groups of people exposed to the annual average concentrations of pollution that are indicated by a network of fixed-site outdoor monitors.

Existing studies on the association of air pollution with different diseases [13,14] recognize the importance of measuring PAPE among individuals. By monitoring activity patterns, it is possible to establish correlations between different populations or levels of socioeconomic status and PAPE. Although there are novel methods to measure and model these exposures, the great variability in PAPE remains a major challenge [15] and provides a compelling case for research on the effects of air pollution on health. Dias and Tchepel [16] suggest that, in order to assess personal exposure, not only the spatial-temporal variability of urban air pollution should be taken into account but also the indoor exposure and the individual time-dependent activities should be measured.

The recent advances in data technology are expected to play major roles in the next decade, permitting easier access and analysis of data [1]. The Internet of Things (IoT), which is defined as the network of various ubiquitous devices that are capable of computation and communication over the Internet, has been gaining recognition in the development of advanced applications in the healthcare sector [17]. This wave of digital innovation is driving the healthcare industry and paving the way for cheaper, smaller and more efficient wearable technologies that monitor health indicators in real time [18].

Health systems around the world are under pressure to come up with economical solutions to existing problems [18]. These sensor technologies, coupled with sophisticated analytics, have the means to improve process efficiency and achieve cost reductions [19]. With this kind of digital innovation, an economical and scalable application can be built to help developing countries to measure PAPE and other health parameters to improve their current health risk assessment systems.

Although there was initially a concern about sharing private information from sensor technologies or other similar devices, the general public has started to accept the use of digital services, even for sensitive information such as health data [20]. In fact, in a consumer survey performed by Price Waterhouse Coopers in 2015, 83% of respondents indicated that they were willing to share data to aid in the diagnosis and treatment of diseases [21]. With this collaborative support from the general public, exploring IoT applications to test sensor-driven projects can greatly facilitate advances in the healthcare industry [22]. Insights gained from the surveillance of vital health information, such as PAPE, can establish a foundation for predictive, preventive, and personalized healthcare systems [18].

In line with these trends in the healthcare industry, this paper seeks to propose a framework for an Air Quality Decision Support System (AQDSS) and to develop an IoT application that measures PAPE based on this framework. The four sections of this paper are organized as follows. The first section provides a literature review of studies on key air pollutants, determination methods for air pollutants, and PAPE estimation techniques as well as the current opportunities and challenges in the field. The second part discusses the methodology, which includes the proposed framework and an IoT application that was tested by means of a case study. This is followed in the third part by the analysis and a discussion of the results of the case study. The fourth part presents conclusions that highlight the study's important contributions and directions for future research.

2. Literature Review

2.1. Key Air Pollutants

The key air pollutants that are currently being monitored by agencies such as the WHO, the EPA in the United States, and the European Environment Agency (EEA) in Europe, are particulate matter (PM_{10} and $PM_{2.5}$), ozone (O_3), nitrogen oxide (NO), nitrogen dioxide (NO_2), carbon monoxide (CO), sulfur dioxide (SO_2), volatile organic compounds (VOC), and benzene (C_6H_6). They are also frequently studied in academic research [23,24]. Although there is a substantial amount of monitoring data available for each of these pollutants. PM_{10} and $PM_{2.5}$ are considered to be the most widely studied air pollutants in the existing environmental risk and health literature. This is because PM poses one of the greatest risks to human health [25].

The indoor environment is a critical domain where an average person spends an estimated 90% of his or her time [26]. Thus, indoor air pollution is more likely to account for total population exposure than pollution from the outdoor environment [27]. While individuals are spending more and more time indoors, an assessment of the health impact of indoor air pollution has not been studied as extensively as the impact of outdoor air pollutants. One of the main reasons for this is the lack of indoor air quality monitoring information [28]. There are primary indoor air pollutants, which are recommended based on the EU (2008) directive for Clean Air and the WHO [28]. They are schematically listed as benzene, formaldehyde, naphthalene, nitrogen dioxide, polycyclic aromatic hydrocarbon, radon, trichloroethylene, and tetrachloroethylene. In the research community of indoor air quality monitoring and assessment, particulate matter, carbon dioxide, carbon monoxide, ozone, nitrogen oxide, formaldehyde, benzene, total volatile organic compound (TVOC), polycyclic aromatic hydrocarbon, and other VOCs have been extensively studied [29,30].

2.2. Determination Method for Air Pollutants

Most workspaces or industrial environments still apply traditional measuring strategies to assess occupational health and safety. These strategies are mainly based on the EPA Compendium of Methods [31] and the International Organization for Standardization (ISO) method, which rely on complex sampling and analysis techniques. These methods, such as Method-10A and IP-3A, require domain experts to prepare diffusive or passive samplers and are frequently replaced with new ones due to the limited equipment lifespan. Subsequently, the collected samplers are separated by gas chromatography and measured by mass-selective detector or multidetector techniques in a remote laboratory [32]. Moreover, to measure multiple pollutants, the equipment for each pollutant has to be prepared or bought from different manufacturers, which can lead to issues about data manipulation and integration. These aforementioned elements of the traditional measuring strategy restrict the sampling time to a short-term basis [33]. As indoor air quality varies from time to time due to changes in working conditions, human activity, and weather conditions, short term sampling cannot cover all kinds of variations. Therefore, long-term monitoring has become a need in the research community and practical applications such as Occupational Safety and Health (OSH) management.

The rapid development of IoT and sensor techniques enables light, low-cost, and real-time pollution monitoring solutions. The integration of IoT and the sensor network in air quality monitoring addresses the aforementioned gaps: short-term monitoring and complex air monitoring solutions. Recent studies on the development of indoor air quality monitoring systems have been undertaken on PM, carbon dioxide (CO_2), CO, and VOC. Moreover, IoT-based indoor air monitoring devices such as Foobot and AirVisual are already commercially available on the market.

2.3. IPAPE Measurement Techniques

As noted previously, there is a growing interest in measuring PAPE at the individual level. At present, there is a wide range of low-cost sensor technologies [34] that can be leveraged to implement large scale monitoring networks by means of complex measurement techniques [35]. PAPE requires tracking of a person's activity patterns to learn the time and location of their exposure to pollution concentrations as well as the duration of exposure and nature of the pollutants. This is necessary to understand the probable effects on health of the exposure [36].

The different PAPE measurement techniques that have been developed in the last decade can be grouped into three categories. The first group is the traditional method in which pollution data are collected from fixed-site outdoor monitors and assigned to the home address of the individual through spatial interpolation techniques. Examples include Land Use Regression (LUR) [37], Inverse Distance Interpolation [38], and the geostatistical Kriging algorithm [39]. Numerical models, such as the Community Multiscale Air Quality (CMAQ) model and the Urban Atmospheric Dispersion model (DAUMOD), were proposed for regional air pollution modeling prediction in previous studies [40]. However, the expensive computational cost and failure to capture pollution variability make them inadequate for the application of modeling in real time in urban areas where there are severe photochemical pollution conditions. Graz Lagrangian Model (GRAL) is another advanced mathematical model that can handle the motion of pollution in buildings and complex terrains [41]. A major drawback of these types of models, however, is the need to have accurate information about emissions, meteorological data, and the structural and geographical figures of the area, which may not always be available in high resolution [42]. While the performance of the spatial interpolation methods may significantly drop in dynamic terrains such as in urban environments, they have still been used widely in recent studies [43,44] of areas where the detailed information needed for complex numerical models (e.g., street-based monitoring) is still unavailable.

In summary, these methods are inadequate, as they do not address the issue of the individual's spatio-temporal PAPE variability [45] and neglect indoor air pollution. Accordingly, this has led researchers to explore new techniques that can provide more accurate measures of PAPE.

The second group of techniques, which is built on the traditional method but addresses the issue of exposure variability, takes into account the activity patterns by tracking an individual's location. It incorporates indoor pollution data based on the amount of time spent indoors. A commonly used indoor pollution measurement method is the indoor/outdoor ratio [37]. Other techniques, such as modeling based on data from vehicle type and emissions have also been proposed [46]. With respect to activity tracking, different tools have been used in studies to track the location and activity patterns of an individual. These include Global Positioning Systems (GPS) [35], public WiFi networks [37], and accelerometers [45]. A common characteristic shared by these activity tracking tools is the use of a mobile device, particularly a smartphone. This mobile technology has proved to be an enabling tool in the health industry with its ability to access data anytime from anywhere [47]. Although this group of PAPE measurement techniques is an improvement from the first group, it still faces the issue of pollution variability and measurement accuracy with its reliance on fixed-site outdoor monitors and indoor/outdoor ratios alone [48].

The last group of techniques stems from the two previously discussed groups but further captures the issue of indoor pollution measurement accuracy and the spatio-temporal resolutions of data from fixed-site outdoor monitors. The periodic measurements by these fixed-site outdoor monitors by nature have low spatial resolution and do not address the issue of variability in pollution concentration [49]. Since the indoor environment has a much greater impact on human health than the outdoor environment [50], it is essential to have solutions that can provide more accurate measures of indoor air pollution instead of employing the traditional method of using the indoor/outdoor ratio.

Personal exposure measurements can be performed directly and indirectly [51]. Passive samplers are widely used in personal sampling, since they have the merits of being light, electricity-free, and wearable. Passive samplers exist for nitrogen dioxide, carbon monoxide, VOC, ozone, sulfur dioxide, and formaldehyde [52]. Due to sampler lifespan, the sampling time usually lasts from a few days up to one week [53].

On the other hand, using a micro-environmental model is an indirect way of assessing personal exposure. In daily life, people move around and are exposed to various levels of pollutants in various locations. The term "micro-environment" is defined as a chunk of air space with a homogeneous pollutant concentration [54]. Such a micro-environment can be an indoor location (bedroom, kitchen, etc.) or workplace location (meeting room, office, printing room, etc.). The spatio-temporal individual time-activity crossing in micro-environments is tracked through questionnaires or time-activity diaries (TADs).

The key to measuring individual pollution exposure is to track an individual's activities in both the space and time dimensions. GPS technology is the ideal technology, and it has been used successfully for this purpose. Some well-designed integrations of GPS devices and portable pollution monitors have been proposed by some studies [35,55] to determine the potential exposure at the individual level. However, in indoor environments, GPS technology does not function as well as it does outdoors.

Therefore, more extensive approaches have been developed such as the use of mobile sensors (i.e., handheld, USB-pluggable smartphone sensors, wearable sensors) to monitor PAPE indoors. Studies [56,57] that have included this group of PAPE measurement techniques have managed to address the most relevant issues of pollution variability by employing mobile sensors. Beacon technology offers a promising solution for indoor location tracking. Furthermore, the use of indoor monitors instead of mobile sensors, which are often used in similar studies, eliminates the inconvenience of carrying a device around. The use of an indoor monitor and e-beacons also enables unobtrusive and low-cost collection of pollution data for multiple individuals, in contrast to a mobile sensor, which only collects data for a single individual.

2.4. Opportunities and Challenges

Although continuous technological advancements have enabled researchers to propose solutions that provide measures of PAPE, the issue about the cost and scalability of such methods remains to be addressed. The most recent approach, as discussed in the third group of PAPE measurement techniques, employs mobile sensors that the individual carries around. Although these mobile sensors are able to provide better spatial resolution of pollution data, the willingness of individuals to carry these sensors is still a challenge, in addition to the cost and scalability issues of the method.

The trade-off between cost and quality of pollution data continues to be a point of discussion among studies. The proposed PAPE measurement techniques that are currently available in the literature are limited to PAPE estimation alone and, therefore, fail to provide a more comprehensive view of the entire AQDSS. Thus, there is an opportunity to further explore the use of existing technologies to enable the development of a more comprehensive PAPE measurement technique that is able to provide a preventive, predictive, and personalized system.

Within similar studies on the measurement of PAPE, there are some proposed conceptual frameworks [35,36] and system architectures [56,58]. However, they are centered primarily on PAPE measurement and the potential health impacts. In this paper, we present a comprehensive framework that encompasses not only PAPE measurement but provides a holistic view of the entire AQDSS. As a proof of concept for this framework, we also develop a low-cost and unobtrusive IoT application for measuring PAPE that addresses the gaps in the currently available solutions.

3. Methodology

3.1. Framework

Figure 1 shows the proposed framework for an AQDSS. There are three key stakeholders identified, namely the individual, the healthcare industry, and the government. The three pillars at the center represent the elements that are directly linked to the government. They include pollution laws, sectorial regulations, and incentives, all of which make up the air quality monitoring policies of the government.

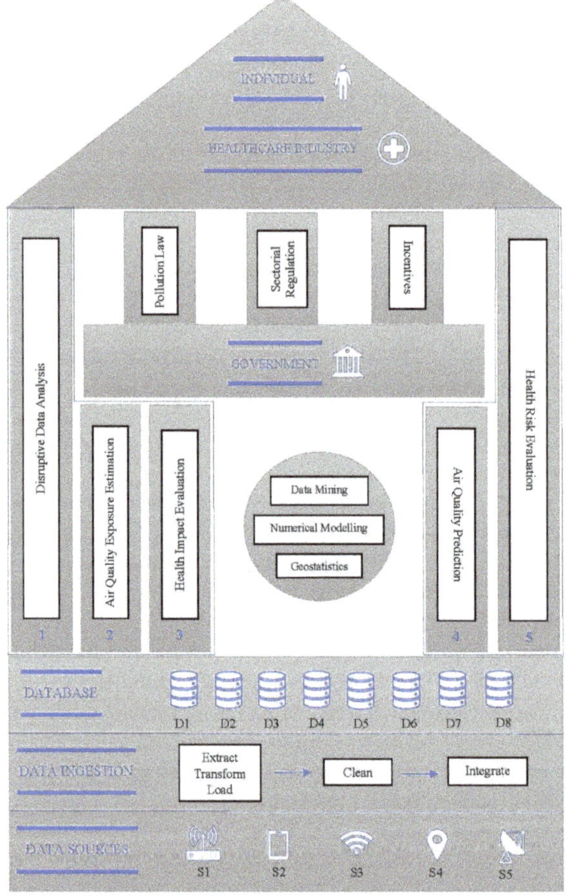

Figure 1. Proposed framework for the Air Quality Decision Support System (AQDSS).

As pointed out previously, these government regulations play an important role, as they largely support and drive policies that enable the measurement and access to air quality data.

The other two stakeholders are the individuals and the healthcare sector, which form the apex of the framework. They are supported by five layers of activities, as illustrated by the pillars on both sides. The first three on the left correspond to the analysis of past data to estimate PAPE and their related health impacts. The two pillars on the right represent future possibilities of forecasting PAPE and the associated health risks.

Although the estimation of PAPE is fundamental to the entire AQDSS, its associations with forecasting and as a predictive health risk assessment system are noteworthy. Air quality forecasting techniques are already being explored in current research [59,60] in environmental modeling literature, where their important contributions to the development of control measures to prevent damage to human health have been highlighted.

This proposed framework can be adopted to aid in the development of an AQDSS and various IoT applications. For instance, consider a mobile application that allows an individual to select the best route to travel from home to work that minimizes the risk to pollution exposure. This could be achieved by employing different modeling techniques to continuously analyze real-time air quality data and forecast PAPE values for each of the possible routes to the destination. Actual PAPE data that are stored in the database can also be used for epidemiological studies and air quality policy development. This could be one of the applications of the proposed framework when all identified pillars are fully employed. In this paper, however, as a proof of concept, we focus mainly on PAPE measurement (pillar 2) in accordance with the system architecture that is illustrated on the base of the framework. In order to manage sensor data in an interoperable way, this implementation considers the Web Service Description Language provided by the Sensor Observation Service v2.1 (SOS) from the SOS-OGC consortium. This standard defines a Web service interface which allows observation queries, sensor metadata, as well as representations of observed features. Furthermore, this standard defines a means to register new sensors and to remove existing ones. Also, it defines operations to insert new sensor observations. The feasibility was assessed through the developed case study.

3.1.1. System Architecture

As indicated in the framework, the base shows the set of activities that are related to the gathering and management of all air quality and personal data. This groundwork is required for the entire system to function. There are five different sources of data. They are the (S1) outdoor pollution monitors, (S2) location tracking application, (S3) indoor pollution monitors, (S4) e-beacons, and (S5) meteorological monitors. S1 and S2 are intended for outdoor pollution modeling, and S3 and S4 are for indoor pollution modeling. We also consider meteorological data, as they are relevant for air quality prediction studies [60,61].

The data are extracted from the mentioned data sources and stored in a database management system. Data mining, numerical modeling, and geostatistics, as shown in the center of the framework are the key activities that support the entire system, as it is a continuous process to discover and analyze spatio-temporal data.

3.1.2. PAPE Measurement

There are essentially three elements to consider when measuring PAPE. They are (1) outdoor pollution, (2) indoor pollution, and (3) the individual's location pattern. With respect to the outdoor pollution, the mobile-phone-based tracking app provides the time and location data of the individual in the outdoor environment. An outdoor pollution map is created by using potentially different strategies, such as

- Numerical-modeling-based dispersion models [62,63],
- Big data, machine-learning-based models [60,64],
- Geostatistic-based techniques, like Kriging [65,66].

Each of these techniques has its specific advantages and limitations, and its consideration in a specific application will support its choice. Actually, the latest family of methods has specific advantages, as it is suitable for working with the fixed network of pollution stations the city has implemented. Indeed, it also deals with the limitations of sparse data, as Data fusion can increase the reliability of data as well as it can contribute to dealing with local effects like street canyons, etc., by using the street granularity-based IoT air quality stations some cities are deploying, such as Airbox in Taipei [67] and Array of Things (AoT) sensor boxes in Chicago [68].

For the duration of time that an individual is outdoors, the corresponding pollution data are estimated by superimposing the developed outdoor pollution map over the collected location pattern data.

For the indoor pollution, the e-beacons indicate the period when the individual is indoors, and the indoor air quality monitors provide the corresponding air quality data, when available. Failing this, outdoor information will be used by default. The integration of personal mobiles and fixed e-beacons located in different indoor micro-environments enables the individual's time-location information to be understood. The corresponding time-location knowledge combined with location-specific indoor air quality information collected from air monitoring devices can provide a detailed picture of personal exposure in the indoor environment.

Both outdoor and indoor data are then integrated, and statistical modeling techniques are employed to either estimate or forecast the individual's PAPE.

3.2. Madrid Case Study

In order to assess the feasibility of the proposed IoT application that measures PAPE and contributes to empowering users because of the relevant figures provided at the personal level, we conducted a case study to analyze significant functionalities.

3.2.1. Study Area

The study area was the City of Madrid, which is the capital of Spain, as well as its largest municipality. It was the first city in Spain to have air quality monitoring stations and has always been at the forefront of the fight against air pollution. In response to the most recent EU directive (Directive 2008/50/EC) regarding the establishment of limits to major air pollutants, the Madrid government has committed to maintaining acceptable pollution levels by continuous air quality monitoring.

The Madrid air pollution monitoring network consists of 24 fixed-site outdoor monitors (Figure 2). The hourly averaged measurements of SO_2, CO, NO, NO_2, PM_{10}, $PM_{2.5}$, C_6H_6, toluene (C_6H_5–CH_3), hexane (C_6H_{14}), propene (C_3H_6), m-xylene, o-xylene, and methane (CH_4) hydrocarbons can be downloaded free of charge from the official open data website of the *Ayuntamiento de Madrid* [69]. Meteorological data, such as temperature, humidity, ultraviolet radiation, pressure, solar radiation, rainfall, precipitation, diffuse solar radiation, global radiation, wind speed, and wind direction can also be accessed through the website of the *Agencia Estatal de Meteorologia* [70].

Figure 2. Air Quality Monitoring Network of Madrid.

3.2.2. Data Collection

In this case study, S1, S2, S3, and S4 data sources were used and meteorological data were excluded (see Figure 1). We had one individual volunteer whose activities were monitored during the study period.

For this particular case, Madrid does not yet implement the street-based pollution monitoring strategy, but based on similar studies [43,44], the research team adopted the geostatistics-based approach, as it becomes linear scalable with time and is suitable for integrating additional data sources. Therefore, outdoor pollution figures were downloaded from the mentioned open data website of Madrid City Hall. For location tracking, we used the mobile app Moves [71] in which time, location, and activity were accessed through an open Application Programming Interface (API). Other similar open source mobile apps are widely available, such as OwnTracks [72], Miataru [73], and Geo2Tag [74].

For the indoor pollution, Foobot indoor monitors were used. One of them was placed in the individual's workplace, as this is where she spends most of her indoor time. The indoor pollution data were retrieved from Foobot's API [75]. The e-beacon devices were placed in proximity to the indoor monitors. They helped us to determine whether the individual was within the indoor vicinity. The e-beacon data were broadcasted through Eddystone, an open-source beacon format, and were retrieved through an app that we developed in Cordova [76]—a free and open-source platform for building mobile applications.

All of these data sources promote the scalability of the proposed IoT application, as most are publicly available without charge. The only costs incurred were for the indoor monitor and e-beacons. E-beacons, however, are low-cost, small enough to attach to any surface, and are finding an increasing number of location-based applications in various industries such as retail and transportation as well as in households [77]. Hence, beacon technology offers a promising solution for indoor location tracking.

All data that were collected from the mentioned data sources were processed as indicated in the Data in Brief Collection documents that were submitted to the journal for this paper. The developed code can be also found in a public repository [78].

The selection of the pollutants used for the PAPE estimation was based primarily on the data provided by the devices, which also agreed with the data on the most common air pollutants that have been widely studied previously [23,24]. Table 1 shows the available pollutants for each of the data sources used.

Table 1. Available Pollutants for each Data Source.

Source	Pollutant	Unit
Indoor Monitor	$PM_{2.5}$	$\mu g/m^3$
	CO_2	ppm
	VOC	ppb
Outdoor Monitor	$PM_{2.5}$	$\mu g/m^3$
	PM_{10}	$\mu g/m^3$
	CO	$\mu g/m^3$
	NO_2	$\mu g/m^3$
	SO_2	$\mu g/m^3$
	O_3	$\mu g/m^3$
	NO_x	$\mu g/m^3$

3.2.3. Outdoor Pollution Modeling

Existing studies of PAPE essentially rely on modeling techniques in which data collected from fixed-site outdoor monitors are used to estimate pollution at specific geographic locations. To create an outdoor pollution map, there are several alternative methods. These include using micro meteorological numerical-based models (WRF, CMAQ, etc.) [79], or machine-learning-based models [64]. However, for the sake of simplicity and considering the computational costs and the number of potential users, we adopted some classical but still cost-effective approaches like the Inverse Distance Weighting (IDW), Simple Kriging, Ordinary Kriging, and Co-Kriging algorithms.

Table 2 shows the formulae and main characteristics of these techniques in which $\hat{z}_0$ is the measured value at the prediction location, λ_i is the weight of the measured value at the ith location, and X_i is the measured value at the ith location. The parameters that were tuned are also indicated in the table.

All three methods estimate the value at a particular location by assigning a weight of the surrounding known values and calculating the weighted sum of the data. These techniques differ mainly in the calculation of the assigned weight λ_i. Kriging, which is a geostatistical method, offers advantages over other interpolation techniques, as it provides an interpolation error estimate, and it is an exact interpolation. The interpolations are based on weights that do not depend on data values [80]. The advantages of the deterministic interpolation technique IDW, on the other hand, are that it is simple, intuitive, and computes the interpolated values quickly [81]. We created the outdoor pollution map by employing these three interpolation techniques in R, an open-source statistical modeling software.

Table 2. Techniques Used for Outdoor Pollution Modeling.

Technique	IDW	Simple Kriging	Ordinary Kriging	Co-Kriging
Formula	$\hat{z}_0 = \sum_{i=0}^{n} \lambda_i X_i$	$\hat{z}_0 - c = \sum_{i=0}^{n} \lambda_i (X_i - c)$	$\hat{z}_0 = \sum_{i=0}^{n} \lambda_i X_i$	$\hat{z}_0 = \sum_{i=0}^{n} \lambda_i X_i + \sum_{j=0}^{n} \beta_j t_j$
Characteristics	The weight, λ_i, depends solely on the distance to the prediction location.	Assumes a constant and known mean c of the samples. The weight, λ_i, depends on the use of a fitted model to the measured points, the distance to the prediction location, and the spatial relationships among the measured values around the prediction location.	Condition that $\sum_{i=0}^{n} \lambda_i = 1$ assumes a constant and unknown mean of the samples. The weight, λ_i, depends on the use of a fitted model to the measured points, the distance to the prediction location, and the spatial relationships among the measured values around the prediction location.	t_j is the secondary regionalized variable which is co-located with the target variable t_j. The weight β_j assigned to t_j varies between 0 to 1.
Parameters	Idp $0.1, 0.3, 0.5, 0.8, 2, 5$	Sill $1, 10, 100, 250, 500, 600, 700, 800, 900$ Range $0.1, 0.5, 1, 10, 20, 50, 80$ Nugget $0.00001, 0.0001, 0.001, 0.01, 0.1, 1, 10, 100, 200, 50$ Beta $0.05, 0.12, 0.2, 0.5, 0.9, 1.5, 3, 1$ Variogram Model: Gaussian, Circular, Exponential		

1. Optimal Parameters and Model Selection

   In order to select the optimal parameters and the best modeling technique for each of the hourly outdoor pollution datasets, a 5-fold cross validation was performed to avoid overfitting. For each of the 24-hourly datasets and each of the three modeling techniques and all combinations of their respective parameters, the selection of optimal values was based on the root-mean-squared-error (RMSE) metric. The dataset was separated into two parts, training and testing, which were used to fit the model and calculate errors, respectively. The parameters and the model that provided the least RMSE were selected.

   For the Simple and Ordinary Kriging techniques, the weights λ_i were derived by fitting a covariance function or variogram. First, a graph of the empirical variogram was plotted and a model was fitted to the points based on this plot. Table 3 shows the different models and functions from which to choose when fitting a model to the empirical variogram. Based on the 5-fold cross validation, the Gaussian Model was selected as the optimal configuration.

2. Outdoor Pollution Map

Similar to [43], an hourly outdoor pollution map was created that was based on the identified optimal parameters and modeling technique for each respective hour. Figure 3 shows an example of the pollution maps based on the $PM_{2.5}$ pollution data on 2017-03-24. It shows that, from midnight to the morning at around 6:00, the highest pollution levels consistently occurred in the southwestern part of the city and moved towards the north with maximum levels that ranged from 8 to 12 $\mu g/m^3$. Concurrently, high pollution levels were also experienced in the northwestern part of the city at midnight and in the northeastern part at 01:00 in the morning.

The selection of time frequency (hourly-based in this case) also impacts the accuracy, depending on how spiky the pollution looks. In Madrid, the pollution sources are strongly related to traffic and then variations are smooth [82]. Therefore, hourly-based frequency is a rather convenient basis for calculations.

Table 3. Available Pollutants for each Data Source.

Model	Function
Circular	$semivar(d) = c_0 + c(1 - \frac{2}{\pi}cos^{-1}(\frac{d}{\alpha}) + \sqrt{1 - \frac{d^2}{\alpha^2}}\ 0 < d \leq \alpha$ $semivar(d) = c_0 + cd > \alpha$ $sermivar(0) = 0$
Spherical	$semivar(d) = c_0 + c(\frac{3d}{2\alpha} - \frac{1}{2}(\frac{d}{\alpha})^3\ 0 < d \leq \alpha$ $semivar(d) = c_0 + cd > \alpha$ $sermivar(0) = 0$
Exponential	$semivar(d) = c_0 + c(1 - e^{\frac{-d}{r}})\ d > 0$ $sermivar(0) = 0$
Gaussian	$semivar(d) = c_0 + c(1 - \exp(\frac{-d^2}{r^2}))\ d > 0$ $sermivar(0) = 0$

d = distance between two locations, c_0 = y-intercept, α = range.

3.2.4. Indoor Pollution Modeling

The main data sources used to model the indoor pollution were the e-beacons and indoor monitor. The timestamps recorded from the e-beacons provide the time when the individual was detected indoors.

In this study, we refer to "indoor" as the work location, since the indoor monitor was only present at the individual's workplace. The "outdoor" environment, on the other hand, refers to any other location outside the workplace. To obtain the corresponding pollution values during these periods, each of these timestamps was matched to the closest timestamp logged from the Foobot device. As illustrated in Figure 4, the pollution values were then aggregated in time periods based on Equation (1) on the assumption that, if the difference between two sequential timestamps recorded on the e-beacons was more than 10 min, the individual was outdoors and a new indoor period would start. For instance, in Figure 4, from 2017-03-24 at 14:37:46 to 2017-03-24 at 14:44:59, the pollution levels were aggregated, since the timestamp immediately following 2017-03-24 at 14:44:59 is 2017-03-24 at 14:59:00 and the difference is longer than 10 min. Therefore, during the period between 2017-03-24 at 14:44:59 and 2017-03-24 at 14:59:00, the individual was outdoors and the new indoor period resumed at 2017-03-24 14:59:00.

Similarly, micro-environments (office, printing room, meeting room) in the workplace could be replicated by deploying e-beacons and air monitoring devices in all available micro-environments.

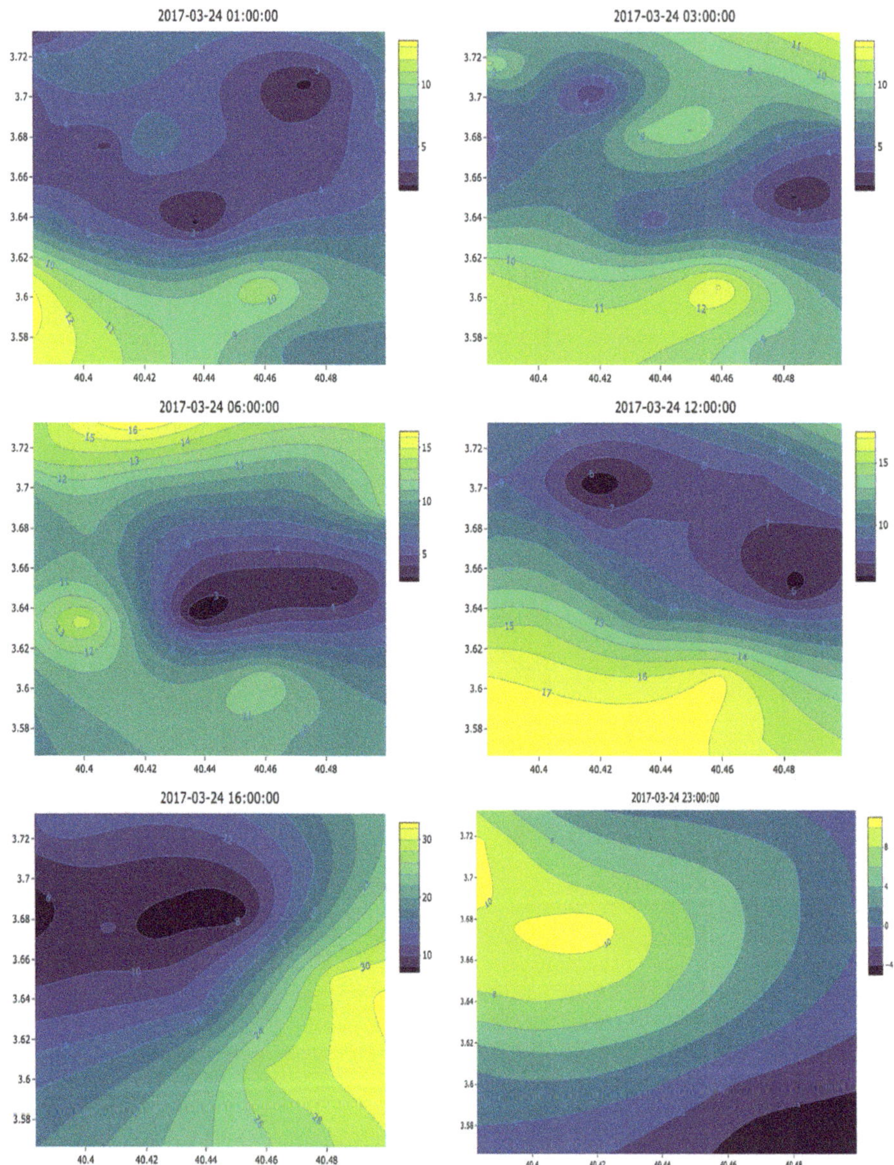

Figure 3. Co-kriging interpolation of PM$_{2.5}$ on 24 March 2017.

The PAPE Exposure(p) in period p inhaled by the individual was calculated by multiplying the pollution value SZ(p) by the respective minute ventilation (VE) value using Equation (2).

$$SZ(p) = \sum_{t=i}^{n} \frac{1}{2}(Z_{t_{i+1}} + Z_{t_i})(t_{i+1} - t_i) \tag{1}$$

where $t_{i+1} - t_i < 10$ mins, and $SZ(p)$ is the fully aggregated pollution value during the period from time t_i to t_n. This period is named p.

$$Exposure(p) = SZ(p) * VE \qquad (2)$$

where $VE \in (VER, VEW, VERT, VEC)$

VER = VE for activity type "run";
VEW = VE for activity type "walk";
VERT = VE for activity types "rest" and "transport";
VEC = VE for activity type "cycle".

Integrated Foobot and Beacon Data

Foobot Time	PM$_{2.5}$	CO$_2$	VOC	Beacon Time
....	....	....	....	....
3/24/2017 13:58:38	5.05	2542.50	701.00	3/24/2017 13:59:01
3/24/2017 13:58:38	5.05	2542.50	701.00	3/24/2017 14:00:01
3/24/2017 13:58:38	5.05	2542.50	701.00	3/24/2017 14:01:02
3/24/2017 14:03:35	4.82	2368.00	653.00	3/24/2017 14:03:02
3/24/2017 14:38:23	3.90	2214.00	611.00	3/24/2017 14:37:46
3/24/2017 14:43:23	3.44	2335.50	644.50	3/24/2017 14:44:59
3/24/2017 14:58:17	2.52	2490.00	687.00	3/24/2017 14:59:00
3/24/2017 14:58:17	2.52	2490.00	687.00	3/24/2017 15:00:00
3/24/2017 15:08:14	3.90	2517.00	694.00	3/24/2017 15:07:02
3/24/2017 15:13:11	4.59	2506.50	691.50	3/24/2017 15:11:02
3/24/2017 15:13:11	4.59	2506.50	691.50	3/24/2017 15:12:02
....	....	....	...	....

Aggregated Indoor Pollution Values

Start Time	End Time	SPM$_{2.5}$	SCO$_2$	SVOC
....	....	....	....	....
3/24/2017 13:16:59	3/24/2017 14:03:02	268.66	94846.21	26177.55
3/24/2017 14:37:46	3/24/2017 14:44:59	28.72	19068.53	5262.14
3/24/2017 14:59:00	3/24/2017 15:58:07	273.88	146118.00	40308.73
3/24/2017 16:09:41	3/24/2017 17:00:49	275.26	152560.10	42077.58
3/24/2017 17:12:51	3/24/2017 17:41:56	253.87	88466.63	24391.58
3/24/2017 17:56:58	3/24/2017 18:32:01	278.09	103618.80	28567.60
3/24/2017 18:46:02	3/24/2017 19:19:03	193.88	80739.33	22272.76
...	....	....	....	....

Figure 4. Aggregation of Indoor Pollution Values.

3.2.5. Indoor and Outdoor Pollution Integration

The individual's location was tracked through the Moves mobile application. The recorded data from this tracking app include the starting and ending times, latitude, longitude, and activity type, as shown in Table 4. To obtain the corresponding pollution values for these periods, the time and location records were matched against the interpolated values from the created outdoor pollution map. The resulting outdoor pollution data were then matched against the aggregated indoor pollution values in Figure 4, in which the outdoor data were replaced by the corresponding indoor data.

Table 5 shows the resulting individual's indoor and outdoor PAPE values for PM$_{2.5}$ with the respective period (i.e., starting and ending times), location (i.e., longitude and latitude), environment type (i.e., indoor or outdoor), activity type (i.e., transport, rest, walk, run, cycle), and minute ventilation (VE). The PAPE values are indicated in its last column "Exposure".

VE (m^3/min) measures the volume of gas inhaled by an individual. It varies with the type of activity. The type of activity or travel mode may have a significant effect on the exposure values [83,84] and, hence, it is important to account for VE. We obtained the VE values from a study done by [85] on human inhalation rates. The types of activities in the tracking app include "transport", "walk", "run", and "cycle" and are based primarily on the speed of movement of the individual. In this study, for the time periods that lack one of these types of activity data, we assumed that the individual was at "rest" (i.e., sleeping, sitting, etc.). Since VE is based primarily on the body movement of the individual, we used the same VE values for both activity types "transport" and "rest".

3.2.6. Practical Application

To illustrate a possible IoT application [86,87] that can be developed using the proposed framework, we identified different travel routes and their corresponding forecasted PAPE values [60] that give the individual an opportunity to select a travel route that minimizes the risk of exposure to pollution.

As an example, we selected an entry in Table 5 for the time period 12:04:30 to 12:23:44 on 24 March 2017, in which the individual was outdoors and in transport mode. During this selected period, by using the starting and ending location data that the tracking app provided, we identified alternative routes using the ggmap package in R.

From this package, the estimated travel time and route locations (i.e., latitude and longitude) were obtained. Then, based on these specific time and location data, the corresponding pollution values were taken from the previously interpolated outdoor pollution values.

Table 4. Data from Tracking App.

Start Time	End Time	Latitude	Longitude	Activity
2017-03-24 00:00:00	2017-03-24 11:55:37	40.4612	−3.7093	Rest
2017-03-24 11:55:37	3/24/2017 11:59:20	40.4592	−3.7106	Walk
2017-03-24 11:59:20	3/24/2017 12:04:30	40.4571	−3.7118	Rest
2017-03-24 12:04:30	3/24/2017 12:23:44	40.4486	−3.7006	Transport
2017-03-24 12:23:44	3/24/2017 19:52:00	40.4400	−3.6894	Rest
2017-03-24 19:52:00	3/24/2017 20:08:40	40.4506	−3.6994	Transport
2017-03-24 20:08:40	3/24/2017 21:13:07	40.4612	−3.7093	Rest
2017-03-24 21:13:07	3/24/2017 21:21:55	40.4594	−3.7105	Walk
2017-03-24 21:21:55	3/24/2017 22:32:59	40.4575	−3.7117	Rest
2017-03-24 22:32:59	3/24/2017 22:42:51	40.4594	−3.7105	Walk
2017-03-24 22:42:51	3/25/2017 00:00:00	40.4612	−3.7093	Rest

Table 5. Integrated Indoor and Outdoor Personal Air Pollution Exposure (PAPE).

Start	End	Latitude	Longitude	$PM_{2.5}$ ($\mu g/m^3$ * min)	Environment	Activity	VE (m^3/min)	Exposure (μg)
2017-03-24 0:00	2017-03-24 11:55	40.461	−3.709	5354.15601	Outdoor	Rest	0.00893	47.81261
2017-03-24 11:55	2017-03-24 11:59	40.459	−3.711	13.9543	Outdoor	Walk	0.01326	0.18503
2017-03-24 11:59	2017-03-24 12:04	40.457	−3.712	20.08333	Outdoor	Rest	0.00893	0.17934
2017-03-24 12:04	2017-03-24 12:23	40.449	−3.701	111.43482	Outdoor	Transport	0.00893	0.99511
2017-03-24 13:16	2017-03-24 14:03	40.43999	−3.68938	268.65655	Indoor	Rest	0.00893	2.3991
2017-03-24 14:03	2017-03-24 14:37	40.44	−3.689	347.06125	Outdoor	Walk	0.01326	4.60203
2017-03-24 14:37	2017-03-24 14:44	40.43999	−3.68938	28.72532	Indoor	Rest	0.00893	0.25652
2017-03-24 14:44	2017-03-24 14:59	40.44	−3.689	141.90761	Outdoor	Walk	0.01326	1.88169
2017-03-24 14:59	2017-03-24 15:58	40.43999	−3.68938	273.87957	Indoor	Rest	0.00893	2.44574
2017-03-24 15:58	2017-03-24 16:09	40.44	−3.689	80.2915	Outdoor	Walk	0.01326	1.06467
2017-03-24 16:09	2017-03-24 17:00	40.43999	−3.68938	275.25632	Indoor	Rest	0.00893	2.45804
2017-03-24 17:00	2017-03-24 17:12	40.44	−3.689	70.15613	Outdoor	Walk	0.01326	0.93027
2017-03-24 17:12	2017-03-24 17:41	40.43999	−3.68938	253.86892	Indoor	Rest	0.00893	2.26705
2017-03-24 17:41	2017-03-24 17:56	40.44	−3.689	63.02457	Outdoor	Walk	0.01326	0.83571
2017-03-24 17:56	2017-03-24 18:32	40.43999	−3.68938	278.09301	Indoor	Rest	0.00893	2.48337
2017-03-24 18:32	2017-03-24 18:46	40.44	−3.689	85.27723	Outdoor	Walk	0.01326	1.13078
2017-03-24 18:46	2017-03-24 19:19	40.43999	−3.68938	193.88363	Indoor	Rest	0.00893	1.73138
2017-03-24 19:52	2017-03-24 20:08	40.451	−3.699	182.9052	Outdoor	Transport	0.00893	1.63334
2017-03-24 20:08	2017-03-24 21:13	40.461	−3.709	626.14768	Outdoor	Rest	0.00893	5.5915
2017-03-24 21:13	2017-03-24 21:21	40.459	−3.711	68.61289	Outdoor	Walk	0.01326	0.90981
2017-03-24 21:21	2017-03-24 22:32	40.457	−3.712	414.90097	Outdoor	Rest	0.00893	3.70507
2017-03-24 22:32	2017-03-24 22:42	40.459	−3.711	51.34302	Outdoor	Walk	0.01326	0.68081
2017-03-24 22:42	2017-03-25 0:00	40.461	−3.709	6.56381	Outdoor	Rest	0.00893	0.05861

4. Results and Discussion

4.1. Outdoor Pollution Model Performance

The adopted modeling technique based on geostatistics [65,66] using hourly-based data [43] from the fixed network of pollution stations can be interpolated by using different techniques, and criteria for technique selection is needed. Therefore, a cross validation with the leave out strategy was adopted. Based on the 5-fold cross validation performed for each of the three modeling techniques, among the 24 hourly datasets of $PM_{2.5}$ captured on 34 March 2017, the Simple Kriging technique proved to be the best model with a selection occurrence of 13, followed by the Ordinary Kriging with 10, Co-kriging with 7, and IDW with 1, out of 24 datasets. Our results agree with previous studies such as [88], where Simple Kriging outperformed Co-Kriging, and [89], in which Simple Kriging turned out to be the best model for estimating NO_2 and PM_{10}.

It can be argued that some local effects like turbulence around buildings, roughness of constructions, and some other aspects impact the accuracy of the estimation. The techniques used in such dimensions can be those related to the integration of meteorological, chemical and transportation numerical modeling (WRF and CMAQ models), with the limitations of being able to precisely estimate the boundary conditions as well as to properly model the city configuration (buildings, trees, surface properties, etc.). When running with high spatial resolution, they produce good results, although the quality is slightly reduced and numerical stability becomes an issue [90]. Another potential contribution could be to use artificial-intelligence-based models to estimate pollution levels. In these fields, the authors have already made significant contributions. Actually, some papers [91] have shown the competitive advantage of these methods over those based on numerical simulations. However, to keep the implementation interoperable and extendable, interpolation was finally adopted, because it can easily be enriched with the data fusion option based on IoT-based, street level pollution sensors.

4.2. Device Performance

To validate the fully aggregated indoor pollution values (SZ(p)) obtained from the indoor monitor and e-beacon devices, they were matched against the pollution data that were measured simultaneously during the study period using a portable air pollution monitoring tool- Atmotube [92] that was carried by the individual.

Figure 5 shows the indoor VOC values measured from the Atmotube and the Foobot monitor on 2017-03-31. It can be seen that there is significant measurement variance between the two devices. Nevertheless, the measured values follow the same trend. There is no consistent Air Quality Index (AQI) provided for comparing the pollution values measured by each device. In agreement with [93], the AQI scales differ across countries, organizations, and devices, and this presents an obstacle for comparison and invalidates its usability, which emphasizes the need for a standardized awareness procedure.

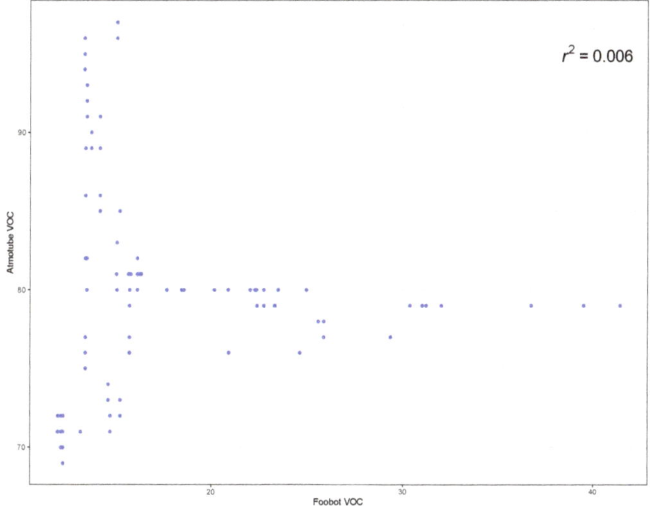

Figure 5. Indoor VOC Values from Atmotube and Foobot for 2017-03-31.

Variance in measurements can be attributed to the differences in calibration and measurement methods that were used by these sensors (see Figure 5). However, this situation partially unveils the observed difficulties in getting people aware of the real importance of pollution, as someone can exhibit different figures for the same pollutants at the same place and point in time. Actually, it is another strong point to have a common framework such as the one proposed in this paper, because it mainly fosters transparency and then allows interpolated or modeled values for outdoor pollution over time at a particular place to be compared with local, privately owned sensors from both outdoor, and indoor locations. From such observations where different local sensors can indeed participate, a better understanding about outliers and commonalities and trends can be derived.

4.3. PAPE Values

Figure 6 illustrates a color map of the average $PM_{2.5}$ levels ($\mu g/m^3$) for one day, in which the range of the specific values is presented on a color scale on the right. The location pins indicate the environment, activity type, time percentage (%), and the respective amount of $PM_{2.5}$ (μg) that the individual was exposed to within the indicated time duration. It can be seen that the individual spent most of the day (62.78 %) outdoors (i.e., outside the workplace) on the northwest side of the city where the highest daily average pollution level of 12 $\mu g/m^3$ was concentrated, and this resulted in a total $PM_{2.5}$ exposure of 52.7 μg.

Figure 7 shows the one day $PM_{2.5}$ exposure levels by activity type. Based on this plot, the individual spent most of the day (88.13%) at rest and was exposed to approximately 70 μg of $PM_{2.5}$ during this period. $PM_{2.5}$ exposure values within a selected time period on the same day are also plotted in Figure 8, which shows that the individual had the highest pollution exposure at 15:32 in the afternoon during this selected period.

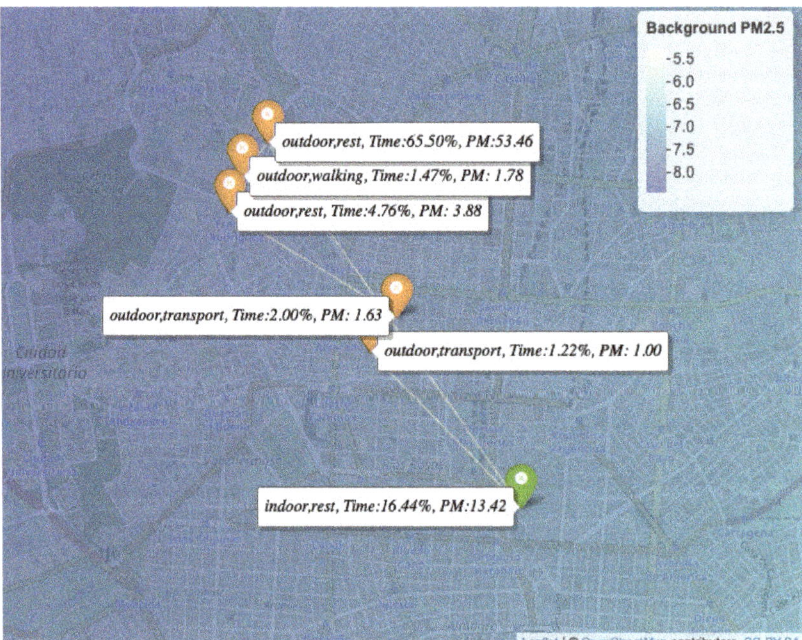

Figure 6. One Day $PM_{2.5}$ Exposure Per Location, Activity Type, and Time Percentage.

From this analysis, the value of people being able to figure out the distribution of the total intensity of pollutants based on their activity becomes evident, as this method can make them aware of the real dimension of the problem and avoid classical myths, like the idea that most of the pollution is acquired outdoors (see Figure 6). While there are similar studies such as in [87], where the authors demonstrated the cleanest air routing algorithm for path navigation by calculating the $PM_{2.5}$ exposure, they mainly focused on pollution acquired outdoors and not indoors.

Since information is the key aspect in having the opportunity to make proper decisions, the advantage of such an integrated framework that is able to integrate not only outdoor conditions but also indoor ones when available becomes more evident. This can also have an impact not only at the individual level by making everyone aware of their exposed pollution levels but at an aggregated level as well, because the public health dimension is impacted when buildings are seen as actionable regarding the indoor conditions. Therefore, KPIs can be adopted by considering the gradient between outdoor and indoor levels per area of occupancy of the buildings. By having systematic monitoring inside, the management dimension can be adopted.

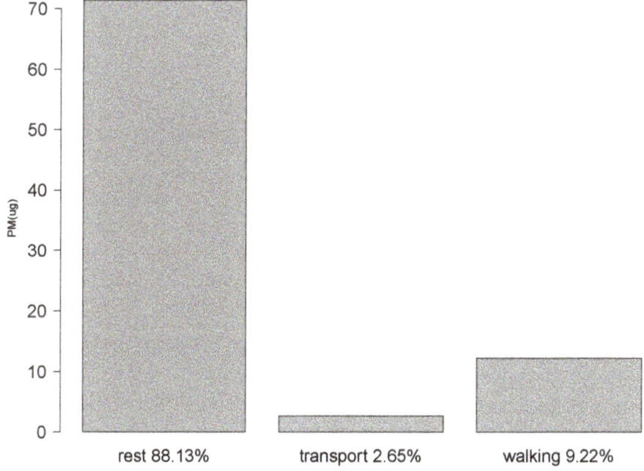

Figure 7. One Day $PM_{2.5}$ Exposure by Activity Type Percentage.

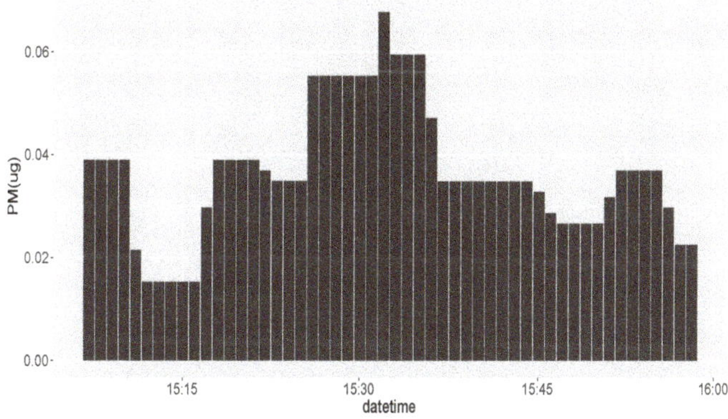

Figure 8. Indoor $PM_{2.5}$ Exposure Values Across Time. (DeltaT = 1 min).

4.4. Alternative Travel Routes

Similar to [87,94], another non-neglectable dimension that is possible to consider is the impact in terms of transportation decisions. Figure 9 shows different routes that one individual can take when moving from one location to another, and the corresponding aggregated pollution and exposure values are provided in Table 6. These values were predicted on the basis of the individual's activity data for 24 March 2017 from 12:04:30 to 12:23:44. The most frequent one adopted by the user was labeled "Actual", while the other potential routes were named A to C.

In this example, the better individual route will is B, as it causes the least amount of PAPE at 0.769 µg, which is 22.75% lower than the actual exposure of 0.995 µg. However, the decision process can be more complex, because there will certainly be some time duration uncertainties, which will consequently result in uncertainty about the total PAPE value of each alternative route.

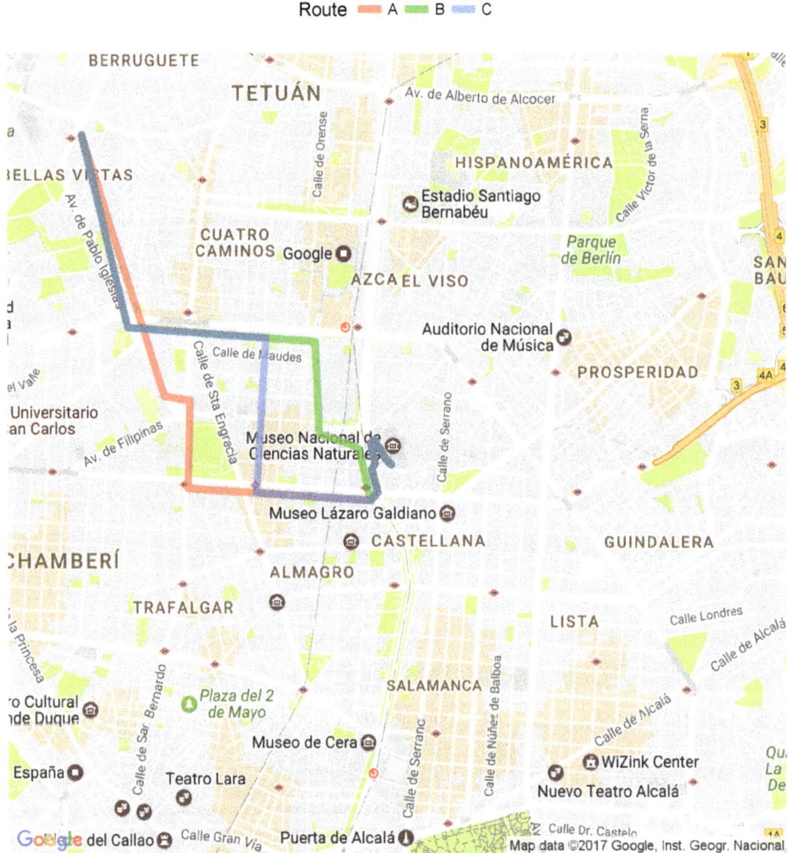

Figure 9. Alternative Travel Routes.

Although most of the tools that give routing solutions for transportation problems are based on duration, some of them have the capability of filtering them out based on pollution exposure outdoors [87,94]. In terms of added value, this contribution enables alternatives to be ranked based on estimated pollution levels both outdoors and indoors, provided that pollution data is also available inside public transportation modes such as trains, buses, and subways. In these cases, as forecast for pollution is needed, machine-learning-based models that infer outdoor pollution values need to be used.

Table 6. Total PM$_{2.5}$ and Exposure Values of the Different Routes..

Route	$PM_{2.5}$ (µg/m^3 * min)	Exposure (µg)
A	89.34	0.80
B	86.08	0.77
C	100.17	0.89
Actual	111.43	0.99

4.5. Limitations

Due to the lack of publicly available air quality information for other indoor areas such as shops, buses, cars, metros, etc., outdoor pollution data from the fixed-site outdoors information must be used in such cases. If there are more available resources, additional monitoring IoT devices in other indoor areas will provide greater accuracy. In most cases, good results demand good inputs, and existing data are replaced whenever better data become available. Quality improvements can be expected from those actions. Smart city empowered data sharing platforms such as IOTA Tangle [95] would boost IoT-based indoor air quality resource availability.

Accuracy for outdoor pollution estimation is another known limitation, both because of the time frequency resolution of available data and because of the interpolation errors. It would be possible to implement Weather Research and Forecasting (WRF) models such as the CMAQ. This decision requires significant effort, not only because of using the appropriate Digital Elevation Model (DEM) required to represent the landscape and building configuration, which is a complex task, but because it requires the boundary conditions to be realistic. This means adopting pressure and wind speed conditions for all surfaces external to the volume of interest. These situations need to be updated regularly throughout the day, as environmental conditions change as well. Indeed, numerical stability conditions must be carefully managed in this case as well.

For future applications, the best solution for environments will come from both the increasing deployment of dense (e.g., street-level) IoT-based air quality sensors and the prosperity of the data sharing platform, which can increase the available data and, consequently, will increase the accuracy.

5. Conclusions

This paper (1) proposed a framework for an AQDSS and (2) developed an IoT application based on this framework. The feasibility of the IoT application in measuring PAPE was evaluated through a case study. In comparison to mobile sensors that were used in previous studies, this IoT application has higher scalability, because it involves minimal cost and intrusion to the individual. This pilot case study also presents evidence that PAPE can be estimated by employing indoor monitors and e-beacon technologies that have not been used previously in similar studies.

Using our proposed framework as a general guideline, the IoT application that we developed can be further extended to include prediction models that will allow an individual to make smart decisions when it comes to PAPE risk. Furthermore, PAPE data obtained from the application can be used in air quality policy development as well as in epidemiological studies to explore the correlations of PAPE with certain diseases.

We faced difficulties during the extraction and integration of data from multiple devices, which highlights the importance of choosing the right technologies to use when developing such IoT applications. There was an observed variance among the different devices, which can be attributed to device calibration and the measurement techniques used. Future research should, therefore, explore these issues and identify emerging technologies that permit seamless data integration and more accurate PAPE measurements.

Author Contributions: The conceptual framework was carried out by B.G. and J.O.-M., while implementation of software was carried out by B.G. Technical writing was carried out mainly by K.A.G.A. and S.S. Funding acquisition was done mainly by J.O.-M.

Funding: Financial support for this research was granted by China Scholarship Council and the European Commission through the EU funded project under the RFCS program with project number 793505 4.0 Lean system integrating workers and processes (WISEST).

Acknowledgments: The authors would like to acknowledge the City Hall of Madrid for the public data on air quality. Indeed, additional acknowledge is due to the Spanish Agencia Estatal de Investigacion, through the research project with code RTI2018-094614-B-I00 into the "Programa Estatal de I+D+i Orientada a los Retos de la Sociedad".

Conflicts of Interest: The authors declare no conflict of interest.

References

1. Initiative, G.R. *Sustainability and Reporting Trends in 2025: Preparing for the Future*; Technical Report; Global Reporting Initiative: Amsterdam, The Netherlands, 2015.
2. EY. *Rapid-Growth Markets*; Technical Report; EY: Oxford, UK, 2014.
3. KPMG. *Ten Emerging Trends in 2017*; Technical Report; KPMG: Zurich, Switzerland, 2017.
4. Bartels, W.; Fogerlberg, T.; Hoballah, A.; Van der Lugt, C.C.A. *Carrots & Sticks: Global Trends in Sustainability Reporting Regulation and Policy*; Technical Report; KPMG: Zurich, Switzerland, 2016.
5. Bollen, J.; Brink, C. Air pollution policy in Europe: Quantifying the interaction with greenhouse gases and climate change policies. *Energy Econ.* **2014**, *46*, 202–215. [CrossRef]
6. Shah, A.S.V.; Langrish, J.P.; Nair, H.; McAllister, D.A.; Hunter, A.L.; Donaldson, K.; Newby, D.E.; Mills, N.L. Global Association of Air Pollution and Heart Failure: A Systematic Review and Meta-Analysis. *Lancet* **2013**, *382*, 1039–1048. [CrossRef]
7. WHO. *Evolution of WHO Air Quality Guidelines: Past, Present and Future*; Technical Report; WHO: Copenhagen, Denmark, 2017.
8. OECD. *Environmental Outlook To 2050: The Consequences of Inaction Key Findings on Health and Environment*; Technical Report; OECD: Paris, France, 2012.
9. WHO. *Household (Indoor) Air Pollution*; WHO: Liverpool, UK, 2017.
10. Fisk, W.J.; Rosenfeld, A.H. Estimates of Improved Productivity and Health from Better Indoor Environments. *Indoor Air* **2004**, *7*, 158–172. [CrossRef]
11. Chen, R.; Chu, C.; Tan, J.; Cao, J.; Song, W.; Xu, X.; Jiang, C.; Ma, W.; Yang, C.; Chen, B.; et al. Ambient air pollution and hospital admission in Shanghai, China. *J. Hazard. Mater.* **2010**, *181*, 234–240. [CrossRef] [PubMed]
12. Micheli, G.J.; Farné, S. Urban railway traffic noise: Looking for the minimum cost for the whole community. *Appl. Acoust.* **2016**, *113*, 121–131. [CrossRef]
13. Kuo, C.Y.; Chan, C.K.; Wu, C.Y.; Phan, D.V.; Chan, C.L. The Short-Term Effects of Ambient Air Pollutants on Childhood Asthma Hospitalization in Taiwan: A National Study. *Int. J. Environ. Res. Public Health* **2019**, *16*. [CrossRef]
14. Saldiva, S.R.D.M.; Barrozo, L.V.; Leone, C.R.; Failla, M.A.; Bonilha, E.D.A.; Bernal, R.T.I.; Oliveira, R.C.D.; Saldiva, P.H.N. Small-Scale Variations in Urban Air Pollution Levels Are Significantly Associated with Premature Births: A Case Study in Sao Paulo, Brazil. *Int. J. Environ. Res. Public Health* **2018**, *15*. [CrossRef] [PubMed]
15. Laumbach, R.J.; Kipen, H.M. Respiratory health effects of air pollution: Update on biomass smoke and traffic pollution. *J. Allergy Clin. Immunol.* **2012**, *129*, 3–11. [CrossRef]
16. Dias, D.; Tchepel, O. Spatial and Temporal Dynamics in Air Pollution Exposure Assessment. *Int. J. Environ. Res. Public Health* **2018**, *15*. [CrossRef]
17. Philip, V.; Suman, V.K.; Menon, V.G.; Dhanya, K.A.A. Review on Latest Internet of Things Based Healthcare Applications. *Int. J. Comput. Sci. Inf. Secury* **2017**, *15*, 248–255.
18. EY. *Health Reimagined: A New Participatory Health Paradigm*; Technical Report; EY: Melbourne, Australia, 2016.
19. Deloitte. *2015 Global Health Care Outlook: Common Goals, Competing Priorities*; Technical Report; Deloitte: London, UK, 2015.
20. Niedermans, F.; Biesdorf, S. *Healthcare's Digital Future*; Technical Report; McKinsey: Munich, Germany, 2014.

21. Barnes, K.; Isgur, B.; Tsouderos, T. *Top Health Industry Issues of 2016: Thriving in the New Health Economy*; Health Research Institute: USA, 2015; pp. 1–18.

22. Reh, G.; Korenda, L.; Boozer, C. *Will Patients and Caregivers Embrace Technology-Enabled Health Care?* Technical Report; Deloitte: London, UK, 2016.

23. Kim, K.H.; Kabir, E.; Kabir, S. A review on the human health impact of airborne particulate matter. *Environ. Int.* **2015**, *74*, 136–143. [CrossRef] [PubMed]

24. Szigeti, T.; Dunster, C.; Cattaneo, A.; Cavallo, D.; Spinazze, A.; Saraga, D.E.; Sakellaris, I.A.; de Kluizenaar, Y.; Cornelissen, E.J.; Hänninen, O.; et al. Oxidative potential and chemical composition of PM2.5 in office buildings across Europe—The OFFICAIR study. *Environ. Int.* **2016**, *92–93*, 324–333. [CrossRef] [PubMed]

25. Guerreiro, C.B.; Foltescu, V.; de Leeuw, F. Air quality status and trends in Europe. *Atmos. Environ.* **2014**, *98*, 376–384. [CrossRef]

26. Abraham, S.; Li, X. Design of A Low-Cost Wireless Indoor Air Quality Sensor Network System. *Int. J. Wirel. Inf. Netw.* **2016**, *23*, 57–65. [CrossRef]

27. WHO. Air Quality Guidelines. Global Update 2005. *Environ. Sci. Pollut. Res.* **2006**, *3*, 23.

28. Penney, D.; Benignus, V.; Kephalopoulos, S.; Kotzias, D.; Kleinman, M.; Verrier, A. *Guidelines for Indoor Air Quality*; Technical Report; WHO: Copenhagen, Denmark, 2010.

29. Anna Mainka, B.K. Assessment of the BTEX concentrations and health risk in urban nursery schools in Gliwice, Poland. *AIMS Environ. Sci.* **2016**, *3*, 858. [CrossRef]

30. Yurdakul, S.; Civan, M.; Ozden, O.; Gaga, E.; Dogeroglu, T.; Tuncel, G. Spatial variation of VOCs and inorganic pollutants in a university building. *Atmos. Pollut. Res.* **2017**, *8*, 1–12. [CrossRef]

31. EPA. *Compendium of Methods for the Determination of Air Pollutants in Indoor Air*; Technical Report; EPA: Springfield, VA, USA, 1990.

32. Ishizaka, T.D.; Kawashima, A.; Hishida, N.; Hamada, N. Measurement of total volatile organic compound (TVOC) in indoor air using passive solvent extraction method. *Air Qual. Atmos. Health* **2019**, *12*, 173–187. [CrossRef]

33. Topping, M. OCCUPATIONAL EXPOSURE LIMITS FOR CHEMICALS. *Occup. Environ. Med.* **2001**, *58*, 138–144. [CrossRef]

34. Kumar, P.; Morawska, L.; Martani, C.; Biskos, G.; Neophytou, M.; Sabatino, S.D.; Bell, M.; Norford, L.; Britter, R. The rise of low-cost sensing for managing air pollution in cities. *Environ. Int.* **2015**, *75*, 199–205. [CrossRef]

35. Reis, S.; Seto, E.; Northcross, A.; Quinn, N.W.; Convertino, M.; Jones, R.L.; Maier, H.R.; Schlink, U.; Steinle, S.; Vieno, M.; et al. Integrating modelling and smart sensors for environmental and human health. *Environ. Model. Softw.* **2015**, *74*, 238–246. [CrossRef]

36. Steinle, S.; Reis, S.; Sabel, C.E. Quantifying human exposure to air pollution—Moving from static monitoring to spatio-temporally resolved personal exposure assessment. *Sci. Total Environ.* **2013**, *443*, 184–193. [CrossRef] [PubMed]

37. Su, J.G.; Jerrett, M.; Meng, Y.Y.; Pickett, M.; Ritz, B. Integrating smart-phone based momentary location tracking with fixed site air quality monitoring for personal exposure assessment. *Sci. Total Environ.* **2015**, *506–507*, 518–526. [CrossRef] [PubMed]

38. Milando, C.W.; Martenies, S.E.; Batterman, S.A. Assessing concentrations and health impacts of air quality management strategies: Framework for Rapid Emissions Scenario and Health impact ESTimation (FRESH-EST). *Environ. Int.* **2016**, *94*, 473–481. [CrossRef] [PubMed]

39. Mercer, L.D.; Szpiro, A.A.; Sheppard, L.; Lindström, J.; Adar, S.D.; Allen, R.W.; Avol, E.L.; Oron, A.P.; Larson, T.; Liu, L.J.S.; et al. Comparing universal kriging and land-use regression for predicting concentrations of gaseous oxides of nitrogen (NOx) for the Multi-Ethnic Study of Atherosclerosis and Air Pollution (MESA Air). *Atmos. Environ.* **2011**, *45*, 4412–4420. [CrossRef] [PubMed]

40. Rojas, A.L.P. Simple atmospheric dispersion model to estimate hourly ground-level nitrogen dioxide and ozone concentrations at urban scale. *Environ. Model. Softw.* **2014**, *59*, 127–134. [CrossRef]

41. Oettl, D. *Documentation of the Lagrangian Particle Model GRAL (Graz Lagrangian Model)*; Amt der Steiermärk: Graz, Austria, 2016.

42. Marjovi, A.; Arfire, A.; Martinoli, A. High Resolution Air Pollution Maps in Urban Environments Using Mobile Sensor Networks. In Proceedings of the 2015 International Conference on Distributed Computing in Sensor Systems, Fortaleza, Brazil, 10–12 June 2015; pp. 11–20. [CrossRef]

43. Gomez-Losada, A.; Santos, F.M.; Gibert, K.; Pires, J.C. A data science approach for spatiotemporal modelling of low and resident air pollution in Madrid (Spain): Implications for epidemiological studies. *Comput. Environ. Urban Syst.* **2019**, *75*, 1–11. [CrossRef]

44. Montero, J.M.; Fernández-Avilés, G. Functional kriging prediction of atmospheric particulate matter concentrations in Madrid, Spain: Is the new monitoring system masking potential public health problems? *J. Clean. Prod.* **2018**, *175*, 283–293. [CrossRef]

45. de Nazelle, A.; Seto, E.; Donaire-Gonzalez, D.; Mendez, M.; Matamala, J.; Nieuwenhuijsen, M.J.; Jerrett, M. Improving estimates of air pollution exposure through ubiquitous sensing technologies. *Environ. Pollut.* **2013**, *176*, 92–99. [CrossRef] [PubMed]

46. Yu, T.C.; Lin, C.C.; Chen, C.C.; Lee, W.L.; Lee, R.G.; Tseng, C.H.; Liu, S.P. Wireless sensor networks for indoor air quality monitoring. *Med. Eng. Phys.* **2013**, *35*, 231–235. [CrossRef] [PubMed]

47. Schlesinger, J.; Burris, S.; Tippmann, C. *Health and Mobility: Realizing the Power of Mobile Technology*; Technical Report; EY: Los Angeles, CA, USA, 2015.

48. Tong, Z.; Chen, Y.; Malkawi, A.; Adamkiewicz, G.; Spengler, J.D. Quantifying the impact of traffic-related air pollution on the indoor air quality of a naturally ventilated building. *Environ. Int.* **2016**, *89–90*, 138–146. [CrossRef] [PubMed]

49. Mead, M.; Popoola, O.; Stewart, G.; Landshoff, P.; Calleja, M.; Hayes, M.; Baldovi, J.; McLeod, M.; Hodgson, T.; Dicks, J.; et al. The use of electrochemical sensors for monitoring urban air quality in low-cost, high-density networks. *Atmos. Environ.* **2013**, *70*, 186–203. [CrossRef]

50. Saad, S.M.; Mohd Saad, A.R.; Kamarudin, A.M.Y.; Zakaria, A.; Shakaff, A.Y.M. Indoor air quality monitoring system using wireless sensor network (WSN) with web interface. In Proceedings of the 2013 International Conference on Electrical, Electronics and System Engineering (ICEESE), Selangor, Malaysia, 4–5 December 2013; pp. 60–64. [CrossRef]

51. Ott, W.R. Concepts of human exposure to air pollution. *Environ. Int.* **1982**, *7*, 179–196. [CrossRef]

52. Nash, D.G.; Leith, D. Use of Passive Diffusion Tubes to Monitor Air Pollutants. *J. Air Waste Manag. Assoc.* **2010**, *60*, 204–209. [CrossRef] [PubMed]

53. Monn, C. Exposure assessment of air pollutants: A review on spatial heterogeneity and indoor/outdoor/personal exposure to suspended particulate matter, nitrogen dioxide and ozone. *Atmos. Environ.* **2001**, *35*, 1–32. [CrossRef]

54. Klepeis, N. Modeling Human Exposure To Air Pollution. Available online: http://citeseerx.ist.psu.edu/viewdoc/download?doi=10.1.1.460.8304&rep=rep1&type=pdf (accessed on 15 February 2017).

55. Gu, Y.; Yim, S. The air quality and health impacts of domestic trans-boundary pollution in various regions of China. *Environ. Int.* **2016**, *97*, 117–124. [CrossRef] [PubMed]

56. Predic, B.; Yan, Z.; Eberle, J.; Stojanovic, D.; Aberer, K. ExposureSense: Integrating Daily Activities with Air Quality Using Mobile Participatory Sensing. In Proceedings of the 2013 IEEE International Conference on Pervasive Computing and Communications Workshops (PERCOM Workshops, San Diego, CA, USA, 18–22 March 2013; pp. 303–305.

57. Steinle, S.; Reis, S.; Sabel, C.E.; Semple, S.; Twigg, M.M.; Braban, C.F.; Leeson, S.R.; Heal, M.R.; Harrison, D.; Lin, C.; et al. Personal exposure monitoring of PM2.5 in indoor and outdoor microenvironments. *Sci. Total Environ.* **2015**, *508*, 383–394. [CrossRef] [PubMed]

58. Liu, R.; Wang, Y.; Shu, M. Internet of Things Healthcare Cloud System Based on IEEE 802.15.4. *J. Appl. Sci.* **2013**, *13*, 1582–1586. [CrossRef]

59. Stojić, A.; Maletić, D.; Stojić, S.S.; Mijić, Z.; Šoštarić, A. Forecasting of VOC emissions from traffic and industry using classification and regression multivariate methods. *Sci. Total Environ.* **2015**, *521–522*, 19–26. [CrossRef] [PubMed]

60. Gong, B.; Ordieres-Mere, J. Prediction of daily maximum ozone threshold exceedances by preprocessing and ensemble artificial intelligence techniques: Case study of Hong Kong. *Environ. Model. Softw.* **2016**, *84*, 290–303. [CrossRef]

61. Diaz-Robles, L.A.; Ortega, J.C.; Fu, J.S.; Reed, G.D.; Chow, J.C.; Watson, J.G.; Moncada-Herrera, J.A. A hybrid ARIMA and artificial neural networks model to forecast particulate matter in urban areas: The case of Temuco, Chile. *Atmos. Environ.* **2008**, *42*, 8331–8340. [CrossRef]

62. Beevers, S.D.; Kitwiroon, N.; Williams, M.L.; Carslaw, D.C. One way coupling of CMAQ and a road source dispersion model for fine scale air pollution predictions. *Atmos. Environ.* **2012**, *59*, 47–58. [CrossRef] [PubMed]

63. Raducan, G. Pollutant dispersion modelling with OSPM in a street canyon from Bucharest. *Roman. Report Phys.* **2008**, *60*.

64. Gong, B.; Ordieres-Mere, J. Reconfiguring existing pollutant monitoring stations by increasing the value of the gathered information. *Environ. Model. Softw.* **2017**, *96*, 106–122. [CrossRef]

65. Farhi, A.; Boyko, V.; Almagor, J.; Benenson, I.; Segre, E.; Rudich, Y.; Stern, E.; Lerner-Geva, L. The possible association between exposure to air pollution and the risk for congenital malformations. *Environ. Res.* **2014**, *135*, 173–180. [CrossRef] [PubMed]

66. Nikzad, N.; Verma, N.; Ziftci, C.; Bales, E.; Quick, N.; Zappi, P.; Patrick, K.; Dasgupta, S.; Krueger, I.; Rosing, T.V.; et al. CitiSense: Improving Geospatial Environmental Assessment of Air Quality Using a Wireless Personal Exposure Monitoring System. In *Proceedings of the Conference on Wireless Health*; ACM: New York, NY, USA, 2012; pp. 11:1–11:8. [CrossRef]

67. Government, T.C. Air Boxes PM2.5. Available online: https://smartcity.taipei/posts/3?locale=en (accessed on 30 April 2019).

68. Catlett, C. Array of Things. Available online: https://www.anl.gov/mcs/array-of-things (accessed on 30 April 2019).

69. Council, M.C.Air Quality: Real-time data. Available online: https://www.madrid.es/portal/site/munimadrid (accessed on 13 February 2017).

70. AEMET. Open data from AEMET. Available online: https://opendata.aemet.es/centrodedescargas/inicio (accessed on 13 February 2017).

71. Welsh, B.; Baird, T.; Zhao, J.; Block-Schachter, D. Web App Design to Implement Travel Behavioral Nudging Using "Moves". In Proceedings of the Transportation Research Board 93rd Annual Meeting, Washington, DC, USA, 12–14 January 2014.

72. OwnTracks. OwnTracks: Your Location Companion. Available online: https://owntracks.org/ (accessed on 13 February 2017).

73. Kirstenpfad, D. Miataru - Be Found: OpenSource Location Tracking. Available online: http://miataru.com/ios/ (accessed on 13 February 2017).

74. Geo2tag. TAG EVERYWHERE(EVERYTHING) WITH GEO2TAG. Available online: http://www.geo2tag.com/ (accessed on 13 February 2017).

75. Loh, M.; Sarigiannis, D.; Gotti, A.; Karakitsios, S.; Pronk, A.; Kuijpers, E.; Annesi-Maesano, I.; Baiz, N.; Madureira, J.; Oliveira Fernandes, E.; et al. How Sensors Might Help Define the External Exposome. *Int. J. Environ. Res. Public Health* **2017**, *14*. [CrossRef] [PubMed]

76. Mahesh, B.R.; Kumar, M.B.; Manoharan, R.; Somasundaram, M.; Karthikeyan, S.P. Portability of mobile applications using PhoneGap: A case study. In Proceedings of the International Conference on Software Engineering and Mobile Application Modelling and Development (ICSEMA 2012), Chennai, India, 19–21 December 2012; pp. 1–6. [CrossRef]

77. Danova, T. These Are Some Of The Top Emerging Business Applications For Beacons. Available online: https://www.businessinsider.com/top-emerging-applications-for-beacons-2014-4 (accessed on 3 March 2017).

78. Gong, B. masak1112/IoT-for-Estimating-Personal-Pollution-Doses 0.1.0. Available online: https://zenodo.org/record/818074#.XWVKMpNKiYU (accessed on 25 June 2017).

79. Vedrenne, M.; Borge, R.; Lumbreras, J.; Rodriguez, M.E. Advancements in the design and validation of an air pollution integrated assessment model for Spain. *Environ. Model. Softw.* **2014**, *57*, 177–191. [CrossRef]

80. Andria, G.; Cavone, G.; Lanzolla, A.M. Modelling study for assessment and forecasting variation of urban air pollution. *Measurement* **2008**, *41*, 222–229. [CrossRef]

81. Li, L.; Losser, T.; Yorke, C.; Piltner, R. Fast Inverse Distance Weighting-Based Spatiotemporal Interpolation: A Web-Based Application of Interpolating Daily Fine Particulate Matter PM2.5 in the Contiguous U.S. Using Parallel Programming and k-d Tree. *Int. J. Environ. Res. Public Health* **2014**, *11*, 9101–9141. [CrossRef]

82. Council, M.C. Web Portal of Air Quality of the Madrid City Council. Available online: https://www.madrid.es/portal/site/munimadrid (accessed on 15 June 2019).

83. de Nazelle, A.; Bode, O.; Orjuela, J.P. Comparison of air pollution exposures in active vs. passive travel modes in European cities: A quantitative review. *Environ. Int.* **2017**, *99*, 151–160. [CrossRef]

84. O'Donoghue, R.; Gill, L.; McKevitt, R.; Broderick, B. Exposure to hydrocarbon concentrations while commuting or exercising in Dublin. *Environ. Int.* **2007**, *33*, 1–8. [CrossRef]

85. Allan, M.; Richardson, G.M.; Jones-Otazo, H. Probability Density Functions Describing 24-Hour Inhalation Rates for Use in Human Health Risk Assessments: An Update and Comparison. *Human Ecol. Risk Assess. Int. J.* **2008**, *14*, 372–391. [CrossRef]

86. Dirks, K.N.; Salmond, J.A.; Talbot, N. Air Pollution Exposure in Walking School Bus Routes: A New Zealand Case Study. *Int. J. Environ. Res. Public Health* **2018**, *15*. [CrossRef] [PubMed]

87. Mahajan, S.; Tang, Y.S.; Wu, D.Y.; Tsai, T.C.; Chen, L.J. CAR: The Cleanest Air Routing Algorithm for Path Navigation with Minimal PM2.5 Exposure on the Move. In Proceedings of the 16th Annual International Conference on Mobile Systems, Applications, and Services, Munich, Germany, 10–15 June 2018; ACM: New York, NY, USA, 2018; pp. 532–532. [CrossRef]

88. Krivoruchko, K. *Using Linear and Non-Linear Kriging Interpolators to Produce Probability Maps*; Environmental Systems Research Institute: Redlands, CA, USA, 2001.

89. Halimi, M.; Farajzadeh, M.; Zarei, Z. Modeling spatial distribution of Tehran air pollutants using geostatistical methods incorporate uncertainty maps. *Pollution* **2016**, *2*, 375–386. [CrossRef]

90. San José, R.; Pérez, J.; González, R. Very High Resolution Urban Simulations with WRF/UCM and CMAQ over European Cities. In *Urban Environment*; Rauch, S., Morrison, G., Norra, S., Schleicher, N., Eds.; Springer: Dordrecht, The Netherlands, 2013; pp. 293–301.

91. Fernando, H.; Mammarella, M.; Grandoni, G.; Fedele, P.; Marco, R.D.; Dimitrova, R.; Hyde, P. Forecasting PM10 in metropolitan areas: Efficacy of neural networks. *Environ. Pollut.* **2012**, *163*, 62–67. [CrossRef]

92. Atmotube. Atmotube Makes You Smarter about the Air You Breathe. Available online: https://atmotube.com/ (accessed on 13 February 2017).

93. Lemes, S. Air Quality Index (AQI)—Comparative Study And Assessment Of An Appropriate Model For B&H. In Proceedings of the 12th Scientific/Research Symposium with International Participation "Metallic And Nonmetallic Materials" MNM 2018, Zenica, Bosnia and Herzegovina, 19–20 April 2018.

94. Luo, J.; Boriboonsomsin, K.; Barth, M. Reducing pedestrians' inhalation of traffic-related air pollution through route choices: Case study in California suburb. *J. Trans. Health* **2018**, *10*, 111–123. [CrossRef]

95. Tangle, I. Meet the Tangle. Available online: https://www.iota.org/research/meet-the-tangle (accessed on 13 February 2017).

International Journal of
Environmental Research and Public Health

Article

The Relationship Between the Actual Level of Air Pollution and Residents' Concern about Air Pollution: Evidence from Shanghai, China

Daxin Dong, Xiaowei Xu *, Wen Xu and Junye Xie

School of Business Administration, Southwestern University of Finance and Economics, Chengdu 611130, China; dongdaxinedu@126.com (D.D.); xxww@smail.swufe.edu.cn (W.X.); johnny@smail.swufe.edu.cn (J.X.)
* Correspondence: xiaoweix@swufe.edu.cn

Received: 5 October 2019; Accepted: 26 November 2019; Published: 28 November 2019

Abstract: This study explored the relationship between the actual level of air pollution and residents' concern about air pollution. The actual air pollution level was measured by the air quality index (AQI) reported by environmental monitoring stations, while residents' concern about air pollution was reflected by the Baidu index using the Internet search engine keywords "Shanghai air quality". On the basis of the daily data of 2068 days for the city of Shanghai in China over the period between 2 December 2013 and 31 July 2019, a vector autoregression (VAR) model was built for empirical analysis. Estimation results provided three interesting findings. (1) Local residents perceived the deprivation of air quality and expressed their concern on air pollution quickly, within the day on which the air quality index rose. (2) A decline in air quality in another major city, such as Beijing, also raised the concern of Shanghai residents about local air quality. (3) A rise in Shanghai residents' concern had a beneficial impact on air quality improvement. This study implied that people really cared much about local air quality, and it was beneficial to inform more residents about the situation of local air quality and the risks associated with air pollution.

Keywords: air pollution; public concern; air quality index; Baidu index; Shanghai

---

1. Introduction

According to the World Health Organization (WHO), 91% of the world's population lives in areas where air pollution exceeds the safety limits [1]. Air pollution negatively affects both human's daily life, such as emotional and physical health, and sustainable economic growth, such as labor productivity and tourism (e.g., [2–5]). However, air pollutants are not always visible, which might lead to the public being unaware of pollution. Prior studies have indicated that the public perception of air pollution might be inconsistent with objective air quality, which was evaluated using scientific indices such as $PM_{2.5}$ and PM_{10} [6]. A potential reason is that individuals' perceived air quality could differ on the basis of their sociodemographic status, including gender, age, education, knowledge, and health status (e.g., [7,8]). Therefore, perhaps there exists a gap between objective and subjective measures of air quality.

There are two important reasons why the relationship between the actual level of air pollution and residents' concern about air pollution should be examined. On the one hand, ignorance or underestimation of the severity of air pollution potentially poses a threat to residents' health since it could increase the probability of long term exposure to air pollutants. For instance, it was reported that 90% of residents in Hong Kong would not stop their outdoor activities in the face of poor air quality [9]. Whether residents' concern about air pollution could correctly reflect the actual pollution level has become a critical question. On the other hand, a high level of public awareness regarding air pollution might contribute to political and social enthusiasm for the enforcement of environmental protection behaviors and policies. Larson [10] reported that the concurrent rise of online platforms in China has become a positive force for environmental data transparency in China. Kay et al. [11] provided further evidence showing that social media are capable of empowering the government to respond to the air pollution problem to ensure social stability. More recently, Lu et al. [12] stressed that public concern about air quality might have a more direct impact than perceived air pollution in influencing people's behaviors and the actions of the community, production sectors, and the government. Therefore, how Chinese residents responded to air pollution should be carefully examined. In particular, more studies are needed to analyze the relationship between the actual level of air pollution and residents' concern about air pollution, given the importance of public concern in shaping public health and environmental regulation practices.

The purpose of this study is to examine whether and to what extent actual air pollution is correlated with residents' concern about air quality in Shanghai, China. Shanghai was chosen as the target city in this study since it is one of the largest and most developed cities in China. Although Shanghai has a high level of economic and social development, its air pollution issue is considerable. According to the Shanghai Environmental Bulletin published by the local government, during the year 2018, air quality was classified as "good" on only 93 days. In 2017 and 2016, the numbers were 58 and 78, respectively. By the end of 2018, Shanghai's air pollution was even reported to be worse than that of Beijing [13], which is well known for its air pollution problem. However, it seems that most attention has been centered on air pollution issues in Beijing in recent years. Public opinion on air pollution from Shanghai residents should be examined since it might maintain pressure on governments to roll out environmental regulations to reduce air pollution effectively and efficiently. From another perspective, understanding Shanghai residents' concern about pollution is valuable in providing residents with useful advice about public health and environmental protection.

In this study, actual air quality was measured by the air quality index (AQI) reported by environmental monitoring stations. Residents' concern about air pollution was measured using the Baidu index for the online search keywords "Shanghai air quality". Baidu index data were provided by Baidu, which is the most popular Internet search engine in China. The index was calculated on the basis of the search volume for a specific search item on a daily basis within a specific region. The Baidu index has been widely used to predict public health issues (e.g., [14–17]) and tourism flows (e.g., [18–20]) in China. Similar to its usage in public health and tourism studies, the Baidu index is also applicable in measuring the degree of public concern about air pollution [12]. This point will be discussed in detail later in the literature review section.

Figure 1 shows the logarithmic values of the actual AQI (blue curve) in Shanghai and the Baidu index (yellow curve) for "Shanghai air quality". The sample period spanned from 2 December 2013 to 31 July 2019, covering 2068 days. The 30-day moving average values of the variables are also displayed to more clearly show varying trends. It can be observed that both the AQI and the Baidu index followed a similar cyclical pattern with apparent fluctuations. The correlation coefficient between logarithmic AQI and the Baidu index was 0.432, indicating a statistically significant positive correlation.

Although Figure 1 provides preliminary visual evidence on the relationship between actual air pollution and residents' concern about air pollution, a basic correlation analysis is not sufficient. To better understand the relationship, the following three research questions are proposed. (1) To what extent and how soon could the actual air pollution level influence public concern about air quality? (2) Does a decline in air quality in another major city, such as Beijing, affect public concern about air quality in Shanghai? (3) Could public concern about air quality and the level of actual air pollution reciprocally influence each other? A vector autoregression (VAR) model was applied to answer these research questions due to its high ability to capture linear interdependencies among multiple variables over time.

It is expected that this study could add to the air pollution literature by examining the reciprocal interactions between public concern about air pollution and the actual degree of air pollution. From a practical perspective, it could help the municipal government of Shanghai better understand the degree of public concern about air quality and better assess the current environmental management practices. This study also provides insights into the spillover effects that the actual air pollution in other major tourist cities in China might have on public concern about air quality in Shanghai.

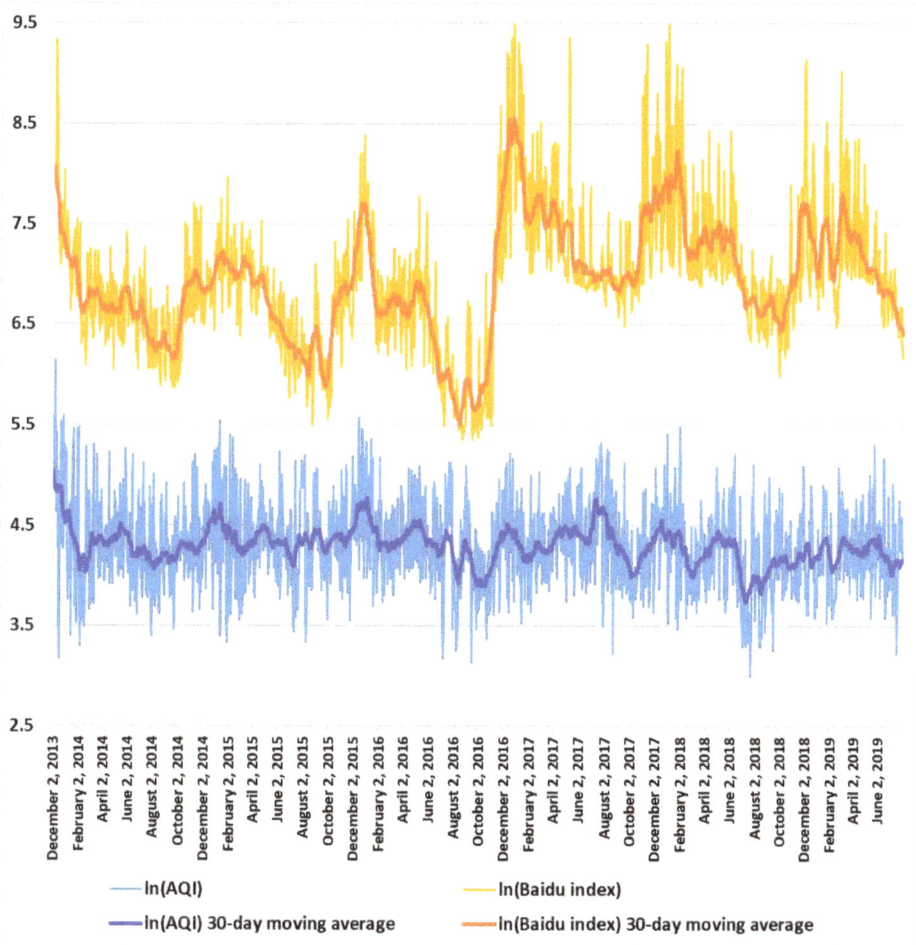

Figure 1. Actual air pollution and residents' concern about air pollution in Shanghai from 2 December 2013 to 31 July 2019.

The rest of this paper is organized as follows. Section 2 presents a literature review and develops the hypotheses. Section 3 describes the empirical model and the data used in the analyses. The estimated results of the empirical model are reported in Section 4. Section 5 discusses the main findings and implications of the results. Section 6 concludes the paper and talks about limitations and directions for future research.

2. Literature Review and Hypothesis Development

2.1. Air Pollution: Actual Level, Perceived Level, and Public Concern

Actual air quality levels are measured by scientific techniques, and they are reported in the forms of different air pollutant indicators such as $PM_{2.5}$ (particulate matter with a size of 2.5 micrometers or less) and SO_2 (sulfur dioxide), or comprehensive indices such as AQI and API (air pollution index) constructed using the air pollutant indicators. Generally speaking, there is not much controversy on the proper measurement of actual air quality level, and scientists can measure it objectively and accurately

People's perception of air quality is subjective. Traditionally, researchers have used questionnaire surveys to collect data on perceived air quality level and people's displeasure with air pollution (e.g., [21–23]). Some of the literature reported that perceived and actual levels of air quality are strongly correlated. For example, Atari et al. [24] conducted a community health survey in the "Chemical Valley" in Sarnia, Ontario, Canada, and they observed a significant correlation between odor annoyance scores and modeled ambient pollution. Peng et al. [22] analyzed the data for over 5000 valid respondents in 62 cities, on the basis of the China Social Survey 2013, and reported a congruence of perceived and actual air pollution. Similar findings were reported by some other studies, such as for Northeast England [25], for Switzerland [26], and for China [27].

However, some of the literature found that perceived air quality is not strongly related to actual air quality. For instance, Egondi et al. [21] focused on two slums in Nairobi, Kenya, on the basis of a cross-sectional study of 5317 individuals. They reported that the majority of respondents were exposed to air pollution. However, the perceived air pollution level was low among residents. This implied the need for promoting public awareness on air pollution. Focusing on the city of Wuhan in China, Guo et al. [7] reported that most persons they surveyed believed that air quality had become worse, though statistics of measured air pollutants over the period of 2010–2014 did not actually show that trend. During the period of 2005–2006, Semenza et al. [23] surveyed the residents of two U.S. cities, Portland and Houston. They reported that the residents' perception of poor air quality was not related to $PM_{2.5}$ or ozone indicators. On the basis of a sample of 200 people interviewed in London in 1999, Williams and Bird [28] reported that public perception of air quality was not a reliable indicator of actual air pollution levels in the investigated areas. The inconsistency between actual and perceived level of air quality was also reported by Berezansky et al. [29] for Haifa, Israel, by Brody et al. [30] for Texas in the U.S., and by Kim et al. [31] for Seoul, South Korea.

A notable recent study was by Lu et al. [12]. They used the correlation-analysis technique to study the relationship between actual $PM_{2.5}$ concentration and public concern about haze in five large Chinese cities for the period between 2013 and 2017, where public concern was measured by the Baidu index. They reported that short term fluctuations in actual pollution level and concern about pollution were strongly and positively correlated, but long term annual trends of these two variables were opposite.

Overall, the literature has not provided a consensus on whether the public could correctly perceive air quality and whether public concern about air pollution concurs with the actual level of air pollution. There are two possible explanations of this disagreement in the literature. First, different studies were based on different samples. Since social and individual characteristics strongly affect people's viewpoints [31–33], it is natural to see that perceptions of some groups are consistent with actual air quality, but some other groups tend to over- or under-estimate the level of air pollution. Second, public perceptions vary over time. Along with the development of information technology, improvements in education, and the propagation of mass media, the public receives increasing levels of information on the current situation and importance of air quality. Therefore, many people might not have perceived and been concerned with air pollution problems in the past, but they have changed their minds in recent years.

In this study, we focus on the city of Shanghai. Shanghai is one of the most important and developed cities in China. In 2018, Shanghai's GDP per capita exceeded 20,000 U.S. dollars, which was close to that of Greece and Slovakia. Shanghai has a large population size of over 24 million residents, close to the population of a middle sized country such as Australia. As the financial center of China, many companies have their headquarters in Shanghai. Shanghai has good educational resources, including 64 universities. As an international city, Shanghai is visited by approximately nine million (person-times) inbound tourists annually. Therefore, on average, Shanghai residents have the knowledge and ability to perceive the importance of clean air and to worry about pollution if they feel that air quality has declined. Accordingly, the first research hypothesis in this study was established:

Hypothesis 1. *Deprivation of air quality in a local area raises the public concern about the air pollution problem.*

Furthermore, since the Internet has allowed information to flow swiftly across regions, news about air pollution in one major city could quickly propagate and might cause mental or even physical responses of residents in another city. In this study, Beijing, the capital of China, was used as a study area because it is notorious for its severe air pollution (it is also important to note that other major tourist cities, including Nanjing and Guangzhou, are also examined in the robustness check section). If Beijing's AQI rose to an unhealthily high value, the residents in Shanghai might also want to check the air quality in their local area after they receive the information about Beijing. Thus, the second research hypothesis was built:

Hypothesis 2. *Deprivation of air quality in another major city raises public concern about the air pollution problem in the local city.*

2.2. Importance of Air Pollution Awareness and Concern

People's concern about air pollution is largely due to the health risks it has [21,34,35]. People's concern about air quality promotes people's avoidance of the polluted environment. For instance, people tend to limit their social activities on polluted days [36]; people are unwilling to buy houses located in polluted areas [29,37,38]; and tourists flow from polluted to less polluted cities [39,40].

Public concern about air pollution generates pressure on the government to implement policies to promote better air quality. Environmental perception and concern also promote more pro-environmental behaviors [41,42]. Tam and Chan [43] further examined the association between environmental concern and behavior across 32 countries (China was not included), and they indicated that the association was weaker in countries with stronger distrust and belief in external control. To better understand the strength of this association in the case of Shanghai, the third research hypothesis was built:

Hypothesis 3. *Public concern about the air pollution problem pushes people or governments to take actions to reduce air pollution.*

2.3. Applicability of the Baidu Search Index to Measure Residents' Concern about Air Pollution

Internet search engines enable quick access to information. Do et al. [44] indicated that tracking online search behavior using relative search volumes was an effective way to gauge public interest. A number of researchers have used the search index as a proxy of issue salience or attention in the public. The majority of search-volume data are extracted from two dominant engines, Google and Baidu. For example, Caputi et al. [45] noticed a significant public interest in heat-not-burn tobacco products on the basis of Google trends data. Do et al. [44] utilized Google trends data to assess public awareness on protected wetlands in South Korea. Mellon [46] found that Google trends data could be used to measure the salience of four issues (fuel prices, economy, immigration, and terrorism) in the United States. Search index data have also been used to predict tourism flows (e.g., [18–20,47]).

Given the fact that the Baidu search engine occupies a dominant market share in China, the Baidu search index has become a useful proxy for Chinese residents' interest and concerns. In Li et al. [16], Baidu search query data were found to be a reliable indicator for monitoring and predicting the HIV/AIDS epidemic in China. Liu et al. [48] used the Baidu index to assess people's awareness of avian influenza A(H7N9) in Zhejiang, China. Using the Baidu index, Yang et al. [49] found that the daily average $PM_{2.5}$ concentration had a weak impact on public awareness of lung cancer risk in China.

Lu et al. [12] used the Baidu index to measure public concern about haze in five large Chinese cities. Given that the use of the search index has been extensively applied in multiple disciplines as a method for surveillance, monitoring, and measuring interest on and concern about specific topics, this study adopted Baidu index data to measure residents' concern about air pollution in China.

3. Methods

3.1. Model

This study is based on time series data relevant to air pollution in Shanghai. VAR analysis, a powerful tool in modeling complex time series, was used in this study. Beyond basic correlation analysis or the ordinary least squares (OLS) regression technique, the VAR model treats all variables as endogenous and contains time lagged variables. Thus, the VAR model can measure the reciprocal reactions among different variables and lagged and persistent effects. The following reduced form VAR model was built:

$$y_t = c + \sum_{i=1}^{p} A_i y_{t-i} + \varepsilon_t, \tag{1}$$

where y_t is a vector of endogenous variables in period t, c is the intercept vector, A_i refers to the autoregressive coefficient matrix that captures system dynamics, and ε_t is the residual term. Lag order p of the model is selected on the basis of some statistical criteria that are discussed later.

Besides the variables of AQI and the Baidu index in Shanghai, the AQI in Beijing was also included in the model as a potential explanatory variable that might affect Shanghai residents' concern about air pollution. Indeed, in modern times, information is highly mobile via different channels. People's views and opinions are impacted by not only situations in their local area but also by things occurring in other districts.

Therefore, in this study, vector $y_t = [AQI_t, AQI_t^{Beijing}, BaiduIndex_t]'$ was used. It contains three variables: AQI_t, daily air quality index in Shanghai; $AQI_t^{Beijing}$, air quality index in Beijing; and $BaiduIndex_t$, the Baidu index for keywords "Shanghai air quality". As usual, all three variables were log transformed to mitigate potential scaling problems. Hence, variable variations are expressed as percentage changes.

3.2. Data for Measuring Actual Air Pollution

Daily AQI data used to measure actual air pollution level in Shanghai are publicly available from the website of the China Air Quality Online Monitoring and Analysis Platform: https://www.aqistudy.cn. AQI is a synthesized index reflecting the degree of pollution in ambient air, calculated using the measured data for several major air pollutants ($PM_{2.5}$, PM_{10}, CO, NO_2, O_3, SO_2) according to the guidelines of the official environmental protection sector. A low AQI value indicates a low degree of air pollution, and a high AQI value implies a high degree of air pollution.

The daily AQI values of 2068 days over the period between 2 December 2013 and 31 July 2019 in Shanghai were retrieved to measure actual air pollution in Shanghai. Additionally, in order to answer the second research question of whether the decline in air quality in another major city, such as Beijing, affects the public concern about air quality in Shanghai, the daily AQI data of Beijing during the same period were also retrieved. Data prior to 2 December 2013 were not included in this study due to data unavailability.

3.3. Data for Measuring Residents' Concern about Air Pollution

The Baidu index was utilized to measure residents' concern about air pollution in Shanghai. The Baidu index data were provided by Baidu, Inc., and they are publicly available from the web page: http://index.baidu.com. Baidu is the most popular Internet search engine that occupies a major market share in China. The Baidu index is calculated on the basis of the search volume for a specific search item on a daily basis at the municipal, provincial, and national levels [16]. A high (low) value of the Baidu index for a certain item indicates that many (few) persons searched for information on the item and cared about the relevant topic. Compared to self-administrated survey methods, the Baidu index has two advantages in measuring the degree of public concern about air pollution. First, it covers a wide range of study samples. Since Baidu dominates the search engine market in China, its Baidu index could reflect the aggregate behaviors of most Chinese Internet users. Second, the Baidu index data are available at a daily frequency for several years. This property enabled us to examine not only the long term trend, but also short term fluctuations of public concern about air pollution.

In this study, the Baidu index was restricted within the district of Shanghai in order to rule out individuals who were not living in Shanghai, but were still interested in Shanghai's air quality. The keywords "Shanghai air quality" ("shanghai kongqi zhiliang" in Chinese) were selected as the searched-for item. Other related search terms, such as "Shanghai haze" ("shanghai wumai" in Chinese) and "Shanghai PM$_{2.5}$" ("shanghai pm2.5" in Chinese) were also tested for robustness checks. Consistent with the sample period for the AQI data, the Baidu index data between 2 December 2013 and 31 July 2019 were exploited.

Table 1 shows the summary statistics of the variables used in the analyses. In the table, both the original level and logarithmic value of variables are shown. In the VAR estimation, the logarithmic values were used to mitigate potential scaling problems.

Table 1. Summary statistics.

Variable		Observations	Mean	Std. Dev.	Minimum	Maximum
AQI	original	2068	81.27	38.95	20	468
	logarithmic	2068	4.30	0.44	3.00	6.15
AQI^{Beijing}	original	2068	108.24	68.01	21	500
	logarithmic	2068	4.51	0.58	3.04	6.21
BaiduIndex	original	2068	1351.33	1375.95	212	15,858
	logarithmic	2068	6.95	0.67	5.36	9.67

4. Results

This section reports the results of the empirical analyses. This section first explains how the optimal lag order for the VAR model was selected. Then, this section reports the core estimation results, focusing on the impulse response figures of different variables and forecast error variance decomposition (FEVD) estimates. After that, several robustness checks on the results are conducted.

4.1. Selection of Optimal Lag Order

Before estimating a VAR model, it is necessary to choose the optimal lag order for the model on the basis of lag order selection statistics. Table 2 reports useful statistics. It shows that different criteria suggest different selections of lag order. The LR (likelihood ratio), FPE (final prediction error), and AIC (Akaike's information criterion) statistics suggested to use eight lags. Differently, the HQIC (Hannan and Quinn information criterion) suggested six lags, and SBIC (Schwarz's Bayesian information criterion) suggested three lags. Since three of these five criteria suggested to use eight lags in the model, eight lags were initially selected for VAR estimation. In the robustness check section, situations using three and six lags are examined, and they demonstrated robust results.

Table 2. Lag order selection statistics for vector autoregression (VAR) model. LR, likelihood ratio; FPE, final prediction error; AIC, Akaike's information criterion; HQIC, Hannan and Quinn information criterion; SBIC, Schwarz's Bayesian information criterion.

lag	LR	FPE	AIC	HQIC	SBIC
0		0.022541	4.72123	4.72424	4.72944
1	4890.7	0.002112	2.35356	2.36559	2.38638
2	177.43	0.001955	2.27609	2.29715	2.33353
3	72.068	0.001904	2.24981	2.27990	2.33188 *
4	62.367	0.001863	2.22826	2.26737	2.33494
5	43.078	0.001841	2.21607	2.26421	2.34737
6	42.971	0.001819	2.20394	2.26110 *	2.35986
7	34.426	0.001804	2.19596	2.26215	2.37649
8	28.796 *	0.001795 *	2.19071 *	2.26593	2.39587
9	11.875	0.001800	2.19369	2.27793	2.42346
10	13.575	0.001804	2.19584	2.28911	2.45023

Note: * indicates optimal lag order selection according to each statistic.

4.2. Estimation Results

In order to generate meaningful estimation results, the whole VAR system must be stable. Figure 2 shows that all eigenvalues were inside the unit circle. This indicates that the established VAR system satisfied the required stability condition, and it could be relied on to analyze interactions among variables.

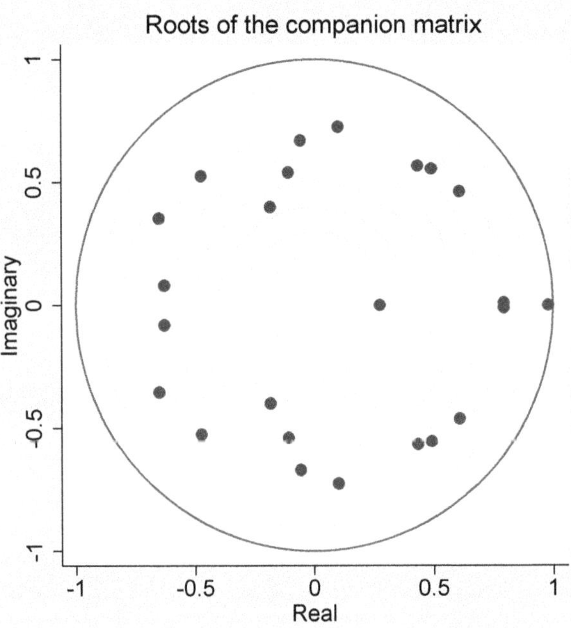

Figure 2. Eigenvalue stability condition.

Since the estimation results of a VAR model are rarely only explained by the estimated coefficients, the analyses primarily focused on impulse response figures (IRFs). Figure 3 shows the impulse response figures based on the estimated coefficients.

The three subfigures in the first row of Figure 3 show the responses of variables to an orthogonalized, one unit, positive shock of (logarithmic) AQI in Shanghai. As shown in Figure 3(i.a), after such an AQI shock, AQI notably increased. As displayed in Figure 3(i.b), the AQI in Beijing fluctuated a little bit, but variation was quite small and not statistically significant. Figure 3(i.c) notably shows that, in response to the AQI shock, the Baidu index rose significantly. To observe the response more clearly, this subfigure was amplified and is displayed in Figure 4(i). It is clear that the AQI shock immediately raised the Baidu index without any time lag. The response was persistent and significantly positive even after ten days. Therefore, Hypothesis 1 is supported. The residents in Shanghai really cared much about the ambient air quality. As air quality deteriorated, Shanghai citizens expressed more concerns about local air pollution, as reflected by the increase in the Baidu index.

The three subfigures in the second row of Figure 3 show the variable responses to a positive shock of (logarithmic) AQI in Beijing. In Figure 3(ii.a), an increase in Shanghai's AQI was observed. This could be explained by the spatial spillover effects of air pollution, as discussed in previous studies reporting the spatial interactions of air pollution among different regions (e.g., [50–52]). Figure 3(ii.b) shows that Beijing's AQI increased persistently. An important finding was that, as demonstrated by Figure 3(ii.c), the Baidu index for Shanghai's air quality rose in response to air pollution in Beijing. This implied that Shanghai residents' concern about local air quality increased after they observed that Beijing's air quality became worse. This graph was amplified and is displayed in Figure 4(ii). As demonstrated in that graph, one day after Beijing's AQI increased, the Baidu index for Shanghai's air quality began to increase. This increase was persistent for five days. The Baidu index returned to its original value after the sixth day. Therefore, Hypothesis 2 is supported. Air pollution problems in another large city indeed tend to increase the residents' concern about air pollution in Shanghai.

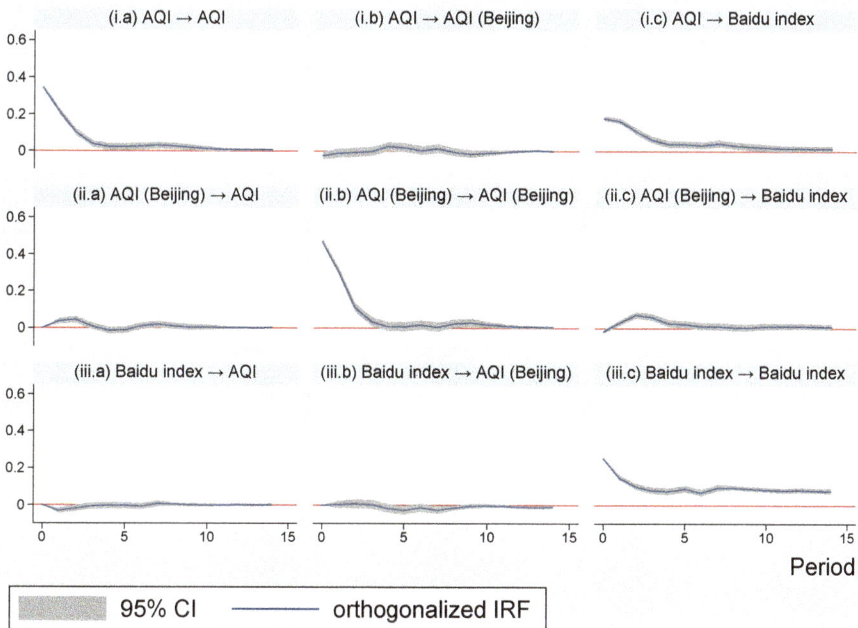

Figure 3. Impulse response figures (IRFs). Note: Each subfigure with the title of "X→Y" demonstrates the response of variable Y to an orthogonalized positive shock of variable X. In other words, X is an impulse variable, and Y is a response variable. One period in the figure denotes one day.

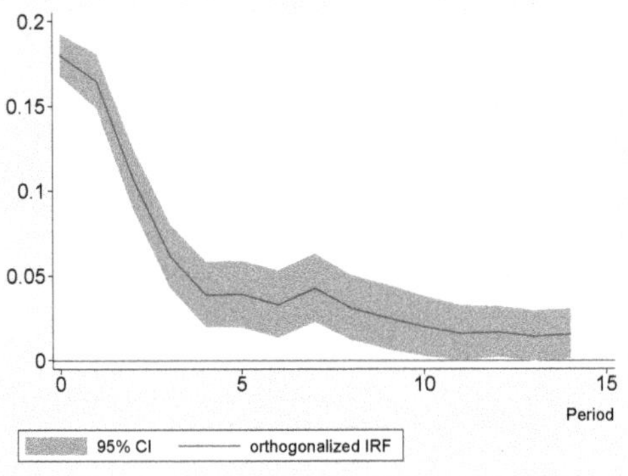

(i) AQI → Baidu index

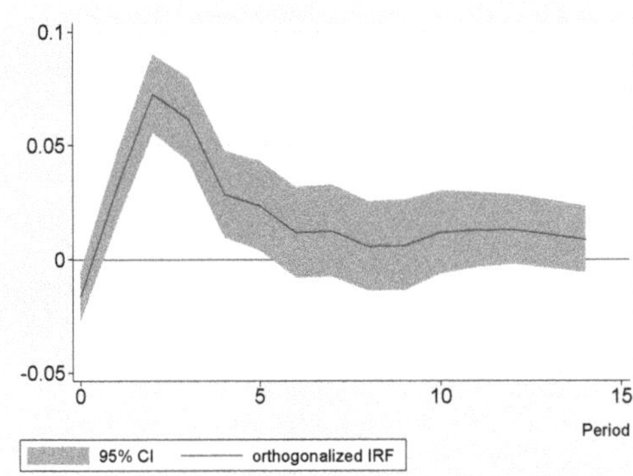

(ii) AQI (Beijing) → Baidu index

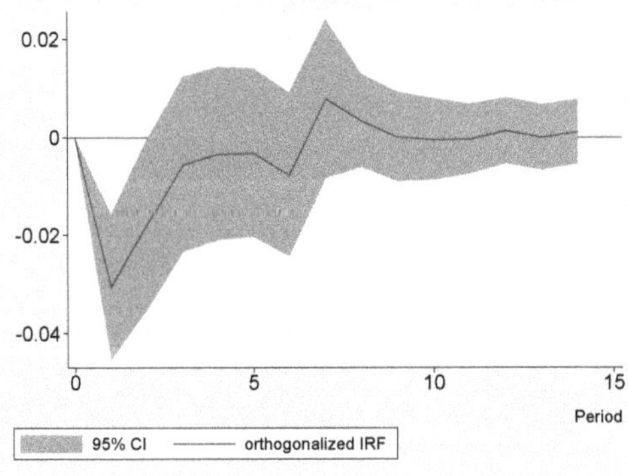

(iii) Baidu index → AQI

Figure 4. Amplified impulse response figures (IRFs) of interest. Note: Each subfigure with the title of "X→Y" demonstrates the response of variable Y to an orthogonalized positive shock of variable X. In other words, X is an impulse variable, and Y is a response variable. One period in the figure denotes one day.

The subfigures in the last row of Figure 3 demonstrated the variable responses to a positive shock of the (logarithmic) Baidu index. Figure 3(iii.a) presents the interesting finding that the AQI in Shanghai actually decreased after the Baidu index increased. In order to observe the details more clearly, this impulse response figure was amplified in Figure 4(iii). As can be seen from the graph, one day after an increase in the Baidu index, the AQI in Shanghai was below its initial level. This phenomenon lasted for two days. After that, the AQI returned back to its initial level. This finding supports Hypothesis 3. If public concern about air pollution intensified, people would take action to ameliorate air quality or, at least, avoid aggravating pollution and wait for the air quality to naturally improve. Figure 3(iii.b) demonstrates that the Baidu index had no significant impact on Beijing's AQI. Figure 3(iii.c) demonstrates the persistency of the increase in the Baidu index.

Table 3 shows FEVD estimates for the Baidu index. As can be seen from the table, generally, within the horizon of fourteen days, local AQI explained around 30%–40% forecast error variance of Shanghai residents' concern about air pollution. The AQI in Beijing explained roughly 5% variance. The rest was explained by the Baidu index itself. Obviously, local air quality was quite important in forecasting fluctuations of the Baidu index. Air quality in another famous city could also partly explain changes in the Baidu index for Shanghai's air quality. These support previous findings that were obtained by observing impulse response figures.

Table 3. Forecast error variance decomposition (FEVD) estimates for the Baidu index.

Forecast Horizon	FEVD of the Baidu Index		
	AQI	AQI (Beijing)	Baidu Index
0	0	0	0
1	0.334	0.003	0.663
2	0.404	0.008	0.588
3	0.406	0.037	0.557
4	0.395	0.054	0.551
5	0.386	0.056	0.558
6	0.374	0.056	0.570
7	0.369	0.055	0.576
8	0.358	0.053	0.589
9	0.347	0.051	0.603
10	0.337	0.049	0.614
11	0.328	0.048	0.624
12	0.320	0.047	0.633
13	0.312	0.047	0.641
14	0.305	0.046	0.649

4.3. Robustness Analyses

In this subsection, we outline several robustness checks that were conducted on previous estimation results. First, whether results were sensitive to the selection of the search engine keyword for the Baidu index was examined. Second, alternative sample periods were considered. Third, alternative selections of lag orders in the VAR model were inspected. Lastly, whether air quality in other cities aside from Beijing affected the public concern about air pollution in Shanghai was further investigated. The impulse response figures are displayed in the subfigures of Figure 5.

4.3.1. Alternative Baidu Index Keyword

In previous analyses, the keywords "Shanghai air quality" ("shanghai kongqi zhiliang" in Chinese) were relied on to derive the Baidu index. Next, another search term, "Shanghai haze ("shanghai wumai" in Chinese), was utilized to get the Baidu index. The results are shown in Figure 5(i.a–i.c). It is apparent that the AQI in Shanghai positively affected the Baidu index; the AQI in Beijing positively affected the Baidu index for Shanghai; and a rise in the Baidu index tended to depress the AQI in Shanghai. Thus, the previous findings of this study remained unchanged. In addition, other search terms such as

"Shanghai $PM_{2.5}$", "air quality", "haze", and "$PM_{2.5}$" were checked, and similar results were generated. To save space, the results using those alternative keywords are not reported here. The Supplementary Material attached to this paper provides additional information to demonstrate the robustness of the study results to the selection of Baidu index keywords. In the Supplementary Material, it is shown that the Baidu index values for different keywords were strongly correlated, indicating that different keywords actually reflected highly consistent online searching behaviors and provided similar information. Moreover, the Supplementary Material demonstrates the IRFs for the keywords "Shanghai $PM_{2.5}$" (which had the lowest correlation coefficient with "Shanghai air quality", compared to other candidate keywords), which were almost the same as the IRFs shown in Figure 3.

4.3.2. Shorter Sample Period

The baseline analyses were based on the sample period between 2 December 2013 and 31 July 2019. It was admitted that the level of the Baidu index was not only determined by the degree of public interest on the specific topic, but also influenced by some other factors such as the changes in the market share of the Baidu search engine, total number of Internet users, and Internet users' habits. The longer the sample period was, the larger the impact those alternative factors might have. As pointed out by Lu et al. [12], the long term annual trend of public concern about air pollution probably had characteristics different from those of short term fluctuations in air pollution. To mitigate this issue and inspect whether the study results were robust to the selection of the sample period, a shorter sample period from 1 January 2017 to 31 July 2019 was considered. Results are presented in Figure 5(ii.a–ii.c), which are similar to those that have previously been derived. Other subsample periods, such as between 1 January 2018 and 31 July 2019, were also checked. Results were analogous, but are not reported here to save space.

4.3.3. Alternative Selection of Variable Lag Order

Previously, eight lags of variables were selected for the VAR model according to the LR, FPE, and AIC statistics. Since the HQIC and SBIC suggested to use different lags, the model was re-estimated on the basis of the alternative selections of lag order. According to the SBIC, three lags might be suitable. The corresponding impulse response figures are demonstrated in Figure 5(iii.a–iii.c). Notably, these new impulse response figures did not shake the previous statements in this study. Other lag orders, such as six, to follow the suggestion by HQIC, were also tested. Similar results were obtained.

4.3.4. VAR Model with AQI in Another City

Previous analyses used a VAR model containing the AQI variable in Beijing. Next, whether results were sensitive to the selection of this specific city was checked. The city of Nanjing was taken instead of Beijing. Nanjing is one of the largest and most important cities in East China. The obtained impulse response figures are demonstrated in Figure 5(iv.a–iv.c). Compared to the baseline results, it was found that situations would be similar if Nanjing rather than Beijing were selected. Moreover, circumstances using the AQI in Guangzhou, which is the largest city in South China, were also inspected. Results were similar, but not reported here.

Overall, robustness checks strengthened the findings of this study. Hypotheses 1–3 were all supported.

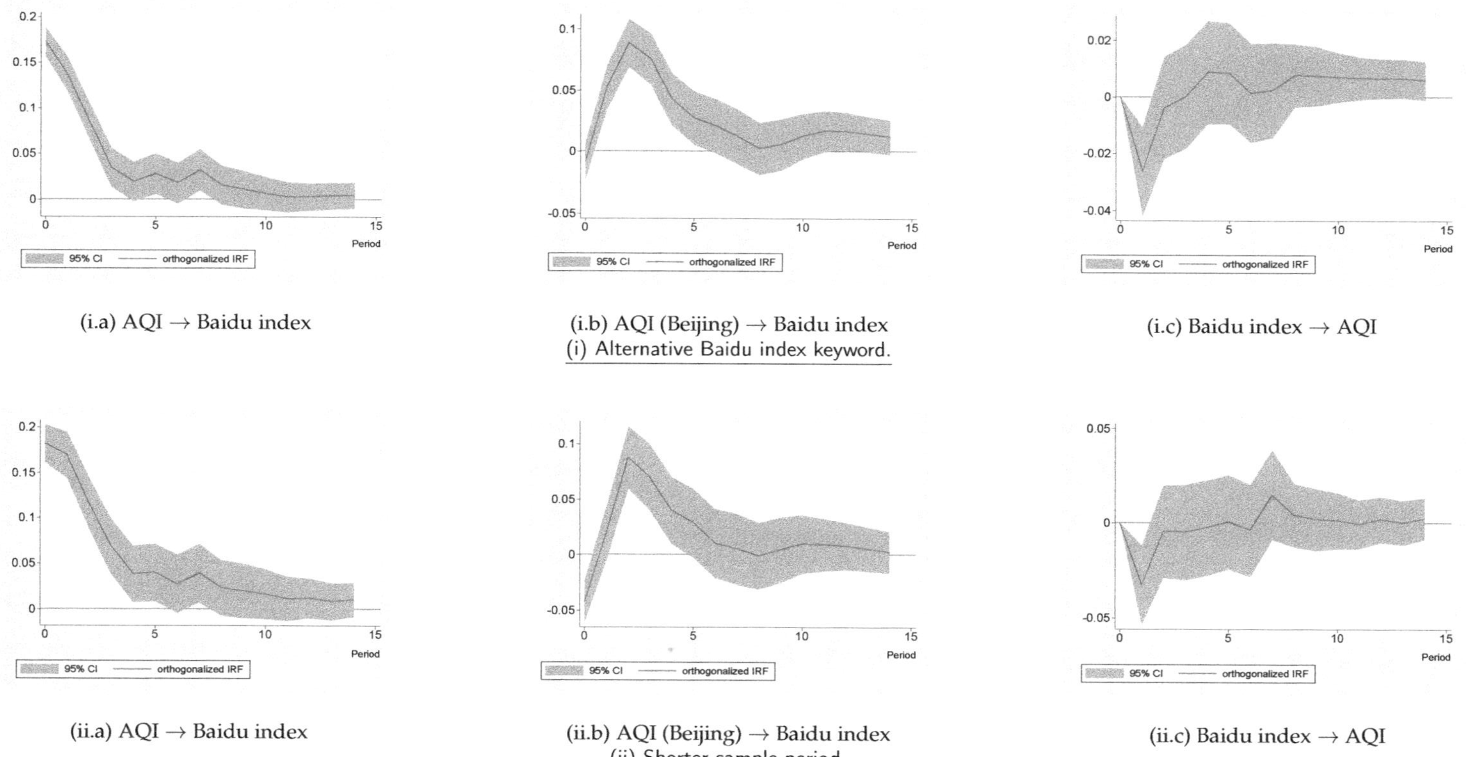

Figure 5. Cont.

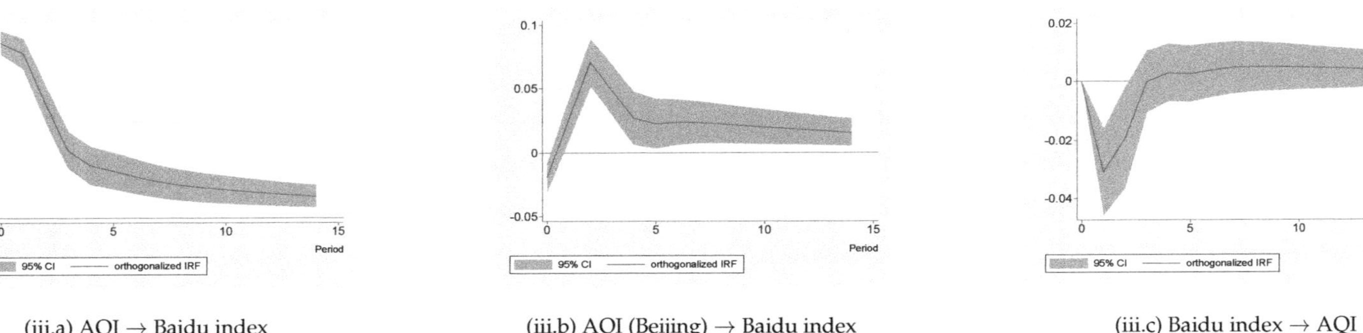

(iii.a) AQI → Baidu index

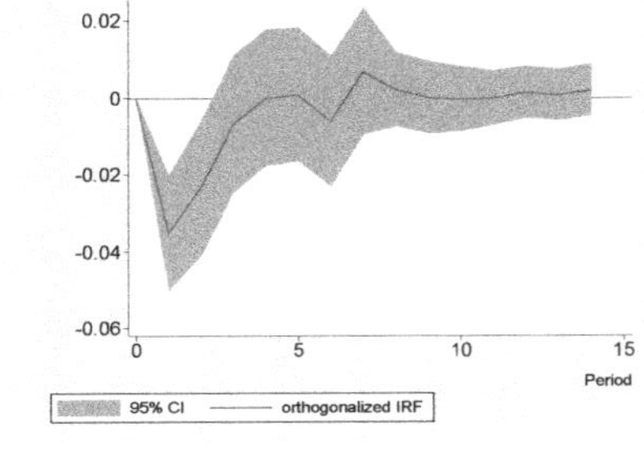

(iii.b) AQI (Beijing) → Baidu index
(iii) Alternative selection of variable lag order.

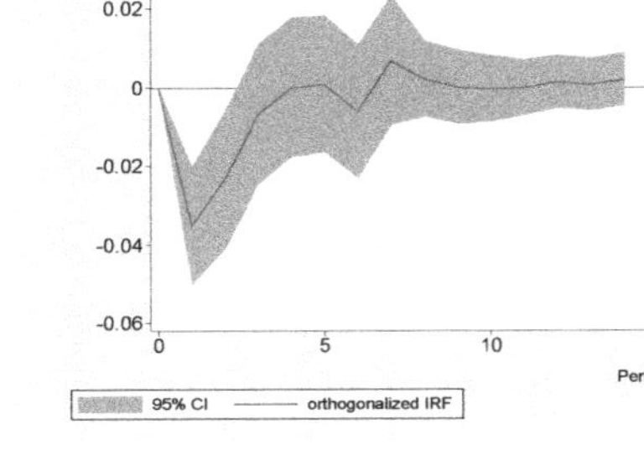

(iii.c) Baidu index → AQI

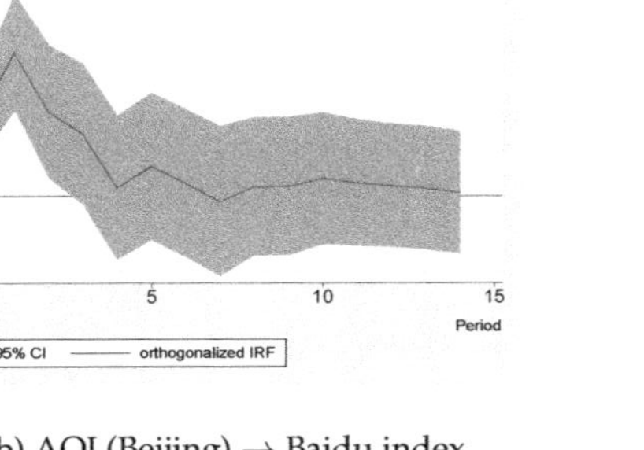

(iv.a) AQI → Baidu index

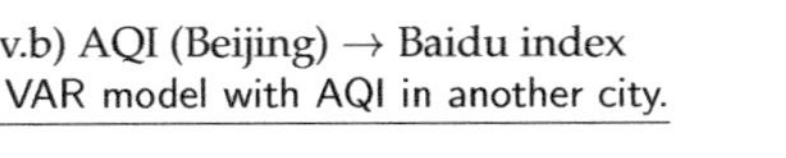

(iv.b) AQI (Beijing) → Baidu index
(iv) VAR model with AQI in another city.

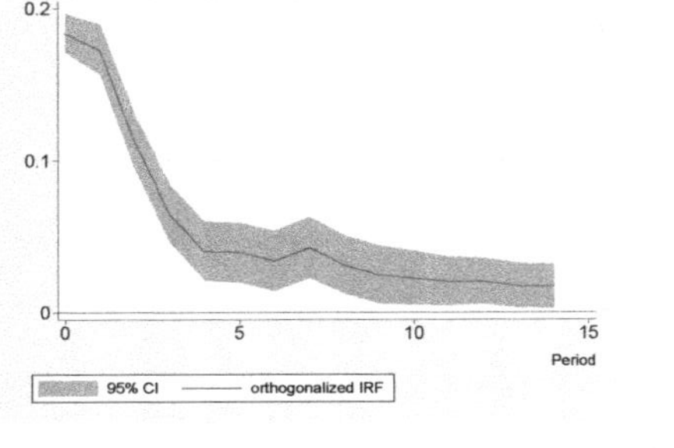

(iv.c) Baidu index → AQI

Figure 5. Robustness analyses: impulse response figures (IRFs). Note: Each subfigure with the title of "X→Y" demonstrates the response of variable Y to an orthogonalized positive shock of variable X. In other words, X is an impulse variable, and Y is a response variable. One period in the figure denotes one day.

5. Discussion and Implications

The estimation results in this study provided three interesting findings. First, local residents perceived the deprivation of air quality and expressed their concern about air pollution quickly, within the day on which the air quality index rose. This supported previous studies that found a strong correlation between perceived and actual level of air quality (e.g., [22,24]). It was implied that air quality is consistently monitored and assessed by Shanghai residents and that they show high awareness of air pollution. This also lends support to the finding by Yan et al. [36] that people who live in richer and more polluted cities are more likely to perceive air pollution. In addition, the concurrent rise of social media platforms in China might also contribute to this strong association. These platforms provide a way to share news and public opinions quickly. Media alerts on AQI would trigger heated debate and discussion on air quality, as well as information seeking behavior.

Second, a decline in air quality in another major city, such as Beijing, also raised the local concern about air quality in Shanghai. This was plausible because air pollutants could be transported by wind, causing pollution to spread over an extensive region within a short time interval. Prior studies have provided evidence that air pollution has a negative spatial spillover impact on neighboring cities' public health [50]. Given the fact that Beijing is 1200 km away from Shanghai, air pollution in Beijing might not directly cause health problems in Shanghai. However, it increases public concerns on air pollution.

Third, a rise in Shanghai residents' concern had a beneficial impact on air quality improvement. On the one hand, this could be explained by prior findings revealing that environmental concerns could promote people's pro-environmental behaviors [41,42]. On the other hand, public concerns about air pollution could force governments to take actions to improve air quality [11,12]. It has been reported that China has curbed industrial emissions, restricted the use of cars on the road, and shut down coal mines in large cities such as Beijing, Shanghai, and Guangzhou [53].

This study contributed to the air pollution literature by empirically examining the reciprocal relationship between public concern about air quality and actual air quality using data on a daily basis. Different from prior studies that relied on survey data to measure perceived air quality, this study utilized a big data based dataset dating back to 2013 to conduct more accurate analyses. Additionally, this study performed VAR analysis rather than only basic correlation analysis, which helped demonstrate the rapidness and persistency of the rise in public concern about air pollution.

From a practical perspective, it was implied that providing timely air quality indices to residents could be a powerful tool to raise public awareness on air pollution to take steps for pollution mitigation. The government could utilize online search engines as a tool for displaying more information and advice explaining how people can minimize their contribution to air pollution during residents' keyword search process. When observing a significant drop in air quality, the government should initiate immediate actions to tackle air pollution by providing information explaining how people can minimize their exposure to polluted air and which air pollution reduction strategies have been implemented in order to prevent excessive concerns on pollution. Furthermore, displaying AQI in other major Chinese cities could be utilized by local governments to promote local residents' awareness of air pollution and, subsequently, support for environmental protection.

6. Conclusions, Limitations, and Future Research

Using Shanghai as a case study, this study empirically examined the interactive relationship between actual level of air pollution and residents' concern about air pollution on the basis of the daily Baidu search index and AQI data. This study highlighted that residents in Shanghai expressed immediate concerns about air pollution as long as the air quality in Shanghai or in other major Chinese cities got worse. The study results also suggested that raising awareness on air pollution would motivate individuals or the government to carry out actions to improve air quality.

In evaluating the significant findings from this study, two major limitations need to be acknowledged. First, this paper focused on the circumstances in only one city because of difficulties in data collection. If we can collect data for a larger number of cities in the future and use the panel VAR approach, we could re-examine the findings based on Shanghai. A dataset covering more regions could also allow us to investigate possible heterogeneities across different regions, which may provide further policy implications. Second, as a preliminary exploration, this study only considered three variables in the VAR model because other variables were not available on a daily frequency. There is no doubt that some other factors may also be influential in the relationship between actual air pollution level and public concern about the problem. In the future, more variables may be introduced into the model as long as the data availability problem is overcome.

Supplementary Materials: The Supplementary Material is available online at http://www.mdpi.com/1660-4601/16/23/4784/s1.

Author Contributions: Conceptualization and funding acquisition, D.D. and X.X.; methodology, data curation, formal analysis, and original draft preparation, D.D.; literature review and review and editing, X.X.; software, validation, and supervision, W.X. and J.X.

Funding: This research was funded by the Fundamental Research Funds for the Central Universities (Grant Nos. JBK1801039 and JBK1809054).

Acknowledgments: The authors are grateful to the Editors and two anonymous referees for their comments and suggestions.

Conflicts of Interest: The authors declare no conflict of interest.

References

1. World Health Organization. Air Pollution. 2019. Available online: https://www.who.int/airpollution/en/ (accessed on 10 September 2019).

2. He, J.; Liu, H.; Salvo, A. Severe Air Pollution and Labor Productivity: Evidence from Industrial Towns in China. *Am. Econ. J. Appl. Econ.* **2019**, *11*, 173–201. doi:10.1257/app.20170286. [CrossRef]

3. Kelly, F.J.; Fussell, J.C. Air pollution and public health: Emerging hazards and improved understanding of risk. *Environ. Geochem. Health* **2015**, *37*, 631–649. doi:10.1007/s10653-015-9720-1. [CrossRef] [PubMed]

4. Poudyal, N.C.; Paudel, B.; Green, G.T. Estimating the Impact of Impaired Visibility on the Demand for Visits to National Parks. *Tour. Econ.* **2013**, *19*, 433–452. doi:10.5367/te.2013.0204. [CrossRef]

5. Zheng, S.; Wang, J.; Sun, C.; Zhang, X.; Kahn, M.E. Air pollution lowers Chinese urbanites' expressed happiness on social media. *Nat. Hum. Behav.* **2019**, *3*, 237–243. doi:10.1038/s41562-018-0521-2. [CrossRef]

6. Sun, Z.; Li, J. Citizens' Satisfaction with Air Quality and Key Factors in China—Using the Anchoring Vignettes Method. *Sustainability* **2019**, *11*. doi:10.3390/su11082206. [CrossRef]

7. Guo, Y.; Liu, F.; Lu, Y.; Mao, Z.; Lu, H.; Wu, Y.; Chu, Y.; Yu, L.; Liu, Y.; Ren, M.; et al. Factors Affecting Parent's Perception on Air Quality—From the Individual to the Community Level. *Int. J. Environ. Res. Public Health* **2016**, *13*, 493. [CrossRef]

8. Reames, T.G.; Bravo, M.A. People, place and pollution: Investigating relationships between air quality perceptions, health concerns, exposure, and individual- and area-level characteristics. *Environ. Int.* **2019**, *122*, 244–255. doi:10.1016/j.envint.2018.11.013. [CrossRef]

9. Mccarthy, S. Low awareness in Hong Kong of health risks from bad air quality, green group says. *South China Morning Post*, 5 November 2018.

10. Larson, C. China's Ma Jun on the Fight To Clean Up Beijing's Dirty Air. *Yale Environment 360*, 10 April 2012.

11. Kay, S.; Zhao, B.; Sui, D. Can Social Media Clear the Air? A Case Study of the Air Pollution Problem in Chinese Cities. *Prof. Geogr.* **2015**, *67*, 351–363. doi:10.1080/00330124.2014.970838. [CrossRef]

12. Lu, Y.; Wang, Y.; Zuo, J.; Jiang, H.; Huang, D.; Rameezdeen, R. Characteristics of public concern on haze in China and its relationship with air quality in urban areas. *Sci. Total Environ.* **2018**, *637–638*, 1597–1606. doi:10.1016/j.scitotenv.2018.04.382. [CrossRef]

13. Howard, E. Shanghai's Air Pollution Is Now Worse than Beijing. 2018. Available online: https://unearthed.greenpeace.org/2018/02/12/shanghai-air-pollution-worse-beijing/ (accessed on 10 September 2019).

14. He, G.; Chen, Y.; Chen, B.; Wang, H.; Shen, L.; Liu, L.; Suolang, D.; Zhang, B.; Ju, G.; Zhang, L.; et al. Using the Baidu Search Index to Predict the Incidence of HIV/AIDS in China. *Sci. Rep.* **2018**, *8*, 9038. doi:10.1038/s41598-018-27413-1. [CrossRef]

15. Li, Z.; Liu, T.; Zhu, G.; Lin, H.; Zhang, Y.; He, J.; Deng, A.; Peng, Z.; Xiao, J.; Rutherford, S.; et al. Dengue Baidu Search Index data can improve the prediction of local dengue epidemic: A case study in Guangzhou, China. *PLoS Negl. Trop. Dis.* **2017**, *11*, e0005354. doi:10.1371/journal.pntd.0005354. [CrossRef] [PubMed]

16. Li, K.; Liu, M.; Feng, Y.; Ning, C.; Ou, W.; Sun, J.; Wei, W.; Liang, H.; Shao, Y. Using Baidu Search Engine to Monitor AIDS Epidemics Inform for Targeted intervention of HIV/AIDS in China. *Sci. Rep.* **2019**, *9*, 320. doi:10.1038/s41598-018-35685-w. [CrossRef] [PubMed]

17. Zhong, S.; Yu, Z.; Zhu, W. Study of the Effects of Air Pollutants on Human Health Based on Baidu Indices of Disease Symptoms and Air Quality Monitoring Data in Beijing, China. *Int. J. Environ. Res. Public Health* **2019**, *16*, 1014. [CrossRef] [PubMed]

18. Huang, X.; Zhang, L.; Ding, Y. The Baidu Index: Uses in predicting tourism flows—A case study of the Forbidden City. *Tour. Manag.* **2017**, *58*, 301–306. doi:10.1016/j.tourman.2016.03.015. [CrossRef]

19. Li, X.; Pan, B.; Law, R.; Huang, X. Forecasting tourism demand with composite search index. *Tour. Manag.* **2017**, *59*, 57–66. doi:10.1016/j.tourman.2016.07.005. [CrossRef]

20. Yang, X.; Pan, B.; Evans, J.A.; Lv, B. Forecasting Chinese tourist volume with search engine data. *Tour. Manag.* **2015**, *46*, 386–397. doi:10.1016/j.tourman.2014.07.019. [CrossRef]

21. Egondi, T.; Kyobutungi, C.; Ng, N.; Muindi, K.; Oti, S.; Vijver, S.V.d.; Ettarh, R.; Rocklöv, J. Community Perceptions of Air Pollution and Related Health Risks in Nairobi Slums. *Int. J. Environ. Res. Public Health* **2013**, *10*, 4851–4868. [CrossRef]

22. Peng, M.; Zhang, H.; Evans, R.D.; Zhong, X.; Yang, K. Actual Air Pollution, Environmental Transparency, and the Perception of Air Pollution in China. *J. Environ. Dev.* **2019**, *28*, 78–105. doi:10.1177/1070496518821713. [CrossRef]

23. Semenza, J.C.; Wilson, D.J.; Parra, J.; Bontempo, B.D.; Hart, M.; Sailor, D.J.; George, L.A. Public perception and behavior change in relationship to hot weather and air pollution. *Environ. Res.* **2008**, *107*, 401–411. doi:10.1016/j.envres.2008.03.005. [CrossRef]

24. Atari, D.O.; Luginaah, I.N.; Fung, K. The Relationship between Odour Annoyance Scores and Modelled Ambient Air Pollution in Sarnia, "Chemical Valley", Ontario. *Int. J. Environ. Res. Public Health* **2009**, *6*. doi:10.3390/ijerph6102655. [CrossRef]

25. Moffatt, S.; Phillimore, P.; Bhopal, R.; Foy, C. 'If this is what it's doing to our washing, what is it doing to our lungs?' Industrial pollution and public understanding in North-East England. *Soc. Sci. Med.* **1995**, *41*, 883–891. doi:10.1016/0277-9536(94)00380-C. [CrossRef]

26. Oglesby, L.; Künzli, N.; Monn, C.; Schindler, C.; Ackermann-Liebrich, U.; Leuenberger, P.; the SAPALDIA Team. Validity of Annoyance Scores for Estimation of Long Term Air Pollution Exposure in Epidemiologic Studies : The Swiss Study on Air Pollution and Lung Diseases in Adults (SAPALDIA). *Am. J. Epidemiol.* **2000**, *152*, 75–83. doi:10.1093/aje/152.1.75. [CrossRef] [PubMed]

27. Pu, S.; Shao, Z.; Fang, M.; Yang, L.; Liu, R.; Bi, J.; Ma, Z. Spatial distribution of the public's risk perception for air pollution: A nationwide study in China. *Sci. Total Environ.* **2019**, *655*, 454–462. doi:10.1016/j.scitotenv.2018.11.232. [CrossRef] [PubMed]

28. Williams, I.D.; Bird, A. Public perceptions of air quality and quality of life in urban and suburban areas of London. *J. Environ. Monit.* **2003**, *5*, 253–259. doi:10.1039/B209473H. [CrossRef]

29. Berezansky, B.; Portnov, B.; Barzilai, B. Objective vs. Perceived Air Pollution as a Factor of Housing Pricing: A Case Study of the Greater Haifa Metropolitan Area. *J. Real Estate Lit.* **2010**, *18*, 99–122. doi:10.5555/reli.18.1.f057w463l258r805. [CrossRef]

30. Brody, S.D.; Peck, B.M.; Highfield, W.E. Examining Localized Patterns of Air Quality Perception in Texas: A Spatial and Statistical Analysis. *Risk Anal.* **2004**, *24*, 1561–1574. doi:10.1111/j.0272-4332.2004.00550.x. [CrossRef]

31. Kim, M.; Yi, O.; Kim, H. The role of differences in individual and community attributes in perceived air quality. *Sci. Total Environ.* **2012**, *425*, 20–26. doi:10.1016/j.scitotenv.2012.03.016. [CrossRef]

32. Deguen, S.; Padilla, M.; Padilla, C.; Kihal-Talantikite, W. Do Individual and Neighborhood Characteristics Influence Perceived Air Quality? *Int. J. Environ. Res. Public Health* **2017**, *14*, 1559. [CrossRef]

33. Liu, X.; Zhu, H.; Hu, Y.; Feng, S.; Chu, Y.; Wu, Y.; Wang, C.; Zhang, Y.; Yuan, Z.; Lu, Y. Public's Health Risk Awareness on Urban Air Pollution in Chinese Megacities: The Cases of Shanghai, Wuhan and Nanchang. *Int. J. Environ. Res. Public Health* **2016**, *13*, 845. [CrossRef]

34. Claeson, A.S.; Lidén, E.; Nordin, M.; Nordin, S. The role of perceived pollution and health risk perception in annoyance and health symptoms: A population-based study of odorous air pollution. *Int. Arch. Occup. Environ. Health* **2013**, *86*, 367–374. doi:10.1007/s00420-012-0770-8. [CrossRef]

35. Zhang, X.; Wargocki, P.; Lian, Z.; Thyregod, C. Effects of exposure to carbon dioxide and bioeffluents on perceived air quality, self-assessed acute health symptoms, and cognitive performance. *Indoor Air* **2017**, *27*, 47–64. doi:10.1111/ina.12284. [CrossRef] [PubMed]

36. Yan, L.; Duarte, F.; Wang, D.; Zheng, S.; Ratti, C. Exploring the effect of air pollution on social activity in China using geotagged social media check-in data. *Cities* **2019**, *91*, 116–125. doi:10.1016/j.cities.2018.11.011. [CrossRef]

37. Shaaf, M.; Rod Erfani, G. Air pollution and the housing market: A neural network approach. *Int. Adv. Econ. Res.* **1996**, *2*, 484–495. doi:10.1007/BF02295473. [CrossRef]

38. Yusuf, A.A.; Resosudarmo, B.P. Does clean air matter in developing countries' megacities? A hedonic price analysis of the Jakarta housing market, Indonesia. *Ecol. Econ.* **2009**, *68*, 1398–1407. doi:10.1016/j.ecolecon.2008.09.011. [CrossRef]

39. Dong, D.; Xu, X.; Yu, H.; Zhao, Y. The Impact of Air Pollution on Domestic Tourism in China: A Spatial Econometric Analysis. *Sustainability* **2019**, *11*. doi:10.3390/su11154148. [CrossRef]

40. Wang, L.; Fang, B.; Law, R. Effect of air quality in the place of origin on outbound tourism demand: Disposable income as a moderator. *Tour. Manag.* **2018**, *68*, 152–161. doi:10.1016/j.tourman.2018.03.007. [CrossRef]

41. Gu, D.; Huang, N.; Zhang, M.; Wang, F. Under the Dome: Air Pollution, Wellbeing, and Pro-Environmental Behaviour Among Beijing Residents. *J. Pac. Rim Psychol.* **2015**, *9*, 65–77. doi:10.1017/prp.2015.10. [CrossRef]

42. Wang, B.Z.; Cheng, Z. Environmental Perceptions, Happiness and Pro-environmental Actions in China. *Soc. Indic. Res.* **2017**, *132*, 357–375. doi:10.1007/s11205-015-1218-9. [CrossRef]

43. Tam, K.P.; Chan, H.W. Environmental concern has a weaker association with pro-environmental behavior in some societies than others: A cross-cultural psychology perspective. *J. Environ. Psychol.* **2017**, *53*, 213–223. doi:10.1016/j.jenvp.2017.09.001. [CrossRef]

44. Do, Y.; Kim, J.Y.; Lineman, M.; Kim, D.K.; Joo, G.J. Using internet search behavior to assess public awareness of protected wetlands. *Conserv. Biol.* **2015**, *29*, 271–279. doi:10.1111/cobi.12419. [CrossRef]

45. Caputi, T.L.; Leas, E.; Dredze, M.; Cohen, J.E.; Ayers, J.W. They're heating up: Internet search query trends reveal significant public interest in heat-not-burn tobacco products. *PLoS ONE* **2017**, *12*, e0185735. doi:10.1371/journal.pone.0185735. [CrossRef] [PubMed]

46. Mellon, J. Internet Search Data and Issue Salience: The Properties of Google Trends as a Measure of Issue Salience. *J. Elections Public Opin. Parties* **2014**, *24*, 45–72. doi:10.1080/17457289.2013.846346. [CrossRef]

47. Park, S.; Lee, J.; Song, W. Short-term forecasting of Japanese tourist inflow to South Korea using Google trends data. *J. Travel Tour. Mark.* **2017**, *34*, 357–368. doi:10.1080/10548408.2016.1170651. [CrossRef]

48. Liu, B.; Wang, Z.; Qi, X.; Zhang, X.; Chen, H. Assessing cyber-user awareness of an emerging infectious disease: Evidence from human infections with avian influenza A H7N9 in Zhejiang, China. *Int. J. Infect. Dis.* **2015**, *40*, 34–36. doi:10.1016/j.ijid.2015.09.017. [CrossRef] [PubMed]

49. Yang, H.; Li, S.; Sun, L.; Zhang, X.; Hou, J.; Wang, Y. Effects of the Ambient Fine Particulate Matter on Public Awareness of Lung Cancer Risk in China: Evidence from the Internet-Based Big Data Platform. *JMIR Public Health Surveill* **2017**, *3*, e64. doi:10.2196/publichealth.8078. [CrossRef] [PubMed]

50. Chen, X.; Shao, S.; Tian, Z.; Xie, Z.; Yin, P. Impacts of air pollution and its spatial spillover effect on public health based on China's big data sample. *J. Clean. Prod.* **2017**, *142*, 915–925. doi:10.1016/j.jclepro.2016.02.119. [CrossRef]

51. Cole, M.A.; Elliott, R.J.R.; Okubo, T.; Zhou, Y. The carbon dioxide emissions of firms: A spatial analysis. *J. Environ. Econ. Manag.* **2013**, *65*, 290–309. doi:10.1016/j.jeem.2012.07.002. [CrossRef]

52. Marbuah, G.; Amuakwa-Mensah, F. Spatial analysis of emissions in Sweden. *Energy Econ.* **2017**, *68*, 383–394. doi:10.1016/j.eneco.2017.10.003. [CrossRef]

53. Greenstone, M. Four years after declaring war on pollution, China is winning. *The New York Times*, 12 March 2018.

International Journal of
*Environmental Research
and Public Health*

Article

Statistical Modeling of the Early-Stage Impact of a New Traffic Policy in Milan, Italy

Paolo Maranzano [1,*], **Alessandro Fassò** [2], **Matteo Pelagatti** [3] and **Manfred Mudelsee** [4]

[1] Department of Statistics and Quantitative Methods (DISMEQ), University of Milano-Bicocca, 20126 Milano, Italy

[2] Department of Management, Information and Production Engineering (DIGIP), University of Bergamo, 24044 Dalmine, Italy; alessandro.fasso@unibg.it

[3] Department of Economics, Management and Statistics (DEMS), University of Milano-Bicocca, 20126 Milan, Italy; matteo.pelagatti@unimib.it

[4] Climate Risk Analysis, 37581 Heckenbeck, Germany; mudelsee@climate-risk-analysis.com

[*] Correspondence: p.maranzano@campus.unimib.it; Tel.: +39-3487-057-117

Received: 29 December 2019; Accepted: 1 February 2020; Published: 8 February 2020

Abstract: Most urban areas of the Po basin in the North of Italy are persistently affected by poor air quality and difficulty in disposing of airborne pollutants. In this context, the municipality of Milan started a multi-year progressive policy based on an extended limited traffic zone (Area B). Starting on 25 February 2019, the first phase partially restricted the circulation of some classes of highly polluting vehicles on the territory, in particular, Euro 0 petrol vehicles and Euro 0 to 3 diesel vehicles, excluding public transport. This is the early-stage of a long term policy that will restrict access to an increasing number of vehicles. The goal of this paper is to evaluate the early-stage impact of this policy on two specific vehicle-generated pollutants: total nitrogen oxides (NO_x) and nitrogen dioxide (NO_2), which are gathered by Lombardy Regional Agency for Environmental Protection (ARPA Lombardia). We use a statistical model for time series intervention analysis based on unobservable components. We use data from 2014 to 2018 for pre-policy model selection and the relatively short period up to September 2019 for early-stage policy assessment. We include weather conditions, socio-economic factors, and a counter-factual, given by the concentration of the same pollutant in other important neighbouring cities. Although the average concentrations reduced after the policy introduction, this paper argues that this could be due to other factors. Considering that the short time window may be not long enough for social adaptation to the new rules, our model does not provide statistical evidence of a positive policy effect for NO_x and NO_2. Instead, in one of the most central monitoring stations, a significant negative impact is found.

Keywords: air pollution; oxides; traffic; state space; milan; area b; cross validation; policy intervention analysis; counter-factual; unobservable components

---

1. Introduction

Air quality monitoring is one of the major challenges that European institutions jointly with national and local administrations are facing in terms of environmental protection. In particular, the 2008 European Air Quality Directive (AQD) 2008/50/EC [1] requires EU Member States to design appropriate air quality plans for zones where the air quality does not comply with the AQD limit values. In the last few decades, European countries implemented various modeling methods to assess the effects of local and regional emission abatement policy options on air quality and human health [2]. On the one side, they include scenario approaches, in which running a chemical-physical simulation model with and without a specific emission source allows for quantifying the impact on air quality levels [3,4]. On the other side, they also include more comprehensive and multidisciplinary approaches, such as Integrated Assessment Models (IAM), which combine simultaneously many features of the economy, society, and scientific findings. These models are based on the combination of multiple mathematical tools and allow for assessing the impact of environmental policies or to improve the air quality control system. Typical tools are the full cost-benefit analyses [5], in which abatement measures, costs, and benefits are expressed in monetary units, optimization, and spatial analysis [6,7].

In areas such as Northern Italy, where the industrial transition in the 1990s reduced coal burning and sulphur concentration, the large majority of environmental studies focus their attention on toxic pollutants that are produced by thermic vehicle engines and house heating plants. These are known to generate serious health effects [8]. Total nitrogen oxides (NO_x), nitrogen dioxide (NO_2), and particulates matters (PM_{10} and $PM_{2.5}$) belong to this class.

According to the above EU rules, governments adopted standards and quantitative limits for pollutant emissions to make economic agents responsible and implement abatement policies. In particular, the maximum concentration for NO_x and NO_2 is set to 40 $\mu g/m^3$ annual average and 200 $\mu g/m^3$ hourly not to be exceeded more than 18 times in a single year. Figure 1 represents the average concentration levels of NO_2 in Europe for the year 2018. The Po basin in Northern Italy stands out as a heavily polluted area with difficulties in pollution management. The negative impact on society is not limited to health only. There is increasing evidence showing that bad air quality in general, and high NO_2 concentrations in particular, impact the economy, including finance [9] and tourism [10].

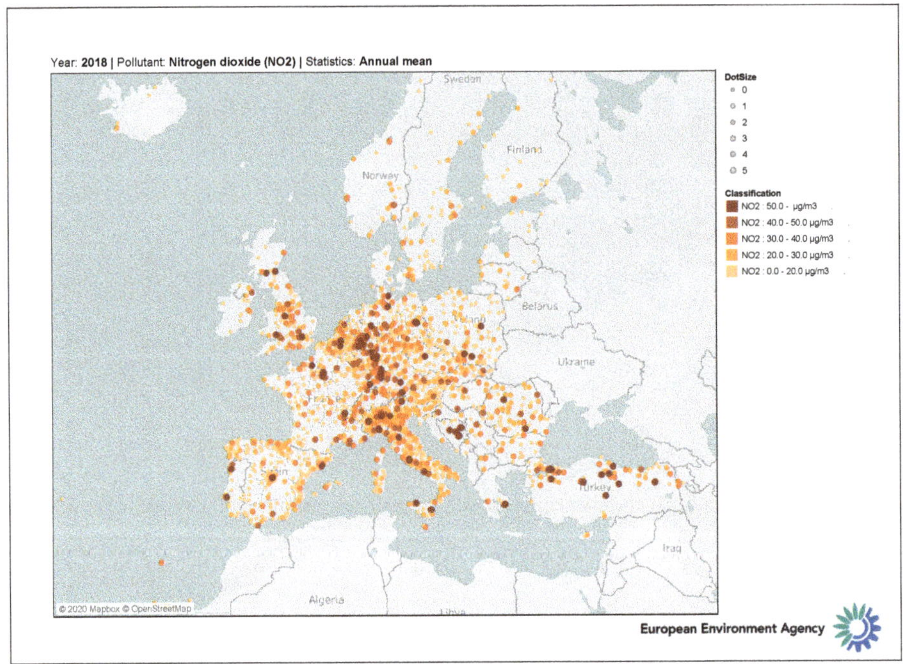

Figure 1. Annual average NO_2 concentrations ($\mu g/m^3$) in Europe during 2018. Levels are expressed in $\mu g/m^3$. Source: European Environmental Agency

The present paper analyses the introduction of the first phase of an air quality control policy in the municipality of Milan, which started on 25 February 2019 and directly acts on traffic rules. The administration defined an extended limited traffic zone, named Area B (https://www.comune.milano.it/aree-tematiche/mobilita/area-b), where the access and circulation for the most polluting vehicles, as well as those longer than 12 meters, have some partial restrictions, enforced by a monitoring system of entrance gates controlling each license plate and imposing a fine on unauthorized vehicles.

The access prohibition concerns Euro 0 petrol vehicles and a large part of Euro 0, 1, 2, and 3 diesel vehicles, with specific exemptions for public transport, itinerant traders, and residents, and it is active from Monday to Friday during business hours (from 7:30 a.m. to 7:30 p.m.), except holidays. According to the municipality of Milan, the share of cars registered in the Milan metropolitan area and involved in the restrictions in 2019 is close to 17%, while the share of freight transport vehicles is around 53% [11]. Area B is a progressive policy divided into various phases, which will concern an increasing number of vehicle classes. In terms of NOx emissions, the administration expects a reduction of 4–5% per year until 2022 and a reduction of 11% between 2023 and 2026. The policy will be fully operative within October 2030.

Area B extends the previously existing limited traffic zone, Area C, which covers just the historical city centre. The physical coverage of the two restriction zones is represented in Figure 2, which highlights the arrangement of both within the city borders. Area B covers almost the entire area of the city, excluding extreme peripheral districts.

Statistical literature on air quality grew up sharply in the last decades. Two main statistical modeling directions have been developed. One has a focus on pollutants concentration and the other on human exposure. Regarding the latter, recent advances are based on crowdsourced data, such as smartphone data modeling [12]. Regarding pollutants concentrations, increasing attention is being given to latent component models; see, as an example [13] and for the problem of misalignment. In particular, the use of the INLA-SPDE approach for misalignment between pollutant concentration and epidemiological data [14] and PCA based methods with missing data [15].

When the territory under study is large and spatial correlation is important, spatio-temporal models are appropriate. See, for example, the multivariate state space approach of Calculli et al. [16], which is capable of handling jointly PM_{10}, NO_2 and weather variables, the approach of Menezes et al. [17] for modeling daily NO_2 trends in Portugal. Moreover, the land-use regression model (LUR) under a state space approach has been used for modeling air pollution in Tehran [18]. Despite this growing spatial literature, time series analysis methods have been recently developed to understand the effect of meteorology on pollutant concentration [19], which will be the main focus of this paper.

The previous Milan limited traffic zone, known as Area C, has already been treated in literature by Fassò [20], who analyzed its introduction through spatio-temporal models, by Invernizzi et al. [21], who considered its impact on black carbon, and by Percoco [22] who considered its effect on traffic. Moreover, similar problems have been studied for London "sulphur-free zone" [23] and the "low emission zone" in Munich [24]. In Fassò [20], the author considered both particulates and nitrogen oxides and observed the presence of a more pronounced permanent reduction of the latter within the restricted area, despite the data showing a strong spatial variability depending on the type of pollutant. This is consistent with the known emissions pattern of particulate matters and nitrogen oxides. The latter are mainly primary gaseous pollutants and can be directly attributed to anthropogenic sources, such as car traffic and house heating. Moreover, from the so-called INEMAR emission inventory [25], in Milan province, 68% of NOx and only 41% of PM10 are due to road traffic. Hence, in this first study of Area B, we will take into account NO_x and NO_2 and postpone the analysis of PM10 and PM2.5 to further research. To adjust for confounding factors, we will consider weather conditions in Milan, the main calendar events, and the concentration levels of oxides observed in neighbouring towns, as in a pseudo-treatment-control approach.

The study aims to identify and quantify variations in pollutant levels due to the above described Area B. Hence, the present paper will try to investigate and test the following two scientific hypotheses:

Hypothesis 1. *The introduction of Area B achieved significant changes in pollution concentration for the city of Milan;*

Hypothesis 2. *The variation occurred homogeneously on the territory and the stations do not show spatial variability of the effects.*

The first hypothesis aims to quantify the impact of the policy on pollution levels measured by several air quality stations scattered around the city and to assess whether this evidence is significantly supported by the data. The impact is evaluated both regarding the statistical significance of the estimates, the absolute magnitudes of the coefficients, and their signs. From the policy maker perspective, the expected coefficients should be negative, indicating a reduction effect on concentrations due to the car traffic restrictions. However, given the complexity of the phenomenon, a change of opposite sign cannot be ruled out either. The second research hypothesis is dedicated to the comparison of the estimates for the considered stations: the effect can be considered homogeneous when the sign and the magnitude of the coefficients for all the stations are similar.

The paper is structured as follows. Section 2 describes the dataset and the methodologies used for the analyses. In particular, we briefly explain the composition of both weather and air quality monitoring systems in Milan, available data sources, and metadata information. Then, we present the methodologies implemented for the preliminary analysis and the state space approach to time series analysis for air quality data. Section 3 reports and discusses the empirical results of the estimated models and their implications. Section 4 concludes the paper discussing the two research hypotheses in light of what emerged from the data analysis and gives some hints for future research developments.

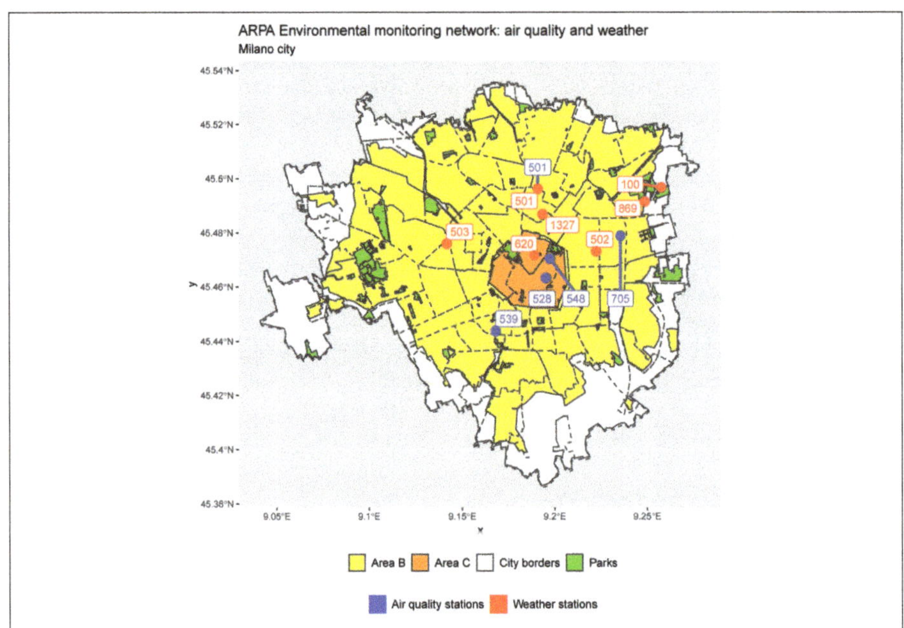

Figure 2. Monitoring system in Milan. Air quality stations (blue points): Marche (501), Verziere (528), Senato (548), Liguria (539) and Città Studi (705). Weather stations (red points): Lambrate (100), Zavattari (503), Brera (620), Feltre (869), Rosellini (1327), Juvara (502), and Marche (501).

2. Materials and Methods

In this section, we present the structure of the ARPA dataset and briefly introduce the methodologies implemented for the analyses. Section 2.1 introduces the data source for the Milan case study and the spatio-temporal structure of the data and provides a brief description of the variables taken into consideration. Section 2.2 designs the preliminary analysis, which introduces a temporal treatment-control experiment to highlighting the differences in concentration levels before and after the policy intervention. Section 2.3 gives a detailed overview of the use of state space models in time series analysis for the study of air quality data, including also a specific subsection for model selection and policy intervention.

2.1. Data

2.1.1. Air Quality and Weather Monitoring Network in Milan

Data on pollution and weather conditions of Lombardy are collected from the Lombardy Regional Agency for Environmental Protection (ARPA Lombardia), which makes a large open data portal fully available to users (https://www.dati.lombardia.it/). The agency manages a diffuse monitoring system distributed among the regional territory and counting on hundreds of monitoring stations collecting intra-daily information on climate and pollution through sensors.

Installed within the borders of Milan are seven weather monitoring stations and five air quality monitoring stations. Air quality stations are classified according to a taxonomy system that identifies the type and function in the network. The stations Liguria (ARPA code 539), Marche (ARPA code 501), Senato (ARPA code 548), and Verziere (ARPA code 528) are urban traffic control units: sensors installed near important roads and intersections in order to accurately quantify the pollution generated by traffic. The station Città Studi (ARPA code 705) is instead of type urban background, that is, the station is located in such a position that the level of pollution is not mainly influenced by specific sources but by the integrated contribution of all the upwind sources at the station with respect to the predominant directions of the winds on the site [8]. The seven weather stations are Marche (ARPA code 501), Lambrate (ARPA code 100), Zavattari (ARPA code 503), Brera (ARPA code 620), Feltre (ARPA code 869), Rosellini (ARPA code 1327), and Juvara (ARPA code 502).

Figure 2 georeferences on the map the exact position of each station and allows for identifying the position with respect to Area B and Area C. Air quality stations are represented as blue points, while weather stations are the red points. Marche station (ARPA code 501), in the upper side of the map, is the only one to collect both weather and pollution data and is represented with a double label, the first one blue and the second red.

The spatial distribution of the stations is not uniform: air quality stations cover northern, eastern, central, and southern parts of the city, leaving the western districts uncovered; climate stations cover in detail the city centre and all the northern neighbours but are not installed in the south.

2.1.2. Temporal Coverage, Pollutants, and Weather Measures

The analysis presented in this paper takes into account daily measures from 1 January 2014 to 30 September 2019, generating an overall sample of 2099 daily observations.

The whole, the monitoring system provides information about many urban pollutants, such as carbon dioxide, particulates, and oxides. All the pollutants are measured as $\mu g/m^3$. As already stated in the Introduction, we focus our attention on concentrations of total nitrogen oxides (NO_x) and nitrogen dioxide (NO_2), which are mainly primary gaseous pollutants, hence considered as proxies of pollution emissions due to human activities, first of all car traffic.

Weather stations provide measures of local temperature ($^\circ$C), rainfall (cumulated mm), humidity (%), global radiation (W/m^2), wind speed (m/s), and wind direction. The wind direction is expressed in clockwise degrees from 0° to 360°; for example, 90° identifies winds going from east to west. To make results easier to interpret, we decide to aggregate the measurements on wind direction and speed by constructing a set of new variables that describe the average speed in the four quadrants of the compass rose. The Northeast quadrant (Q_{NE}) corresponds to degrees between 0 and 90, the Southeast quadrant (Q_{SE}) to degrees from 90 to 180, the Southwest quadrant (Q_{SW}) to degrees from 180 to 270 and the Northwest quadrant (Q_{NW}) to the remaining values lying between 270 and 360 degrees.

These measures will be used in the modeling part to capture local weather conditions specific to the city of Milan. Instead of using the data referring to the weather station closest to each air quality station, we preferred to aggregate each of the climate variables through a daily city average valid for each pollution station. In this way, the subsequent models will be fully comparable guaranteeing the maximum possible spatial coverage.

2.1.3. Anthropogenic Activities

Human activities, and therefore the quality of the air we breathe, are often affected by calendar events that are recorded based on national, local, and religious holidays and weekends. Calendar effects are captured by a set of covariates, which identify the weekends and the main Italian holidays, both religious and secular. Holidays are collected in a dummy variable named *Holidays*, while the weekends are contained in a dummy variable named *WeekEnd*. Specific effects related to the behavior of people can be observed when holidays coincide with the weekend; therefore, we considered two terms of interaction between the two dummies. The interaction terms include those holidays that fall on Saturday, denoted as *Saturday:Holiday*, and those on which they fall on Sunday, which is *Sunday:Holiday*.

For a correct assessment of the effects of the traffic policy on pollutants concentrations in Milan, it is necessary to purify the estimates from any external weather or socio-economic effects overlapping with the policy and which may hence alter policy effects. This operation is accomplished by introducing a counter-factual term into the models represented by the pollution levels observed in other cities surrounding Milan. We considered seven important urban centres located in the Lombardy Po Valley area, which show socio-demographic and economic characteristics and weather conditions similar to Milan, but which cannot be directly affected by the limited traffic zone. These urban centres are Bergamo (East), Brescia (far East), Cremona (far Southeast), Lodi (Southeast), Pavia (South), Saronno (North) and Treviglio (East). As reported in Figure 3, the considered candidates cover a large territory surrounding Milan in all the directions while maintaining a sufficient distance to be considered independent in terms of traffic.

2.2. Methods: Average and Median Difference before and after the Policy

Figure 4 shows the temporal evolution of yearly average and median concentrations in the period preceding and following the entry into force of the policy for each control units located in Milan. According to the figure, starting from 2015, the city of Milan recorded a generalized reduction of concentration levels especially in peripheral areas, such as Marche and Liguria. Observed mean values for 2019 present a further reduction of concentrations rather apparently anomalous and significant. The comparison between the levels of NO_x and NO_2 pairs for each station shows obvious common trends between the two pollutants both considering the annual average and median values. Averages and medians follow similar temporal patterns, but focusing on nitrogen oxides sensors, it is possible to note that the medians are significantly smaller than the averages, highlighting the heavy tailed characteristic of the distribution (positive asymmetry) and the presence of extreme values. Following these facts, an interesting question to investigate is if, and how much, the greater difference observed in 2019 can be attributed to traffic restrictions, or if it is due to a general de-carbonization trend that the city is experiencing, or to weather variations not considered yet.

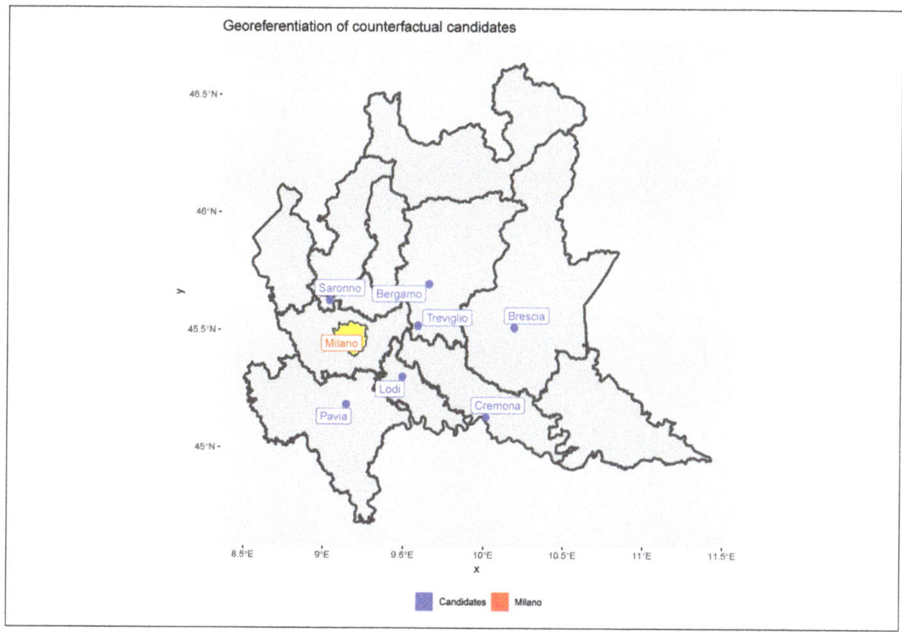

Figure 3. Georeferentiation of counter-factual candidates. Geographical positioning of the counter-factual candidates with respect to Milan.

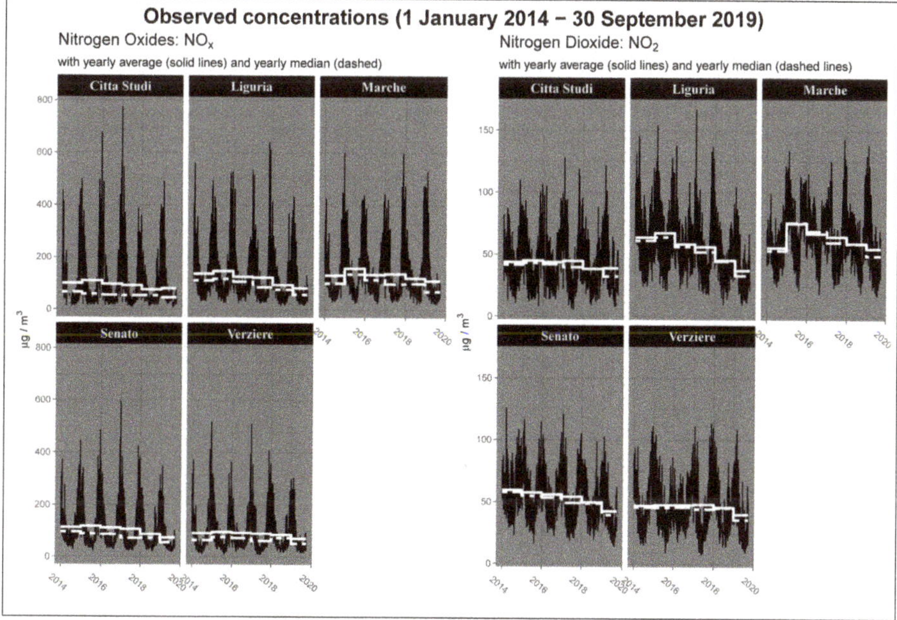

Figure 4. Pollutant levels in Milan ($\mu g/m^3$). Observed concentrations levels of NO_x and NO_2 between 2014 and 2019 with yearly average and median values. Values are expressed as $\mu g/m^3$.

Before investigating the factors and causes that may have generated these sharp reductions, we perform a preliminary analysis of the concentration levels pre-and-post policy, in order to quantify the changes observed in 2019 both in Milan and in the other centres. Since air quality data present outliers and heavy-tail distributions given by extreme events, the only use of average values for central tendency estimation can provide misleading results. Therefore, we compare the central values obtained both considering the sample mean and the sample median, which is notoriously a more robust indicator if outliers occur [26,27].

The comparison is performed through the computation of two statistics based on the difference of central tendency indicators. The first statistic computes the difference between the average of the observations gathered after the policy intervention and the average of observations referring to the sub-period 2014–2018. The second statistics consists of computing the difference between median concentration levels observed in 2019 and before that year. The difference in average concentrations is denoted by d_{AVG}, whereas the difference in median concentrations is denoted as d_{MED}. Since both sub-periods are treated as independent of each other, from the statistical perspective, the statistics are assimilable to unpaired samples statistics.

Both statistics use the observations collected between 25 February and 30 September of each year, with a total length of 214 days. Approaches of this type can be framed in a context of treatment-control analysis, in which the data referred to the year 2019 constitute the treatment group, while the observations collected between 2014 and 2018 compose the control group. Control data refer to a 5-year-period; therefore, the concentrations measured on the same calendar day are aggregated into a single representative value calculated as the daily average concentration of the period 2014 to 2018. Denoting as c_{ij} the observed pollutant concentration during the day j, where $j = 25\ February, ..., 30\ September$, of the year i, where $i = 2014, ..., 2018$, the average for a generic calendar day j is computed as $c_j = \frac{\sum_{i=2014}^{2018} c_{ij}}{5}$.

Let $U = \{u_j, \quad j = 1, 2, ..., 214\}$ be the treatment observations and $V = \{v_j, \quad j = 1, 2, ..., 214\}$ the control observations, the difference of averages is defined as $d_{AVG} = AVG(U) - AVG(V)$ and the difference of medians is calculated as $d_{MED} = MED(U) - MED(V)$, where $AVG(.)$ is the temporal sample mean and $MED(.)$ is the temporal sample median.

2.3. Methods: Time Series Modeling Using a State Space Approach

In this section, we discuss the time series models used to identify the policy effect, the estimation algorithms, and the related inference. Firstly, we introduce a brief description of the basic structural model (BSM) using a state space approach for time series analysis and the estimation algorithm based on the Kalman filter [28,29]. Then, we present a three-step procedure used to select the most representative model in terms of predictive power and quality of fit. As a last step, we explain how the policy intervention is included in the models and how it should be interpreted.

2.3.1. Basic Structural Model for Air Quality Data

According to their physical characteristics, air pollution concentrations time series are often characterized by seasonality, high persistence [30,31], strong right skewness with uni-modal distribution, and scale invariance [32]. Therefore, we analyze the concentrations using the basic structural model [33,34] augmented by deterministic regressors for weather conditions and socio-economic features.

BSM is defined as a simple unobservable components model composed by local linear trend (LLT), stochastic seasonality, and irregular (white noise) component. LLT describes both the temporal evolution of the series level and its slope, while the seasonal component aims to capture cyclical behaviors given by natural and anthropogenic phenomena. We modeled the seasonal component using a trigonometric form for daily data, hence with period $s = 365$, and considering only a few harmonics given the very regular and almost deterministic behavior of the series. This fact avoids the risk of a model over-parametrization.

Let $\{y_1, y_2, ..., y_n\}$ be the time series of the observed pollution concentrations in logarithmic scale, the state space form of BSM without regressors is composed by the following equations:

$$y_t = \mu_t + \gamma_t + \varepsilon_t , \tag{1}$$

where $\varepsilon_t \sim N(0, \sigma_\varepsilon^2)$ is the measurement error and

$$LLT\ (Level): \quad \mu_t = \mu_{t-1} + \beta_{t-1} + \eta_t , \qquad \eta_t \sim WN(0, \sigma_\eta) , \tag{2}$$

$$LLT\ (Slope): \quad \beta_t = \beta_{t-1} + \zeta_t , \qquad \zeta_t \sim WN(0, \sigma_\zeta) , \tag{3}$$

$$Stochastic\ seasonality: \quad \gamma_t = \sum_{j=1}^{k} \gamma_{j,t} , \tag{4}$$

where $k \leq \lfloor \frac{s}{2} \rfloor$ is the number of included harmonics and $\gamma_{j,t}$ is the non-stationary stochastic cycle

$$\begin{bmatrix} \gamma_{j,t} \\ \gamma_{j,t}^* \end{bmatrix} = \begin{bmatrix} cos(2\pi j/s) & sin(2\pi j/s) \\ -sin(2\pi j/s) & cos(2\pi j/s) \end{bmatrix} \begin{bmatrix} \gamma_{j,t-1} \\ \gamma_{j,t-1}^* \end{bmatrix} + \begin{bmatrix} \omega_{j,t} \\ \omega_{j,t}^* \end{bmatrix} , \tag{5}$$

$\omega_t \sim WN(0, \sigma_\omega^2)$ and $\omega_t^* \sim WN(0, \sigma_\omega^2)$ are white-noise processes with mean zero and variance σ_ω^2.

Equation (1) is called measurement equation and describes the evolution of the observed series as the sum of the underlying components, while Equations (2)–(4) are named transition equations. Equations (2) and (3) compose the LLT and describe respectively the unobservable processes of the level and the slope, whereas Equation (4) describes the trigonometric seasonality evolution. Weather, socio-economic factors, and policy intervention will be included in the models adding a set of deterministic components to the measurement Equation (1). Since the BSM with Gaussian errors belongs to the class of Gaussian linear models, the estimation step has been performed using the Kalman Filter algorithm, an iterative procedure, which allows estimating simultaneously the unobservable components and the model's parameters by maximizing the Gaussian likelihood function.

When dealing with Gaussian linear state space models, the parameters estimated using a maximum likelihood (ML) approach inherit the asymptotic properties of ML estimators [29]. The distribution of the MLE is asymptotically approximated using a Gaussian distribution, which allows deriving the usual asymptotic confidence intervals and t-tests for significance. Assuming a significance level of 5%, the estimates are statistically significant if the standardized value lies outside of the interval $[-1.96, 1.96]$, obtained using the quantiles of a Standard Normal distribution. Moreover, since the dependent variable is expressed in logarithmic scale, the coefficients have to be interpreted as relative increases or decreases in concentration levels due to a unitary increase in the explanatory variable.

2.3.2. Three-Step Model Selection

We now propose a three-step procedure for model selection, which considers multiple rules based on cross-validation, information criteria, and stepwise regression. To avoid estimation bias due to the policy introduction, all the steps are computed using only the observations before the introduction of Area B that is, from 1 January 2014 to 24 February 2019.

Step 1 is designed for selecting the most predictive seasonal component, defined in Equation (4), comparing different model specifications, which consider a varying number of harmonics k for the trigonometric function. Specifically, we fit 10 alternative models for each station: in each of them, the trigonometric seasonality is modeled by an increasing number of harmonics ranging from $k = 1$ to $k = 10$. The use of an increasing number of sinusoids, in our case up to 10, allows the modeling of complex seasonality with strong variations within short periods, but at the same time increases the model complexity.

Once the seasonal component has been selected, step 2 introduces in Equation (1) a counter-factual term x_t able to capture weather and socio-economic factors common to the Po basin and affecting the air quality of Milan. In our approach, the counter-factual candidates are the time series introduced in Section 2.1.3 and which refer to the measurements of pollutant concentrations in seven important cities around Milan. The new measurement Equation can be written as follows:

$$y_t = \mu_t + \gamma_t + \theta x_t + \varepsilon_t, \tag{6}$$

where x_t is the logarithm of the counter-factual time series and θ is its coefficient, μ_t follows Equations (2) and (3), and γ_t follows the specification obtained by step 1. The expected sign of θ is positive: higher levels of air pollution should correspond to high values in nearby cities due to similar conditions.

In step 3, we identify the best subset of calendar events and weather covariates, capturing residual variations not yet covered by the counter-factual or by the latent components. These residual variations are estimated by the smoothed observation disturbances from Equation (6) that is $\hat{\varepsilon}_t$, and describe residual patterns that have not been explained by the persistence of series, the seasonality or characteristics common to nearby territories of the region.

Relevant weather and calendar covariates are selected through a backward-forward stepwise regression algorithm, which uses as a starting model the auxiliary linear regression expressed in Equation (7). The equation represents the full model which sets up the smoothed observation errors $\hat{\varepsilon}_t$ as dependent variable and the weather conditions and calendar events as covariates:

$$\begin{aligned}
\hat{\varepsilon}_t = {} & \tau_1 Holidays + \tau_2 WeekEnd + \tau_3 Saturday : Holidays + \tau_4 Sunday : Holidays \\
& + \tau_5 Temperature + \tau_6 Rainfall + \tau_7 Radiation + \tau_8 Humidity \\
& + \tau_9 WindSpeedQ_{\mathrm{NE}} + \tau_{10} WindSpeedQ_{\mathrm{NW}} + \tau_{11} WindSpeedQ_{\mathrm{SW}} + \tau_{12} WindSpeedQ_{\mathrm{SE}} \\
& + e_t,
\end{aligned} \tag{7}$$

The stepwise regression is set up twice for each station: in one case, it selects the model according to the Akaike's Information Criterion (AIC), while, in the other, it uses the Bayesian Information Criterion (BIC). The algorithm starts estimating the full model and computes the AIC or the BIC. Iteratively, it drops out the predictors one at a time; at each step, it computes the new information criterion and considers whether the criterion is improved by adding back in a variable removed at a previous step. The procedure ends when the reintroduction of each omitted variable does not improve the information criteria.

In the first two steps, we select the seasonal component and the counter-factual term by fitting and comparing alternative models based on Equations from (1) to (4) according to their predictive power and their ability to adapt adequately to the observed data. The first principle, which tests the predictive power of the models, relies on the minimization of the cross-validated mean square forecasting error (MSFE) evaluated for up to 10-step-ahead forecast horizon, that is, $\hat{y}_{t+h} \quad \forall\, h = 1, 2, \ldots, 10$, while the second compares the models in terms of estimation quality. The latter computes both corrected Akaike's Information Criteria (AICc) and BIC intending to select the model that minimizes both. To identify a unique model for all the stations located in Milan, we proceed to a global comparison, both graphical and analytical, of the two blocks of indicators, giving attention to the overall performances and not focusing only on individual outputs.

According to the cross-validation principle for time series [35,36], we split the full time series into two subsets: a training set for model estimation and a test set for evaluating the out-of-sample forecast performances. The training set includes all the measurements until 24 February 2018, while the test set contains observations relative to the sub-period 25 February 2018–24 February 2019, for a total count of 365 out-of-sample observations. The exclusion of observations after the start of the traffic restrictions makes it possible to obtain unbiased estimates of the policy effects avoiding overlapping with other unidentified factors. Before starting the iterative loop, the model to evaluate is estimated just on time using the observations included in the original training set. At the end of the estimation, the cross-validation algorithm is iteratively implemented as follows. For each iteration, the algorithm extracts the first ten observations available in the test set, generating a forecasting set, and computes three quantities: the 1-to-10 step-ahead forecasts that is $\hat{y}_{t+h}$ $\forall h = 1, ..., 10$, the forecast errors $\hat{y}_{t+h} - y_{t+h}$ and the quadratic forecast errors $(\hat{y}_{t+h} - y_{t+h})^2$. The first out-of-sample observation is discarded and the set of forecasting observations is updated right-shifting the forecast horizon by 1 unit and adding the new observation. These operations are repeated for a number of times equal to the length of the test set, in our cases 365 times. The algorithm returns the output of 365 different sequences of 1–10 step-ahead forecasts; for each step-ahead $h = 1, 2, ..., 10$, the MSFE is calculated as

$$MSFE(h) = \frac{\sum_{j=1}^{365} (\hat{y}_{t+h} - y_{t+h})^2}{365}.$$

2.3.3. Policy Intervention Analysis

The introduction of new rules or limitations to individual behaviors can lead to the co-existence of multiple effects with different structure, such as simultaneous immediate changes and adaptive changes that take a long time before visible effects occur. Take into consideration that this fact leads to implement intervention analysis, which includes both permanent and transitory effects. Further details and examples of ARMA-like transfer function applied to intervention analysis are available in Pelagatti ([29]).

The policy intervention is modeled through the combination of two individual effects: (1) a permanent effect, estimated by δ_1 that measures the level shift of pollutant concentrations given by the treatment and modeled as a step dummy, which is D_{1t}, which assumes a value equal to 1 starting from 25 February 2019; (2) a transitory effect, estimated by δ_0 and evolving according to a first-order difference dynamics of the type

$$w_t = \lambda w_{t-1} + \delta_0 D_{2,t}, \tag{8}$$

where $D_{2,t}$ is a impulse dummy, which assumes value equal 1 for 25 February 2019 and 0 otherwise and λ measures the persistence of the transitory effect. The sum of the two effects returns the total effect, which expresses the estimated overall reduction or increase in air pollutant levels generated by the policy. The measurement equation after the three-step model selection and augmented by the policy intervention is eventually expressed as follows:

$$y_t = \mu_t + \gamma_t + \theta x_t + Z_t \Phi + \delta_1 D_{1t} + w_t + \varepsilon_t , \tag{9}$$

where y_t is the logarithm of pollution concentrations in one of the stations in Milan, x_t is the logarithm of pollution concentrations in the optimal counter-factual station, μ_t is the LLT evolving according to Equations (2) and (3), γ_t is the optimal seasonal component selected in step one, Z_t is a matrix containing the set of optimal subset of weather and calendar covariates selected in step 3, and Φ is the associate vector of coefficients.

2.3.4. Software

All the statistical computations and figures have been carried out using the statistical software R [37]. For state space models estimation, the *KFAS* package [38] was used. Cross-validation, forecasting, and model selection codes have been developed by the authors. The graphic elaborations were obtained by using the packages *ggplot2* [39] and *sf* [40].

3. Results

In this section, we present and comment on the empirical results relating both to the differences between pre-and-post policy averages and medians and to the policy intervention analysis for the Milan Area B case study. Section 3.1 shows the variations in concentration levels of NO_x and NO_2 for all the stations installed in Milan and for the other seven cities around it. Section 3.2 presents the model selection results, the values of the selection criteria, and the final model specifications. Section 3.3 reports the empirical estimates of the policy effects obtained through the basic structural model augmented by the policy intervention.

3.1. Average and Median Differences

Empirical differences of concentrations levels for all the considered stations are presented in Table 1. For both nitrogen oxides and nitrogen dioxide, using the difference of mean and median respectively, it reports the estimates of the difference between the average (the median) concentrations for the year 2019 and the average (the median) concentrations of the same period for the years 2014–2018.

Table 1. Differences between the average concentration level of the sub-period 2014–2018 and the treatment period 2019. Differences are expressed in $\mu g/m^3$.

Station Name	Nitrogen Oxides		Nitrogen Dioxide	
	d_{AVG}	d_{MED}	d_{AVG}	d_{MED}
Milano city stations				
Città Studi	0.94	−0.43	−2.63	−4.26
Liguria	−25.59	−26.91	−19.49	−20.71
Marche	−17.29	−22.00	−11.53	−13.37
Senato	−14.99	−13.84	−10.46	−9.73
Verziere	−2.66	−5.00	−4.35	−5.78
Other urban centres in Lombardy				
Bergamo	−11.56	−8.90	−5.11	−4.07
Brescia	−4.61	−6.39	−3.46	−4.12
Cremona	3.47	1.10	1.53	0.85
Lodi	−5.99	−7.62	-1.85	−2.57
Pavia	−14.75	−17.77	−7.75	−9.28
Saronno	−7.96	−7.84	−8.66	−8.87
Treviglio	0.06	−3.71	2.25	0.39

The estimates highlight large negative differences in oxides concentrations between 2019 and the period 2014–2018, both in the metropolitan area of Milan and in almost all the surrounding towns. Particularly heavy reductions, and similar to those in Milan, were recorded in the cities of Bergamo (East) and Pavia (South). The simultaneous abatement inside and outside Milan confirms the presence of a general decreasing trend in the aggregate levels of pollutants for the Lombardy Po basin as already indicated by the previous figures.

The differences registered in Milan are relevant both in suburban districts, such as the stations Liguria (West) and Marche (North), and in the historical centre at the Senato station. For those monitoring stations, the reductions are larger than 16 $\mu g/m^3$ for NO_x and 10 $\mu g/m^3$ for NO_2. In general, the differences between the averages and between the medians are quite similar, but in many stations, the reductions for the medians are stronger than the average differences. This fact is related to the skewed and non-symmetric characteristics of the distribution involved, as shown also in Figure 4. The above considerations on average and median pollution abatement are valid for both pollutants, in fact, the stations where the greatest differences are recorded for nitrogen oxides are the same for nitrogen dioxide.

These preliminary results do not allow for identifying the causes of the reductions and to state if they depend on common causes related to the environment and climatic factors or if they have been generated by the introduction of the new policy in Milan. The next section will attempt to investigate the variations through the modeling of possible environmental and anthropogenic factors able to influence the air quality of the city.

3.2. Model Selection

3.2.1. Step 1: Detection of the Seasonal Components

Results relative to the first step of model selection are summarized in Figures 5 and 6, which show the evaluation criteria for all the stations. For each station, the plots are organized in paired-panels; the left panel represents the 10-steps-ahead MSFE as a function of the forecast horizon and the number of harmonics modeling the seasonality (scale colour); the right panel shows the AICc–BIC pairs for each model. The optimal number of harmonics to model the seasonality is identified as the one that evaluates the minimum pair of AICc and BIC and returns the lowest MSFE curve. Both the estimates for NO_x and NO_2 for the city of Milan agree unanimously in the selection of the model in which the seasonality is composed by a single harmonic ($k = 1$); therefore, it can be rewritten as

$$Optimal\ seasonal: \quad \gamma_t = \gamma_{1,t}, \tag{10}$$

where $\gamma_{1,t}$ is

$$\begin{bmatrix} \gamma_{1,t} \\ \gamma_{1,t}^* \end{bmatrix} = \begin{bmatrix} cos(2\pi/365) & sin(2\pi/365) \\ -sin(2\pi/365) & cos(2\pi/365) \end{bmatrix} \begin{bmatrix} \gamma_{1,t} \\ \gamma_{1,t}^* \end{bmatrix} + \begin{bmatrix} \omega_{1,t} \\ \omega_{1,t}^* \end{bmatrix}, \tag{11}$$

$\omega_t \sim WN(0, \sigma_\omega^2)$ and $\omega_t^* \sim WN(0, \sigma_\omega^2)$.

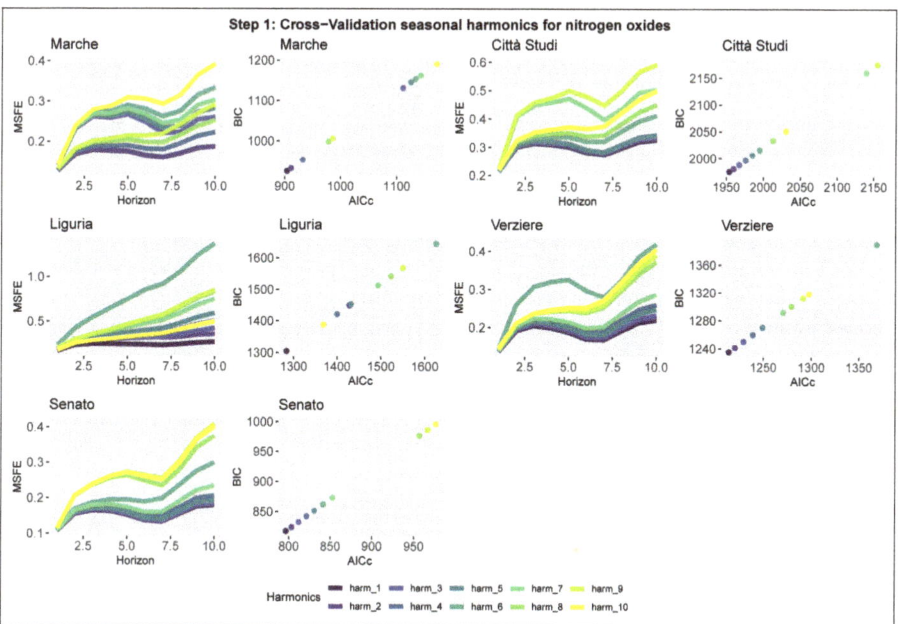

Figure 5. Model selection-Step 1-NO_x. Seasonal component selection for the five nitrogen oxides stations in Milan. Left panel: 10-steps-ahead MSFE in log-scale as function of the number of harmonics. Right panel: AICc and BIC pairs for each model.

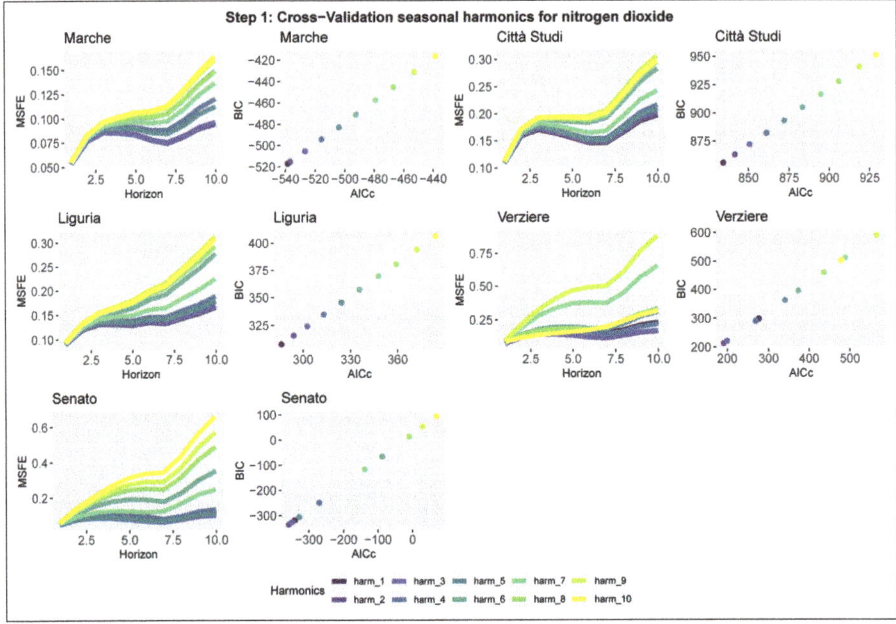

Figure 6. Model selection-Step 1-NO$_2$. Seasonal component selection for the 5 nitrogen dioxide stations in Milan. Left panel: 10-steps-ahead MSFE in log-scale as function of the number of harmonics. Right panel: AICc and BIC pairs for each model.

3.2.2. Step 2: Detection of the Counter-Factual Component

After selecting the seasonal component, we proceed to the selection of the counter-factual term. Estimates are summarized in Figures 7 and 8, which show the results for NO$_x$ and NO$_2$. The plots are graphically organized like those related to step one, with the difference that the MSFEs and the AICc-BIC pairs are functions of one of the seven counter-factual candidates instead of the number of harmonics. The selection criteria follow the same rules used for the previous step.

The search for the optimal counter-factual term requires greater attention and detail than in the previous step as the minimizers are not unique. According to the plots, there is a restricted set of stations that are good candidates for the counter-factual role. The set includes the following cities: Treviglio, Pavia, Saronno, and Cremona. In particular, Pavia's station achieves one of the best forecast and fitting performances for almost all the stations in Area B for both NO$_2$ and NO$_x$. Based on this last consideration, we select as the counter-factual term for future models the air quality monitoring station of Pavia, located South to Milan. Therefore, the final specification of the basic structural model will include as counter-factual term the logarithm of the concentrations in Pavia, $x_t = log(Pavia_t)$.

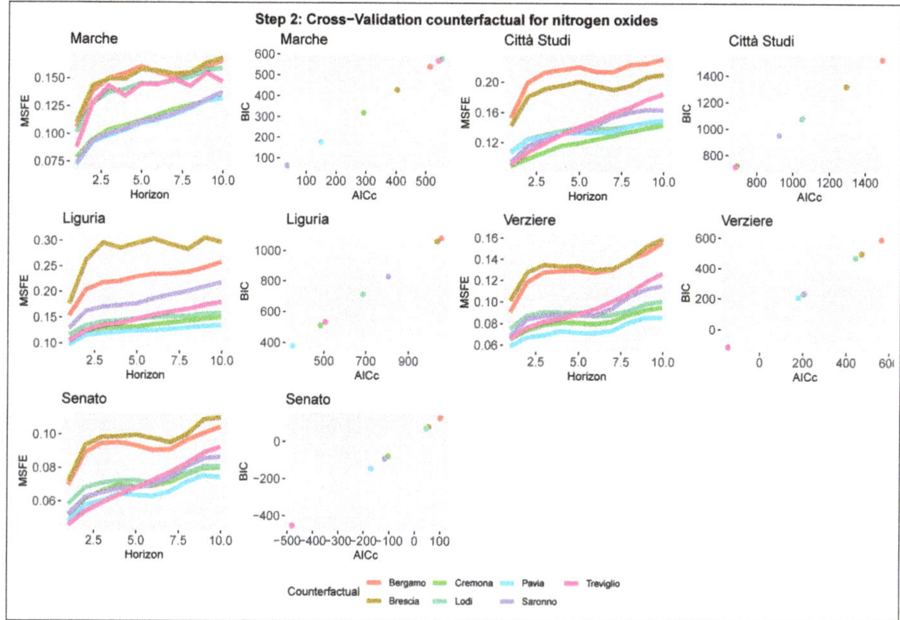

Figure 7. Model selection-Step 2-NO$_x$. Counter-factual term selection for the five nitrogen oxides stations in Milan. Left panel: 10-steps-ahead MSFE in log-scale as function of the candidate. Right panel: AICc and BIC pairs for each model.

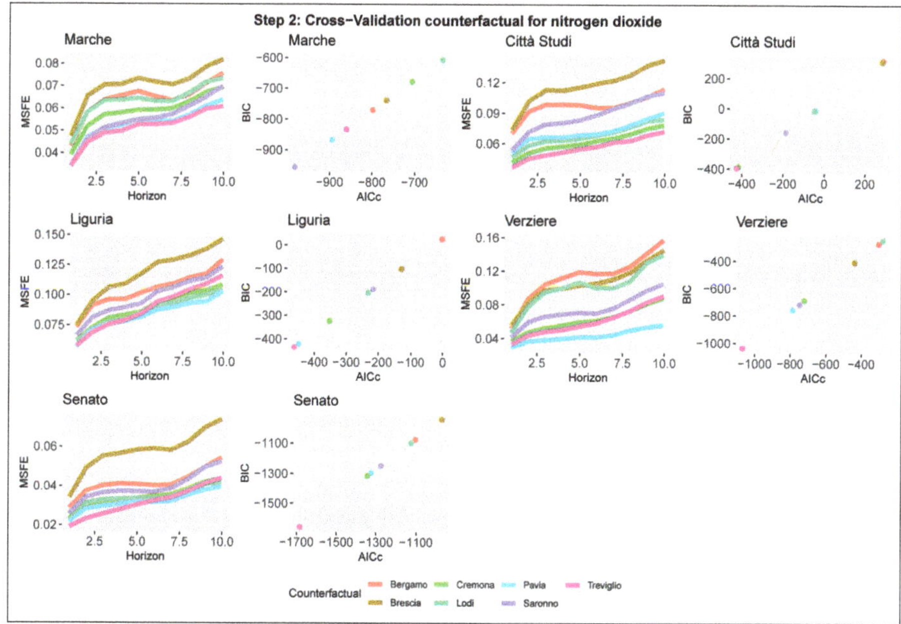

Figure 8. Model selection-Step 2-NO$_2$. Counter-factual term selection for the five nitrogen dioxide stations in Milan. Left panel: 10-steps-ahead MSFE in log-scale as function of the candidate. Right panel: AICc and BIC pairs for each model.

3.2.3. Step 3: Detection of the Weather and Calendar Factors

The last step of model selection aims to select the optimal subset of local weather and calendar regressors after having selected the optimal unobservable components, common weather, and socio-economic factors, captured by the counter-factual. For each station and pollutant, the best models are reported in Tables 2 and 3.

Table 2. Model selection-Step 3-NO_x: Best subset of covariates using backward-forward stepwise algorithms for NO_x.

	Marche		Verziere		Senato		Liguria		Citta Studi	
	AIC	*BIC*	*AIC*	*BIC*	*AIC*	*BIC*	*AIC*	*BIC*	*AIC*	*BIC*
Holidays	✓	✓	✓	✓	✓	✓	✓	✓	✓	✓
Week-End	✓	✓	✓	✓	✓	✓	✓	✓	✓	✓
Saturday:Holidays	✓		✓		✓		✓		✓	
Sunday:Holidays	✓	✓	✓	✓	✓	✓	✓	✓	✓	✓
Wind speed Q_{NE}										
Wind speed Q_{SE}							✓			
Wind speed Q_{SW}	✓	✓	✓	✓	✓	✓	✓	✓	✓	✓
Wind speed Q_{NW}	✓	✓	✓				✓		✓	✓
Temperature										
Rainfall	✓		✓		✓				✓	
Global radiation			✓		✓					
Humidity										

Note: symbol ✓ indicates that the regressor is selected within the best subset of covariates.

Table 3. Model selection-Step 3-NO_2: Best subset of covariates using backward-forward stepwise algorithms for NO_2.

	Marche		Verziere		Senato		Liguria		Citta Studi	
	AIC	*BIC*	*AIC*	*BIC*	*AIC*	*BIC*	*AIC*	*BIC*	*AIC*	*BIC*
Holidays	✓	✓	✓	✓	✓	✓	✓	✓	✓	✓
Week-End	✓	✓	✓	✓	✓	✓	✓	✓	✓	✓
Saturday:Holidays	✓		✓		✓		✓	✓	✓	
Sunday:Holidays	✓	✓	✓	✓	✓	✓	✓		✓	✓
Wind speed Q_{NE}										
Wind speed Q_{SE}										
Wind speed Q_{SW}	✓	✓	✓	✓	✓	✓	✓	✓	✓	✓
Wind speed Q_{NW}	✓								✓	
Temperature										
Rainfall			✓		✓					
Global radiation	✓		✓		✓				✓	
Humidity	✓								✓	

Note: symbol ✓ indicates that the regressor is selected within the best subset of covariates.

As expected, BIC-based models, being more parsimonious, retain fewer covariates than AIC-based models. Following this fact, we will use the BIC-selected models, but we now discuss some details about the selection process. Concerning the calendar, both criteria include in almost all cases the holidays, week-end, and Sunday holidays effects. AIC suggests adding also the interaction term between Sunday and holidays. Even if the interaction between Saturday and holidays is not always included, the final model will take into account the full set of calendar events and their interactions.

Regarding weather covariates, except for the wind speed, none of the others is included within the final models. Winds blowing from Southwest (Q_{SW}) are always selected and those coming from Northwest (Q_{NW}) are often included, and hence kept in the final model. Moreover, temperature, rainfall, solar radiation, and humidity are considered only by the AIC. This fact can be explained by the presence of the counter-factual, which captures not only common characteristics in terms of human behaviour and air quality conditions but also homogeneous climatic conditions common to all the areas considered.

3.2.4. Final Model Specification

Based on the results of the three-step model selection procedure, the final specification of the BSM augmented by the policy intervention can be expressed using the following model:

$$\begin{aligned} Measurement: \quad y_t = {} & \mu_t + \gamma_t + \theta x_t + \phi_1 Holidays + \phi_2 WeekEnd \\ & + \phi_3 Saturday : Holidays + \phi_4 Sunday : Holidays \\ & + \phi_5 WindSpeedQ_{SW} + \phi_6 WindSpeedQ_{NW} \\ & + \delta_1 D_{1t} + w_t + \varepsilon_t \, , \end{aligned} \tag{12}$$

where $\varepsilon_t \sim N(0, \sigma_\varepsilon^2)$

$$Seasonal\ component: \quad \gamma_t = \gamma_{1,t} \, , \tag{13}$$

$$Level: \quad \mu_t = \mu_{t-1} + \beta_{t-1} + \eta_t \qquad \eta_t \sim WN(0, \sigma_\eta^2) \, , \tag{14}$$

$$Slope: \quad \beta_t = \beta_{t-1} + \zeta_t \qquad \zeta_t \sim WN(0, \sigma_\zeta^2) \, , \tag{15}$$

$$Transitory\ policy: \quad w_t = \lambda w_{t-1} + \delta_0 D_{2,t} \, , \tag{16}$$

where y_t is the logarithm of pollution concentrations in one of the stations located in Milan and $x_t = log(Pavia_t)$ is the logarithm of pollution concentrations in Pavia.

3.3. Basic Structural Model and Policy Intervention

In this section, we show the numerical results obtained using state space modeling to estimate both permanent and transitory effects generated by the introduction of Area B controlling for local weather conditions, anthropogenic effects, and common areal trends.

The maximum likelihood estimates of the coefficients and the components' variances for the five air quality monitoring stations installed in Milan are reported in Tables 4 and 5. The results appear to be coherent both for nitrogen oxides and nitrogen dioxide. First, the models identify statistically significant and positive coefficients for the counter-factual term, highlighting its capability to capture socio-economic and climatic factors common to neighbouring areas and coherent with the expected sign. Second, weekends and holidays exert a negative effect on concentration levels probably due to the reduction in the movements and productive activities of the city in those days. Their interactions are almost everywhere not statistically significant but with a positive sign and always less than the sum of the individual effects of the weekend and holidays. This fact underlines how the holiday weekends enjoy more contained effects of emission reductions compared to generic weekends of the year. Third, as to be expected, winds blowing from the West (Q_{SW} and Q_{NW}) greatly reduce the amount of pollutants all over the city with peaks over 40% to 50%.

Table 4. ML estimates of BSM parameters and variances for NO_x.

	Parameter	*Marche*	*Citta Studi*	*Liguria*	*Verziere*	*Senato*
log(Pavia)	θ	0.51 ***	0.93 ***	0.73 ***	0.66 ***	0.57 ***
		(0.01)	(0.02)	(0.02)	0.02	(0.01)
Holidays	ϕ_1	−0.06 ***	−0.02	−0.05 *	−0.10 ***	−0.09 ***
		(0.03)	(0.04)	(0.03)	(0.03)	(0.03)
WeekEnd	ϕ_2	−0.11 ***	−0.09 ***	−0.09 ***	−0.17 ***	−0.15 ***
		(0.01)	(0.02)	(0.01)	(0.01)	(0.01)
Saturday:Holidays	ϕ_3	0.10	0.17	0.04	0.10	0.14 ***
		(0.08)	(0.13)	(0.10)	(0.14)	(0.05)
Sunday:Holidays	ϕ_4	0.09 **	0.09	0.12 *	0.08	0.10 ***
		(0.05)	(0.08)	(0.06)	(0.05)	(0.03)
WindSpeed Q_{SW}	ϕ_5	−0.44 ***	−0.20 ***	−0.56 ***	−0.30 ***	−0.27 ***
		(0.01)	(0.02)	(0.02)	(0.02)	(0.01)
WindSpeed Q_{NW}	ϕ_6	−0.31v***	−0.27 ***	−0.12 ***	−0.20 ***	−0.12 ***
		(0.01)	(0.02)	(0.01)	(0.01)	(0.01)
Level variance	σ^2_η	0.0047	0.0065	0.0037	0.0038	0.0027
Slope variance	σ^2_ζ	0.0000	0.0000	0.0000	0.0000	0.0000
Seasonality variance	σ^2_ω	0.0000	0.0000	0.0000	0.0000	0.0000
Error variance	σ^2_ε	0.0298	0.0745	0.0437	0.0370	0.0308

Note 1: values in parenthesis are standard errors. *Note 2*: * $p < 0.10$, ** $p < 0.05$, *** $p < 0.01$.

Table 5. ML estimates of BSM parameters and variances for NO_2.

	Parameter	*Marche*	*Citta Studi*	*Liguria*	*Verziere*	*Senato*
log(Pavia)	θ	0.36 ***	0.85 ***	0.69 ***	0.65 ***	0.55 ***
		(0.01)	(0.02)	(0.02)	(0.02)	(0.01)
Holidays	ϕ_1	−0.06 ***	−0.08 ***	−0.08 ***	−0.10 ***	−0.08 **
		(0.02)	(0.03)	(0.02)	(0.02)	(0.03)
Week-end	ϕ_2	−0.06 ***	−0.10 ***	−0.09 ***	−0.14 ***	−0.11 ***
		(0.01)	(0.01)	(0.01)	(0.01)	(0.01)
Saturday:Holidays	ϕ_3	0.06	0.18 ***	0.06	0.14 **	−0.09 ***
		(0.05)	(0.08)	0.07	(0.06)	(0.02)
Sunday:Holidays	ϕ_4	0.08 ***	0.11 ***	0.11 ***	0.06	0.10 ***
		(0.03)	(0.05)	(0.04)	(0.03)	(0.03)
WindSpeed Q_{SW}	ϕ_5	−0.35 ***	−0.13 ***	−0.42 ***	−0.23 ***	−0.18 ***
		(0.01)	(0.01)	(0.01)	(0.01)	(0.01)
WindSpeed Q_{NW}	ϕ_6	−0.17 ***	−0.16 ***	−0.08 ***	−0.13 ***	−0.08 ***
		(0.01)	(0.01)	(0.01)	(0.01)	(0.01)
Level variance	σ^2_η	0.0051	0.0065	0.0046	0.0044	0.0024
Slope variance	σ^2_ζ	0.0000	0.0000	0.0000	0.0000	0.0000
Seasonality variance	σ^2_ω	0.0000	0.0000	0.0000	0.0000	0.0000
Errors variance	σ^2_ε	0.0093	0.0282	0.0182	0.0154	0.0114

Note 1: values in parenthesis are standard errors. *Note 2*: ** $p < 0.05$, *** $p < 0.01$.

The short-term impacts adjusted for common anthropic and weather factors are summarized in Table 6, which shows the estimated permanent and transitory effects for each station in Milan expressed in logarithmic scale, hence interpretable as relative variations in concentrations levels. None of these two coefficients identifies an improvement of the considered pollutant concentrations. Moreover, the permanent effect (δ_1) is always positive and in some cases moderately statistically significant. This means that, compared to the generally decreasing areal trend, Milan air quality went worst. It is worth observing that the most significant results are obtained at the Senato station, which is located in the already existing Area C, hence already subject to some car traffic restrictions.

Such a result could be justified by the presence of multiple causes. As a first justification, we are approaching the initial phase of a progressive policy and the time elapsed since its outset may be too short to assess any significant impacts on pollutant levels. This fact is consistent with the forecasts expected by the municipality of Milan about the reductions in nitrogen oxide levels; in fact, the first significant reductions should be observed starting from 2022 [11]. Furthermore, since we are dealing with limitations to human behavior and social perception of new norms, it is not always clear how agents adapt to changes. The deterioration in the air quality of the centre could be linked to new traffic congestions in that area or to a panic shock of drivers, who need time to understand the functioning of the restrictions and adapt their behavior, exactly as in situations mismanagement of individual and organizational changes [41,42]. Eventually, the recent climate changes and the extreme weather conditions that affected the Po valley, such as temperatures higher than the seasonal average and extreme atmospheric events, could increase the noise present in the data and thus mask the real repercussions of the limitations.

Table 6. Estimated permanent and transitory effects in log scale on NO_x and NO_2 for each station.

Stations	Effect	Nitrogen Oxides			Nitrogen Dioxide		
		Estimate	S.E.	*t*-Statistic	Estimate	S.E.	*t*-Statistic
Senato	Perm. eff. δ_1	0.38	0.19	2.03 **	0.29	0.12	2.40 **
	Trans. eff. δ_0	−0.12	0.16	−0.76	−0.14	0.11	−1.25
Verziere	Perm. eff. δ_1	0.26	0.21	1.27	0.22	0.15	1.50
	Trans. eff. δ_0	−0.01	0.18	−0.05	−0.06	0.14	−0.40
Liguria	Perm. eff. δ_1	0.12	0.22	0.54	0.20	0.16	1.25
	Trans. eff. δ_0	−0.02	0.19	−0.10	−0.10	0.15	−0.67
Marche	Perm. eff. δ_1	0.15	0.19	0.80	0.23	0.13	1.82 *
	Trans. eff. δ_0	−0.11	0.17	−0.63	−0.19	0.13	1.56
Citta Studi	Perm. eff. δ_1	0.35	0.29	1.19	0.25	0.19	1.30
	Trans. eff. δ_0	0.08	0.25	0.30	0.02	0.18	0.10

Note 1: * $p < 0.10$, ** $p < 0.05$.

4. Conclusions

This paper analyzed the early-stage effects on air quality of the new traffic policy in Milan, the so-called Area B. The concentrations of nitrogen oxides (NO_x) and nitrogen dioxide (NO_2), which are mainly primary pollutants, have been considered as proxies of pollution emissions.

The first hypothesis in the introduction inquires about the presence of a significant effect on the air quality of the city. As a first point, the preliminary results show that concentrations during spring and summer 2019 are lower than during the same seasons in the previous five years, hinting for a reduction effect due to the policy. On the other side, a similar reducing trend has been observed in various neighbouring cities around Milan, which belong to a homogeneous meteorological, social, and economical cluster within the Po valley. Their similar behavior is used here as an areal common trend capturing both weather and anthropogenic components. Our approach, which adjusts for local weather conditions and the areal common trend, does not provide a further reduction effect for any station comparing to this trend. Instead, in Senato station, which is inside the historical city centre and was already covered by Area C, the estimates provide a strong, but moderately statistically significant, increase for both considered pollutants. This is coherent with the fact that the restriction introduced is very limited as it concerns just some classes of old vehicles, which are a small percentage of the entire vehicle pool, both in terms of number of cars and emissions.

Since the first research hypothesis is confirmed just to a minor extent and with an opposite sign with respect to what was expected, the second research hypothesis, concerning the homogeneity of the possible effects, assumes now only a technical scope. It is confirmed just for what concerns the positive direction of the changes, but not for their significance. In fact, among all the estimated permanent effects, only Senato station is significant at 5%. Moreover, the estimated transitory effects are always not significant at any confidence level.

The above facts hint that, compared to the common trend of the considered area, Milan air quality is improving slowly, and, in this sense, the first phase of Area B seems to have a negative effect on air quality. Due to the limited scope of this first phase and its progressiveness, it is not unexpected to find a limited or a zero effect. Nonetheless, the negative effect needs some more explanations.

Although finding the ultimate motivation for this is not the aim of this paper, a discussion follows. Firstly, the statistically significant increase found is limited in space and is located inside the previously introduced restricted Area C. It may be possible that this further restriction increased congestion of public transport buses, which are often very old vehicles, or to the aforementioned adaptation shocks. This could explain only a part of the results. In fact, this first point is also related to the other sources of nitrogen oxides. According to INEMAR [25], road traffic is about 68% of the total emissions. Hence, a transition to house heating green techniques slower in Milan comparing the other considered cities could have an influence on this result. Moreover, also the other stations experienced a comparative deterioration of air quality and the second-worst station is Città Studi, which is an urban background station, hence with limited relation to local traffic congestion. Second, the increase due to road traffic may have temporal dynamics. Since the traffic restriction is limited to business hours, there may be an increase in congestion early in the morning and in the late evening, affecting the daily average.

In conclusion, although environmental protection policies are in general a fundamental step for sustainability improvement, in some cases, they may not be sufficient or their implementation may be misleading. In our case, we considered only the early-stage of a policy, which is progressive in time. Hence, the results of this paper may be regarded as physiological, provided that they characterize only the initial part of the policy implementation and are improved soon. It follows a recommendation to the municipal government to develop the policy more strongly.

Additional research could be developed in the future. In particular, the effect on traffic congestion inside Area C could be investigated further using historical data related to the vehicle movements crossing the access points. Moreover, the use of a multivariate approach, which includes other pollutants such as PM_{10} and $PM_{2.5}$, and spatio-temporal modeling could highlight hidden effects, which are not visible considering the single stations. Eventually, the extension to hourly data could consider both the presence of intra-daily effects and explaining the spatial dynamics related to traffic.

Author Contributions: Conceptualization, P.M., A.F., M.P. and M.M.; methodology and formal analysis, P.M., A.F. and M.P.; software, P.M. and M.P.; data curation and original draft preparation, P.M.; literature review, P.M. and A.F.; review and editing and validation and supervision P.M., A.F., M.M. and M.P. All authors have read and agreed to the published version of the manuscript.

Funding: This research received no external funding.

Acknowledgments: The authors are grateful to the editors and three anonymous referees for their comments and suggestions.

Conflicts of Interest: The authors declare no conflict of interest.

References

1. European Parliament. Directive 2008/50/EC of the European Parliament and of the Council of 21 May 2008 on ambient air quality and cleaner air for Europe. *Off. J. Eur. Union* **2008**, *29*, 169–212.
2. Thunis, P.; Miranda, A.; Baldasano, J.M.; Blond, N.; Douros, J.; Graff, A.; Janssen, S.; Juda-Rezler, K.; Karvosenoja, N.; Maffeis, G.; et al. Overview of current regional and local scale air quality modeling practices: Assessment and planning tools in the EU. *Environ. Sci. Policy* **2016**, *65*, 13–21. [CrossRef]
3. Cuvelier, C.; Thunis, P.; Vautard, R.; Amann, M.; Bessagnet, B.; Bedogni, M.; Berkowicz, R.; Brandt, J.; Brocheton, F.; Builtjes, P.; et al. CityDelta: A model intercomparison study to explore the impact of emission reductions in European cities in 2010. *Atmos. Environ.* **2007**, *41*, 189–207. [CrossRef]
4. Vautard, R.; Builtjes, P.H.J.; Thunis, P.; Cuvelier, C.; Bedogni, M.; Bessagnet, B.; Honoré, C.; Moussiopoulos, N.; Pirovano, G.; Schaap, M.; et al. Evaluation and intercomparison of Ozone and PM10 simulations by

several chemistry transport models over four European cities within the CityDelta project. *Atmos. Environ.* **2007**, *41*, 173–188. [CrossRef]

5. Gao, J.; Yuan, Z.; Liu, X.; Xia, X.; Huang, X.; Dong, Z. Improving air pollution control policy in China—A perspective based on cost–benefit analysis. *Sci. Total Environ.* **2016**, *543*, 307–314. [CrossRef]

6. Carnevale, C.; Finzi, G.; Pisoni, E.; Volta, M.; Guariso, G.; Gianfreda, R.; Maffeis, G.; Thunis, P.; White, L.; Triacchini, G. An integrated assessment tool to define effective air quality policies at regional scale. *Environ. Model. Softw.* **2012**, *38*, 306–315. [CrossRef]

7. Vlachokostas, C.; Achillas, C.; Moussiopoulos; Hourdakis, E.; Tsilingiridis, G.; Ntziachristos, L.; Banias, G.; Stavrakakis, N.; Sidiropoulos, C. Decision support system for the evaluation of urban air pollution control options: Application for particulate pollution in Thessaloniki, Greece. *Sci. Total Environ.* **2009**, *407*, 5937–5948. [CrossRef]

8. European Environmental Agency. *Air Quality in Europe-2019 Report*; EEA Technical Report 10/2019; European Environment Agency: Copenhagen, Denmark, 2019. Available online: https://www.eea.europa.eu/publications/air-quality-in-europe-2019 (accessed on 7 February 2020).

9. Liu, Z.; Zhao, L.; Wang, C.; Yang, Y.; Xue, J.; Bo, X.; Li, D.; Liu, D. An Actuarial Pricing Method for Air Quality Index Options. *Int. J. Environ. Res. Public Health* **2019**, *16*, 4882. [CrossRef]

10. Xu, X.; Dong, D.; Wang, Y.; Wang, S. The Impacts of Different Air Pollutants on Domestic and Inbound Tourism in China. *Int. J. Environ. Res. Public Health* **2019**, *16*, 5127. [CrossRef]

11. Municipality of Milan. *Delibera Giunta Comunale n. 1366/2018 del 02/08/2018*; Municipality of Milan: Milan, Italy, 2018. Available online: https://www.comune.milano.it/aree-tematiche/mobilita/area-b#collapse_article_u7LTldfp1A6d (accessed on 7 February 2020).

12. Finazzi, F.; Paci, L. Kernel-based estimation of individual location densities from smartphone data. *Stat. Model.* **2019**. [CrossRef]

13. Fassò, A.; Porcu, E. Latent Variables and Space-Time Models for Environmental Problems. *Stoch Environ. Res. Risk Assess* **2015**, *29*, 323–324. [CrossRef]

14. Cameletti, M.; Gómez-Rubio, V.; Blangiardo, M. Bayesian modeling for spatially misaligned health and air pollution data through the INLA-SPDE approach. *Spat. Stat.* **2019**, *31*, 100353. [CrossRef]

15. Vu, P.T.; Larson, T.V.; Szpiro, A.A. Probabilistic Predictive Principal Component Analysis for Spatially-Misaligned and High-Dimensional Air Pollution Data with Missing Observations. *arXiv* **2019**, arXiv:1905.00393.

16. Calculli, C.; Fassò, A.; Finazzi, F.; Pollice, A.; Turnone, A. Maximum likelihood estimation of the multivariate hidden dynamic geostatistical model with application to air quality in Apulia, Italy. *Environmetrics* **2015**, *26*, 406–417. [CrossRef]

17. Menezes Piairo, H.; Garcia-Soidan, P.; Sousa, I. Spatial-temporal modellization of the NO2 concentration data through geostatistical tools. *Stat. Methods Appl.* **2015**, *25*. [CrossRef]

18. Taghavi-Shahri, S.M.; Fassò, A.; Mahaki, B.; Amini, H. Concurrent spatiotemporal daily land use regression modeling and missing data imputation of fine particulate matter using distributed space-time Expectation Maximization. *Atmos. Environ.* **2019**, 117202. [CrossRef]

19. Chen, L.; Guo, B.; Huang, J.; He, J.; Wang, H.; Zhang, S.; Chen, S.X. Assessing air-quality in Beijing-Tianjin-Hebei region: The method and mixed tales of PM2. 5 and O3. *Atmos. Environ.* **2018**, *193*, 290–301. [CrossRef]

20. Fassò, A. Statistical assessment of air quality interventions. *Stoch. Environ. Res. Risk Assess.* **2013**, *27*, 1651–1660. [CrossRef]

21. Invernizzi, G.; Ruprecht, A.; Mazza, R.; De Marco, C.; Močnik, G.; Sioutas, C.; Westerdahl, D. Measurement of black carbon concentration as an indicator of air quality benefits of traffic restriction policies within the ecopass zone in Milan, Italy. *Atmos. Environ.* **2011**, *45*, 3522–3527. [CrossRef]

22. Percoco, M. The effect of road pricing on traffic composition: Evidence from a natural experiment in Milan, Italy. *Transp. Policy* **2014**, *31*, 55–60. [CrossRef]

23. Jones, A.M.; Harrison, R.M.; Barratt, B.; Fuller, G. A large reduction in airborne particle number concentrations at the time of the introduction of "sulphur free" diesel and the London Low Emission Zone. *Atmos. Environ.* **2012**, *50*, 129–138. [CrossRef]

24. Qadir, R.; Abbaszade, G.; Schnelle-Kreis, J.; Chow, J.; Zimmermann, R. Concentrations and source contributions of particulate organic matter before and after implementation of a low emission zone in Munich, Germany. *Environ. Pollut.* **2013**, *175*, 158–167. [CrossRef] [PubMed]
25. INEMAR ARPA Lombardia Settore Aria. *Emission Inventory: 2014 Emission in Lombardy Region—Final Data*; ARPA Lombardia Settore Aria: Milan, Italy, 2018. Available online: https://www.inemar.eu/xwiki/bin/view/InemarDatiWeb/Emission_Inventory (accessed on 7 February 2020).
26. Mudelsee, M.; Alkio, M. Quantifying effects in two-sample environmental experiments using bootstrap confidence intervals. *Environ. Model. Softw.* **2007**, *22*, 84–96. [CrossRef]
27. Mudelsee, M. *Climate Time Series Analysis-Classical Statistical and Bootstrap Methods*, 2nd ed.; Springer: Cham, Switzerland, 2014.
28. Durbin, J.; Koopman, S.J. *Time Series Analysis by State Space Methods*; Oxford University Press: Oxford, UK, 2012.
29. Pelagatti, M.M. *Time series modeling with unobserved components*; Chapman and Hall/CRC: Boca Raton, FL, USA, 2015.
30. Windsor, H.; Toumi, R. Scaling and persistence of UK pollution. *Atmos. Environ.* **2001**, *35*, 4545–4556. [CrossRef]
31. Chelani, A.B. Statistical characteristics of ambient PM2. 5 concentration at a traffic site in Delhi: source identification using persistence analysis and nonparametric wind regression. *Aerosol Air Qual. Res.* **2013**, *13*, 1768–1778. [CrossRef]
32. Lee, C.K. Multifractal characteristics in air pollutant concentration time series. *Water Air Soil Pollut.* **2002**, *135*, 389–409. [CrossRef]
33. Harvey, A.C.; Todd, P. Forecasting economic time series with structural and Box-Jenkins models: A case study. *J. Bus. Econ. Stat.* **1983**, *1*, 299–307.
34. Harvey, A.; Koopman, S. Structural time series models. *Wiley Stats Ref. Stat. Ref. Online* **2014**. [CrossRef]
35. Bergmeir, C.; Hyndman, R.J.; Koo, B. A note on the validity of cross-validation for evaluating time series prediction. *Monash Univ. Dep. Econom. Bus. Stat. Work. Paper* **2015**, *10*, 15. [CrossRef]
36. Bergmeir, C.; Hyndman, R.J.; Koo, B. A note on the validity of cross-validation for evaluating autoregressive time series prediction. *Comput. Stat. Data Anal.* **2018**, *120*, 70–83. [CrossRef]
37. R Core Team. *R: A Language and Environment for Statistical Computing*; R Foundation for Statistical Computing: Vienna, Austria, 2019.
38. Helske, J. KFAS: Exponential Family State Space Models in R. *J. Stat. Softw.* **2017**, *78*, 1–39.10.18637/jss.v078.i10. [CrossRef]
39. Wickham, H. *ggplot2: Elegant Graphics for Data Analysis*; Springer: New York, NY, USA, 2016.
40. Pebesma, E. Simple Features for R: Standardized Support for Spatial Vector Data. *R J.* **2018**, *10*, 439–446.10.32614/RJ-2018-009. [CrossRef]
41. Perlman, D.; Takacs, G.J. The 10 Stages of Change: To cope with change. *Nurs. Manag.* **1990**, *21*, 33–38. [CrossRef]
42. Elrod, P.D.; Tippett, D.D. The "death valley" of change. *J. Organ. Change Manag.* **2002**, *15*, 273–291. [CrossRef]

International Journal of
Environmental Research and Public Health

Article

Long-Term Assessment of Air Quality and Identification of Aerosol Sources at Setúbal, Portugal

Alexandra Viana Silva [1], Cristina M. Oliveira [2], Nuno Canha [1,3,*], Ana Isabel Miranda [3] and Susana Marta Almeida [1]

1. Centro de Ciências e Tecnologias Nucleares (C2TN), Instituto Superior Técnico, Universidade de Lisboa, Estrada Nacional 10, Km 139.7, 2695-066 Bobadela LRS, Portugal; alexandra.silva@kevolution.org (A.V.S.); smarta@ctn.tecnico.ulisboa.pt (S.M.A.)
2. Centro de Química Estrutural (CQE), Faculdade de Ciências da Universidade de Lisboa, Campo Grande, 1749-016 Lisboa, Portugal; cmoliveira@fc.ul.pt
3. Centro de Estudos do Ambiente e do Mar (CESAM), Departamento de Ambiente e Ordenamento, Universidade de Aveiro, 3810-193 Aveiro, Portugal; miranda@ua.pt
* Correspondence: nunocanha@ctn.tecnico.ulisboa.pt

Received: 22 June 2020; Accepted: 26 July 2020; Published: 28 July 2020

Abstract: Understanding air pollution in urban areas is crucial to identify mitigation actions that may improve air quality and, consequently, minimize human exposure to air pollutants and their impact. This study aimed to assess the temporal evolution of the air quality in the city of Setúbal (Portugal) during a time period of 10 years (2003–2012), by evaluating seasonal trends of air pollutants (PM_{10}, $PM_{2.5}$, O_3, NO, NO_2 and NO_x) measured in nine monitoring stations. In order to identify emission sources of particulate matter, $PM_{2.5}$ and $PM_{2.5-10}$ were characterized in two different areas (urban traffic and industrial) in winter and summer and, afterwards, source apportionment was performed by means of Positive Matrix Factorization. Overall, the air quality has been improving over the years with a decreasing trend of air pollutant concentration, with the exception of O_3. Despite this improvement, levels of PM_{10}, O_3 and nitrogen oxides still do not fully comply with the requirements of European legislation, as well as with the guideline values of the World Health Organization (WHO). The main anthropogenic sources contributing to local PM levels were traffic, industry and wood burning, which should be addressed by specific mitigation measures in order to minimize their impact on the local air quality.

Keywords: air pollutants; particulate matter; monitoring; seasonality; chemical characterization; source apportionment

---

1. Introduction

Air pollution is considered one of the main environmental problems that countries face nowadays taking into account its adverse effects on human health and on the environment [1]. Ambient air pollution alone kills around three million people every year, mainly from noncommunicable diseases. Only one person in ten lives in a city that complies with WHO Air quality guidelines [2].

Policies implemented at regional and national level targeting the limitation of emissions have led to acceptable air quality levels across Europe regarding some air pollutants [3], but others still raise concern, such as particulate matter, nitrogen dioxide and ozone [4].

Air pollutants are emitted by natural and anthropogenic sources; they may either be released directly (primary pollutants) or formed in the atmosphere (as secondary pollutants); they may be formed and transported over long distances or produced locally. Effective measures to decrease the impacts of air pollution require a good understanding of its sources, how pollutants are transported and transformed in the atmosphere, and how they affect humans, ecosystems, the climate, and subsequently society and the economy [5]. A regular monitoring program of air pollutants is therefore a crucial tool of successful environmental management.

Setúbal (Portugal) includes an area of high population density, anthropogenic industrial activities, traffic and protected natural areas. Moreover, the region also exhibits high levels of pollution. Several studies have already shown that local activities have a significant impact on the air quality due to: (i) emissions of air pollutants originating in industrial processes [6–8], (ii) dust fugitive emissions from the harbors [9,10], and (iii) intense traffic of heavy duty vehicles [11].

The present study provides a long-term assessment (ten years) of the air quality in Setúbal and aims to identify the main sources of air pollution. Evaluation of the temporal variability of several air pollutants (PM_{10}, $PM_{2.5}$, O_3, NO, NO_2 and NO_x) was performed and a comprehensive characterization of particulate matter levels was conducted, along with the identification of pollution sources, using the Positive Matrix Factorization (PMF) model.

2. Materials and Methods

2.1. Study Area

This study was conducted in Setúbal (south-west Portugal), a coastal city sited where the river Sado flows into the Atlantic Ocean. Setúbal district covers an area of 230 km^2 and it has a total population of 135,000 inhabitants [12]. The city of Setúbal is located next to two protected natural areas (Sado Estuary Reserve and Arrábida Park, which belong to the protected area Natura 2000 network) and with one of the most important industrial areas in the country. This industrial area includes: (i) different types of large industry (such as the production of fresh and dry baker's yeast, a slaughterhouse, a powerplant, and fertilizers, pesticides, cement and chemical industries), (ii) harbors, and (iii) heavy duty vehicle traffic due to transport of raw materials and products to the harbors and industries.

2.2. Monitoring of Air Pollutants by Air Quality Monitoring Networks

Air quality data were obtained from three different air monitoring networks, namely from the Portuguese Environment Agency (APA-QUALAR, with 4 stations), the EDP company (two stations) and the SECIL company (three stations), for the period 2003–2012. Moreover, a field campaign was organized in this study to conduct PM sampling where two monitoring stations were established ("Industrial Mitrena" and "Quebedo").

These monitoring stations were classified as rural background, urban background, suburban background, suburban industrial, urban traffic and suburban traffic. Figure 1 presents the location of the air quality monitoring stations and Table 1 provides a description of each monitoring station, with details of the monitored pollutants and measuring period.

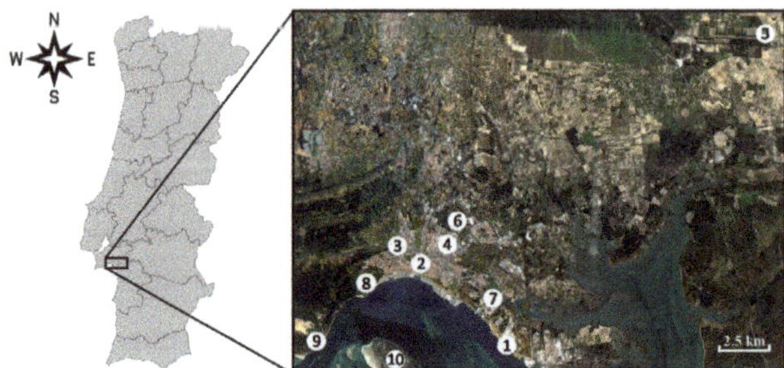

Figure 1. Location of the monitoring stations of the Setúbal area (Portugal): urban background (3 and 4), urban traffic (2), suburban traffic (6), rural background (5), suburban background (8 and 10) and suburban industrial (1, 7 and 9).

Table 1. Description of the monitoring stations located in Setúbal region (regular monitoring stations and field study stations).

Network	#ID	Stations	Coordinates		Altitude (m)	Type	Pollutants Monitored	Sampling Period for Each Pollutant
			Latitude	Longitude				
This study	1	INDUSTRIAL MITRENA	38°29′48″N	08°49′58″W	–	Suburban Industrial	$PM_{2.5-10}$ and $PM_{2.5}$ mass and chemical composition	PM: 2011
								chemical constituents: 2011
	2	QUEBEDO	38°31′27″N	08°53′39″W	16	Urban Traffic	$PM_{2.5-10}$ and $PM_{2.5}$ mass and chemical composition	PM: 2011
								chemical constituents: 2011
APA–QUALAR	2	QUEBEDO	38°31′27″N	08°53′39″W	16	Urban Traffic	NO, NO_2, NO_X, PM_{10}	NO, NO_2: 2003–2012
								NO_X: 2005–2012
								PM_{10}: 2004–2011
	3	ARCOS	38°31′45″N	08°53′39″W	2	Urban Background	$NO, NO_2, NO_X, O_3, PM_{10}$	NO, NO_2, O_3: 2003–2012
								NOx: 2004–2012
								PM_{10}: 2009–2012
	4	CAMARINHA	38°31′50″N	08°52′23″W	15	Urban Background	$NO, NO_2, NO_X, PM_{2.5}, PM_{10}, O_3$	NO, NO_2, O_3, PM_{10}: 2003
								NOx: 2005–2010
								$PM_{2.5}$: 2008–2010
	5	FERNANDO PÓ	38°38′08″N	08°41′26″W	57	Rural Background	$NO, NO_2, NO_X, O_3, PM_{2.5}, PM_{10}$	2008–2011
EDP	6	SUBESTAÇÃO	38°32′08″N	08°51′44″W	30	Suburban Traffic	$NO, NO_2, NO_X, PM_{10}, O_3$	NO, NO_2, O_3, PM_{10}: 2004–2010
								NOx: 2010
	7	PRAIAS SADO	38°31′05″N	08°50′15″W	–	Suburban Industrial	PM_{10}	PM_{10}: 2009–2010
SECIL	8	SÃO FILIPE	38°30′55″N	08°54′40″W	110	Suburban Background	NO, NO_2, NO_X, O_3	NO, NO_2: 2004–2012
								NOx, O_3: 2009–2012
	9	HOSO	38°29′31″N	08°56′02″W	–	Suburban Industrial	$NO, NO_2, NO_X, PM_{2.5}, PM_{10}, O_3$	$NO, NO_2, NO_X, O_3, PM_{2.5}$: 2009–2012
								PM_{10}: 2010–2012
	10	TRÓIA	38°28′45″N	08°53′20″W	3	Suburban Background	$NO, NO_2, NO_X, PM_{2.5}, PM_{10}, O_3$	2009–2012

These monitoring stations provided hourly data for the air pollutants NO, NO_2, NOx, PM_{10}, $PM_{2.5}$ and O_3. Not all monitoring stations provided data for all pollutants nor for all the study period. Different measuring/monitoring methods were applied according to the air pollutant, namely beta-attenuation for PM_{10} and $PM_{2.5}$, chemiluminescence for NO, NO_2 and NOx, and UV photometry for O_3, as defined by European legislation [13].

2.3. Particulate Matter Sampling and Characterisation

2.3.1. Sampling Sites and Methodology

PM sampling was done simultaneously in two different sampling sites within the study area and during two seasons of 2011 (winter: from 17 to 31 January–15 sampling days; summer: from 19 August to 2 September–14 sampling days). One sampling site was located at in urban traffic station (Quebedo) and the other sampling site was located in an industrial site (Mitrena). Figure 1 and Table 1 include these sampling sites, their location and other information.

Coarse and fine particulate matter were sampled using low volume Gent collectors (University of Gent, Gent, Belgium) equipped with a Stacked Filter Unit (SFU) and a PM_{10} pre impactor stage. The SFU carried two 47 mm Nuclepore®polycarbonate filters, one in each of its two different stages. Air flow rate was set to 15–16 L·min^{-1}, allowing the collection of coarse particles in the first stage (particles with aerodynamic diameter (AD) between 2.5 and 10 μm-$PM_{2.5-10}$, using a 8 μm pore size filter) and of fine particles in the second stage (particles with AD < 2.5 μm-$PM_{2.5}$, 0.4 μm pore size filter) [14]. Filter sampling was conducted during periods of 12 h (day and night periods).

2.3.2. Gravimetric Analysis

PM loads in filters were measured by gravimetry in a controlled clean room (class 10,000), with the following conditions: (20 ± 1) °C and relative humidity of (50 ± 5) %, after 48 h equilibrium. Nuclepore filters were weighted on an UMT5 Comparator balance (Mettler Toledo GmbH, 2000, Greifensee, Switzerland), an ultra-micro balance with a 0.1 μg resolution. Filter weight was measured before and after sampling and each final weight was accepted as the average of three measurements only if the variability between them was less than 5%.

2.3.3. Chemical Analysis

Sampled filters were cut into two halves, with each one being used for a specific technique for a specific chemical analysis: (i) chemical elements were quantified by Instrumental Neutron Activation Analysis using the k_0 methodology (k_0-INAA); and (ii) water soluble ions were assessed by Ion Chromatography (IC).

Regarding k_0-INAA, after being rolled up and put in a clean thin aluminum foil, each half filter was irradiated for a period of 5h in a Portuguese Research Reactor, using a thermal neutron flux of 1.03×10^{13} cm $^{-2}$·s^{-1}, as established in the procedure described elsewhere [15]. After being removed from the aluminum foil, irradiated samples were stored in polyethylene containers and two gamma spectra were measured using a hyper-pure germanium detector (the first measured three days after irradiation and the second after four weeks).

For application of the k_0 methodology, comparators were co-irradiated with samples, namely, 0.1% Au–Al discs. This methodology allowed the quantification of 13 chemical elements, namely As, Ce, Co, Cr, Fe, K, La, Na, Sb, Sc, Se, Sm and Zn. Blank filters were processed as regular samples and their concentrations were subtracted from the sampled filters. Quality control was done with the analysis of the reference material NIST-SRM 1633a (Coal Fly Ash) simultaneously with the samples and evaluation was performed using established procedures [16,17].

Regarding IC, a total of three anions (Cl^-, NO_3^- and SO_4^{2-}) and five cations (Na^+, NH_4^+, K^+, Mg^{2+} and Ca^{2+}) were assessed using an established methodology [18]. For this, sampled and blank filters were extracted (using 5 mL of ultrapure water in an ultrasonic bath (Branson 3200, Brookfield,

Connecticut, USA) for 45 min) and, afterwards, extracts were filtered using a pre-washed Whatman 41®
filter (Whatman International Ltd, Maidstone, England). The extract liquid filtered was then analysed
by IC, using a Dionex® DX500 system with an isocratic pump IP20, a conductivity detector (CD20)
equipped with Peaknet® software (Dionex Corporation, Sunnyvale, CA, USA). For the anionic mode,
the used chromatograph had an anion guard column IonPack AG14 (Dionex Corporation, Sunnyvale,
CA, USA), an analytical column IonPack AS14 and an anion suppressor ASRSR Ultra 4 mm. Using a
flow rate of 1.2 mL·min^{-1}, the eluent was a 3.5 mmol·dm^{-3} Na_2CO_3 + 1 mmol·dm^{-3} $NaHCO_3$ buffer
solution. For analysis of cations, the chromatograph had a CSRS 300-II-4mm cation suppressor, a guard
column Ion Pack CG12 and an Ion Pack CS12 column. Using a flow rate of 1.0 mL min^{-1}, the eluent
was a methane-sulfonic acid (MSA) 20 mM solution. The used injection volume was 100 μL and 25 μL
for anions and cations, respectively.

Measurements were conducted after the chromatograph daily calibration using calibrators
with mass concentrations fit to apply the linear regression model. All data were subtracted by the
blanks values.

2.4. Meteorological Data

Hourly meteorological data (wind direction, wind speed, precipitation, temperature and relative
humidity) were measured by two automatic weather stations, one located in the monitoring station
"Subestação" (operated by EDP, registering data from January 2004 to December 2009) and the other
located in the monitoring station "HOSO" (operated by SECIL and registering data from January 2010
to December 2012). For the PM sampling campaigns in 2011, an automatic weather station was used
to gather meteorological data during the sampling periods, located in the suburban industrial site
("Mitrena"). The wind rose and pollution dispersion maps were created using the Openair project
software [19].

2.5. Air Quality Index

Air Quality Indexes provide for the public an easy way to understand the levels of air pollutants in
their area and to gain insights regarding their associated effects on health. Ultimately, this information
aims to raise the awareness of citizens towards air quality and thus to change their behavior or take
mitigation measures in order to minimize their exposure. Several indexes are currently used worldwide
(for example, the European Air Quality Index in Europe [20], the Air Quality Index in USA [21] or the
Air Quality Index in China [22]), but no general methodology has been adopted [23]. The Portuguese
Environment Agency also uses an index in order to provide information about air quality to the citizens
(the QualAr Index [24]).

In order to provide an understanding of the temporal evolution of air quality in the study area,
an Air Quality Index was calculated based on the 10 years' analysis of pollutants (2003–2012). This index
was defined as described in Table 2, with a total of five categories, ranging from "Very good" to "Very
poor". Three main air pollutants were considered (NO_2, O_3 and PM_{10}), but if available two additional
pollutants were also used (SO_2 and CO). The pollutant with the worst index class, in terms of highest
concentrations, was responsible for the global Air Quality Index.

Table 2. Air Quality Index categories and their pollutants range (values in μg·m^{-3}).

Index Class	Mandatory Pollutant			Auxiliary Pollutant	
	NO_2 (1 h)	O_3 (1 h)	PM_{10} (24 h)	SO_2 (1 h)	CO (8 h)
Very good	[0–99]	[0–59]	[0–19]	[0–139]	[0–4999]
Good	[100–139]	[60–119]	[20–34]	[140–209]	[5000–6999]
Moderate	[140–199]	[120–179]	[35–49]	[210–349]	[7000–8499]
Poor	[200–399]	[180–239]	[50–119]	[350–499]	[8500–9999]
Very poor	≥400	≥240	≥120	≥500	≥10,000

2.6. Statistical Analysis and Source Apportionment

Statistical analysis was performed using the STATISTICA software version 13 (StatSoft Europe GmbH, Hamburg, Germany). For the analysis of the results variance, non-parametric statistics at a significance level of 0.050 were selected. Mann-Whitney tests were used to assess significant differences between datasets.

The Positive Matrix Factorization (PMF) model was used to identify the pollution sources contributing to PM levels [25]. PMF is a popular receptor model used for source apportionment studies, which decomposes the data matrix into two sub-data matrixes (factor profiles and factor contributions) without prior knowledge of the profiles of pollution sources [26]. PMF was applied to the datasets of PM sampled in "Quebedo" and "Mitrena". Data below the limit of detection (LoD) were replaced by LoD/2 and the uncertainties were set to 5/6 of the LoD.

3. Results and Discussion

3.1. Meteorological Data

A brief summary of the meteorological data measured during the period 2004–2012 is provided in Table S1 (supplementary material). The average monthly temperature ranged from 12 °C in January/February to 27 °C in July/August. The average temperature and relative humidity (RH) in Setúbal, during the period 2004–2012, was 16.4 °C and 70.3%, respectively. The average annual accumulated precipitation was 900 mm. Rainfall was more frequent during autumn and winter.

The wind patterns in the study area varied according to the location of the meteorological station and the season, as shown in Figure 2.

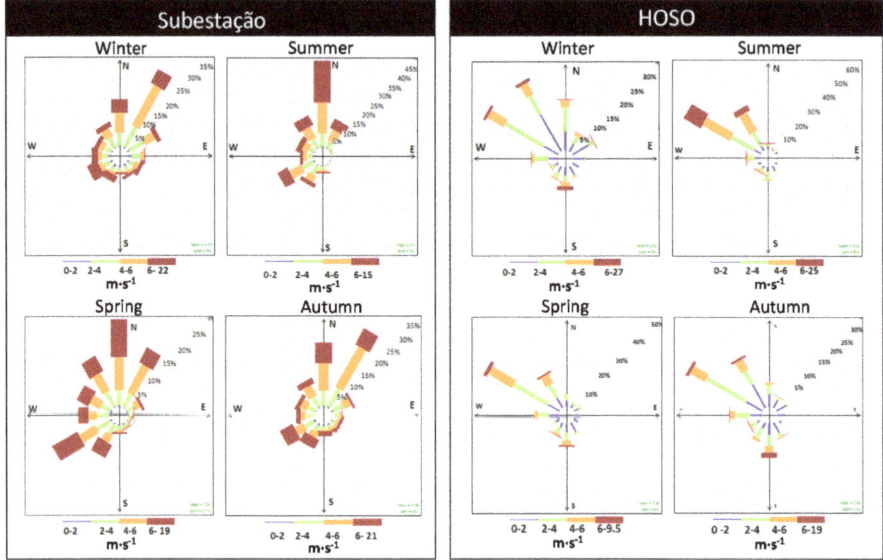

Figure 2. Seasonal wind roses at weather stations Subestação (**left**, during 2004–2009) and HOSO (**right**, during 2010–2012).

At Subestação, the main wind directions were predominantly from N and SW in summer and spring, while in winter and autumn they were from NNE. At HOSO, the predominant wind directions were from NNW and WNW. Overall, winds measured in HOSO were weaker than those registered at Subestação. This fact may be explained by HOSO's location at the bottom of the mountain chains of Arrábida, which may protect the station from prevailing north and northwest winds.

As shown in Figure 3, for the PM sampling campaigns, the prevailing winds were from NNE in winter and from NNW and WSW in summer. In winter period, a mean temperature of 12 °C and a mean relative humidity of 74% were registered, while in summer a mean temperature of 22 °C and a mean relative humidity of 61% were registered.

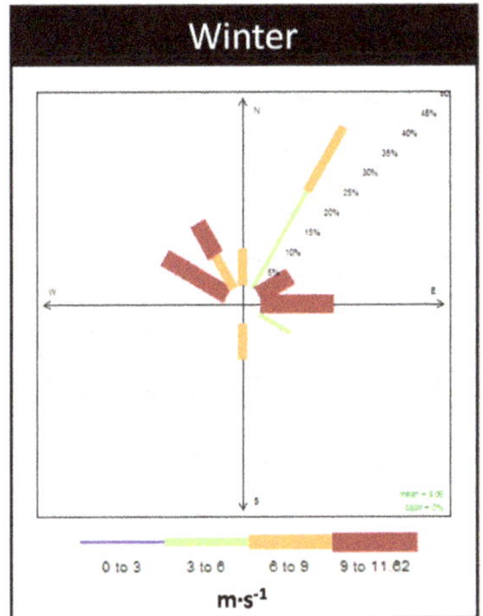

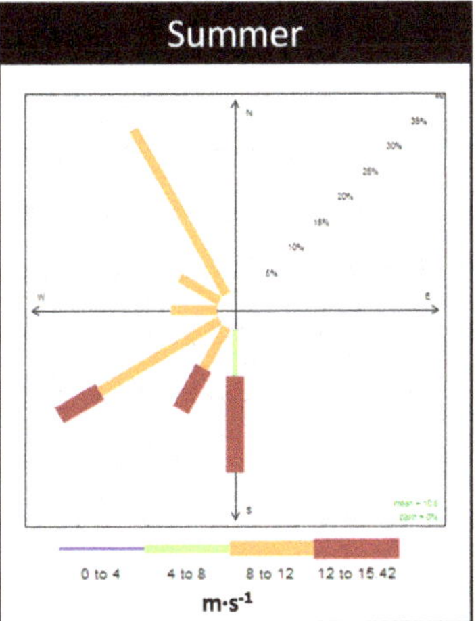

Figure 3. Wind roses during the PM sampling campaign in winter (**left**) and summer (**right**) seasons.

3.2. Air Quality Index (AQI)

Figure 4 provides the temporal evolution of the AQI in the study area, from 2003 to 2012. Overall, it is possible to observe a clear trend of better air quality indexes along the years, with the index "Good" increasing from 150 days per year in 2004 to more than 250 days per year in 2012. The index "Poor" showed a higher peak in 2005 with around 80 days per year and, afterwards, a decreasing trend with around 10 days per year in 2012. During the studied period, almost no days registered a "Very Poor" index, with the exception of a few days in the first four years. However, the number of days with a "Very good" index showed a slight increase over the years, with 2012 having around 20 days.

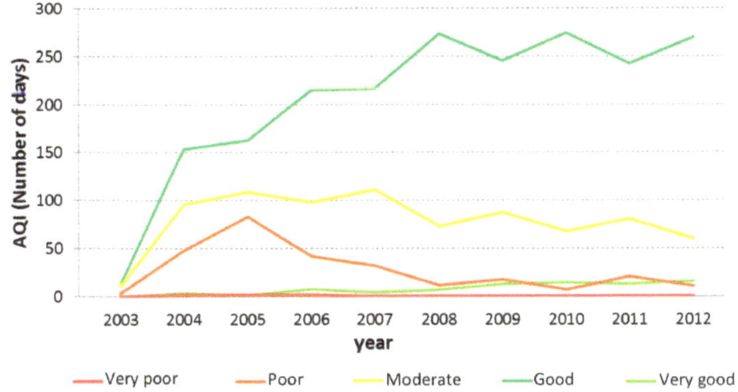

Figure 4. Air Quality Index registered in Setúbal area from 2003 to 2012.

One possible cause for this improvement in the Air Quality Index, especially after 2007, is the reduction of energy consumption, a consequence of the world economic crisis that affected the country, a trend that has already been observed in other Portuguese cities, like Lisbon and Porto, for pollutants such as PM_{10} and NO_2 [27]. Adding to this factor, the geographic position of the study area, which is influenced by clean air masses from the Atlantic Ocean [28], may potentiate this improvement since it contributes to good dispersion conditions of pollutants from local industrial and urban sources.

3.3. Temporal Patterns of Air Pollutants

3.3.1. Annual Trends

Figures 5–7 present the annual variability of $PM_{2.5}$ and PM_{10}, O_3 and NO, NO_2 and NO_X, respectively, during the studied period. When applicable, the number of exceedances taking in account the limit values and air quality guidelines established in the standards defined in Table S2 (supplementary material) are also included.

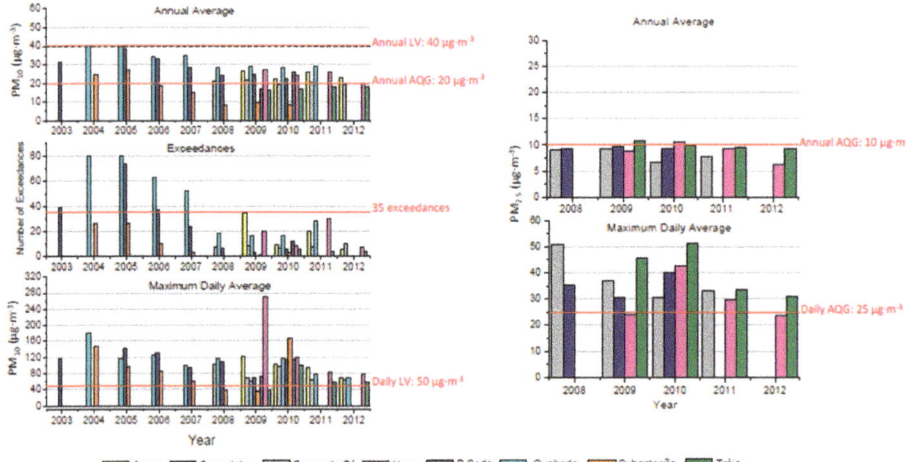

Figure 5. Annual average, exceedances and maximum daily average for PM_{10} (**left**) and $PM_{2.5}$ (**right**) during the period 2003–2012 for Sétubal area. Red line stands for the established limit values (LV) or air quality guidelines (AQG).

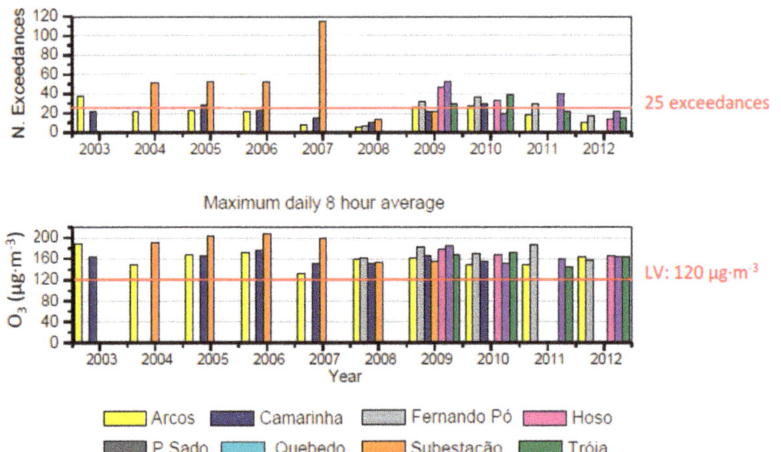

Figure 6. Maximum daily eight hours concentrations of ozone and number of exceedances during the period 2003–2012 at Setúbal area. Red line stands for the established limit values (LV).

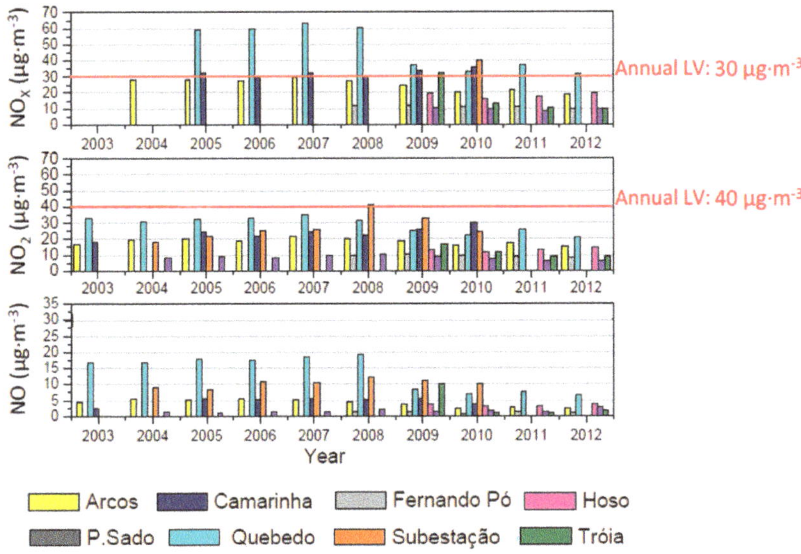

Figure 7. Annual concentrations of nitrogen compounds during the period 2003–2012 for Setúbal area. Red line stands for the established limit values (LV).

The urban traffic station located in Quebedo presented the highest annual PM$_{10}$ concentration with a mean value of 34 µg·m^{-3} for the period 2004–2012, whereas the average PM$_{10}$ concentration in all the studied stations was 25 µg·m^{-3}. The annual averaged PM$_{10}$ concentrations were always below the limit value of 40 µg·m^{-3}, established by EU Directive 2008/50/EC. The annual threshold of 35 exceedances regarding the limit of 50 µg·m^{-3} during a 24 h period, established by EU legislation, was surpassed only in two monitoring stations: the urban background monitoring station "Camarinha" (in 2003, 2005 and 2006) and the urban traffic monitoring station "Quebedo" (from 2004 to 2007). After 2007, no monitoring stations surpassed the annual limit of 35 exceedances regarding the established daily PM$_{10}$ concentration of 50 µg·m^{-3}. However, occasionally, very high daily levels of PM$_{10}$ were measured in several monitoring stations, for instance, the suburban industrial "HOSO" reached a daily concentration slightly above 260 µg·m^{-3} in 2009. Figure 5 shows a decreasing annual trend of the PM$_{10}$ levels from 2005 to 2008.

Considering the annual guide value of 20 µg·m^{-3} for PM$_{10}$ levels recommended by the World Health Organization (WHO) [29], exceedances were observed in all the studied years, for at least one of the studied monitoring stations. The suburban background monitoring station "Tróia" was the only one that presented annual values always below this guideline value (for a total of four years with data). Monitoring stations "Camarinha", "Quebedo", "Arcos" and "P. Sado" always presented annual values above 20 µg·m^{-3} for all monitored years. Moreover, 71%, 60% and 75% of the monitored years above the mentioned threshold were found in the monitoring stations "Subestação", "Fernando Pó" and "HOSO", respectively. In 2011, a study focused on the analysis of trends of air quality in Europe from 2002 to 2011 [30] and revealed that 33% of the urban population in EU-27 lived in areas where the daily limit value for PM$_{10}$ was exceeded and 88% of urban dwellers were exposed to PM$_{10}$ levels that exceeded the WHO AQG for the protection of human health.

Regarding fine particulate matter (PM$_{2.5}$), the annual limit value of 25 µg·m^{-3} defined by EU legislation was not reached in any of the studied monitoring stations. However, regarding the WHO AQG that establishes a guideline value of 10 µg·m^{-3} for the annual concentration, only two monitoring stations surpassed this value, namely "Tróia" in 2009 and "HOSO" in 2010. Regarding the maximum daily average of 25 µg·m^{-3} defined also by the WHO AQG, all monitoring stations presented higher values in all studied years (2008 to 2012), except the monitoring station "HOSO" in 2009. It is relevant to highlight the trend of a decreasing inter-annual variability of PM_{10} levels, while the levels of $PM_{2.5}$ show a stable profile during the studied years.

Figure 6 presents the variability of ozone levels and its exceedances of the established limit value of 120 $\mu g \cdot m^{-3}$ (8 h) during the studied period. The highest concentrations were monitored in the suburban traffic monitoring station "Subestação", with a total of 115 exceedances in 2007 surpassing the annual limit of 25. Regarding the eight hours mean, all monitoring stations presented values above the threshold of 120 $\mu g \cdot m^{-3}$ at least once per year. For the period between 2002 and 2011, in Europe 14% of the urban population was exposed to O_3 levels above the EU target value for protecting human health [30]. Higher O_3 concentrations are more pronounced in Mediterranean countries from southern Europe, due to the more favorable meteorological conditions for its formation such as higher biogenic emissions in summertime, higher insolation, lower deposition under hot and dry conditions and intensive recirculation of air masses [31,32].

Figure 7 presents the levels of nitrogen compounds during the studied period. The WHO AQG for NO_2 is similar to the European annual limit value of 40 $\mu g \cdot m^{-3}$, which was surpassed only once in 2008 in the monitoring station "Subestação". For NOx, the monitoring station "Quebedo" registered annual mean levels of almost twice the annual limit value from 2005 to 2008, and after 2008 measured levels decreased to values slightly higher than the annual limit. The monitoring stations "Camarinha" (in 2005, 2007, 2009 and 2010), "Tróia" (in 2009) and "Subestação" (in 2010) also registered values above this threshold. For NO, the annual levels at all monitoring stations were always below the value of 20 $\mu g \cdot m^{-3}$. The decrease in NOx compared to NO_2 suggests that the proportion of NO_2 in NOx in ambient air has increased. This can be explained by the fact that in the older diesel engines approximately 95% of NOx emissions were NO and only 5% were NO_2. However, in the new diesel passenger cars, both engine size and exhaust after treatments (e.g., catalytic converters) increased the level of NO_2 emissions [33].

Overall, the annual variability of the pollutants shows a decreasing trend, except for ozone. This decrease in pollutant levels is probably due to the implementation of cleaner technologies in the industry, the development of less polluting vehicles and the impact of the economic crisis that promoted the decrease of production and closure of some industrial units in the study area [27].

3.3.2. Monthly Trends

The seasonal variability of pollutants concentrations may provide inputs regarding the processes leading to their production. Figure 8 presents the mean monthly levels of the pollutants measured in the monitoring stations between 2003 and 2012.

Ozone concentrations present a clear seasonal trend with high levels during summer. This season has ideal weather conditions for the formation of this atmospheric oxidant: warm temperatures, sunlight and high emissions of precursor pollutants (NO_x and volatile organic compounds - VOC) that lead to high levels of ozone [32].

The monthly variation of NO, NO_2 and NO_x concentrations followed the opposite trend with lower levels during the summer. The stronger vertical atmospheric mixing in summer helps the dispersion and mixture of pollutants, which contributes to lower NO levels [34]. The apparent NO_2 seasonal variation was probably due to NO_2 depletion during the tropospheric O_3 formation, which is higher in summer [35,36]. It is also important to highlight the high peak of NO_x levels in March registered at the monitoring station "Subestação". For PM, the monthly average concentrations did not present a clear trend.

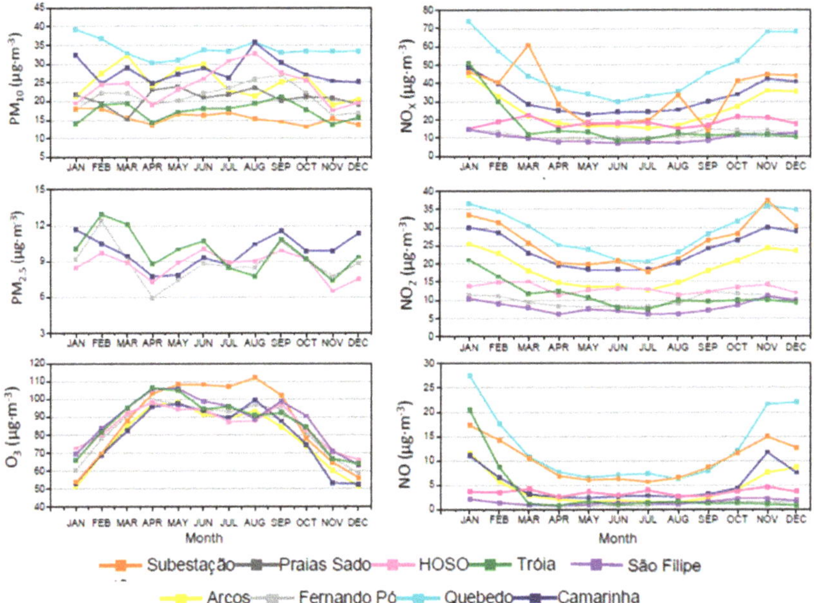

Figure 8. Mean monthly levels of PM_{10}, $PM_{2.5}$, O_3, NO, NO_2 and NO_x registered in the monitoring stations during the period 2003–2012 for Setúbal area.

3.3.3. Hourly Trends

The daily patterns of the studied pollutants are presented in Figure 9. Regarding particulate matter, it is possible to observe different daily profiles depending on the type of monitoring station. "Quebedo" monitoring station is defined as an urban traffic type and it showed higher concentrations of PM_{10} during vehicle peak hours, which highlights traffic influence (mainly due to resuspension [37]). However, it is also possible to observe a similar trend in the stations "Camarinha" (urban background) and "Fernando Pó" (rural background), which highlights traffic contribution to their PM levels, despite being classified as background stations.

Overall, in the monitoring stations influenced by traffic, it is possible to observe two daily peaks during weekdays for PM_{10} and $PM_{2.5}$ concentrations: the first between 7:00 and 10:00 and the second higher one observed between 20:00 and 1:00. This behavior reflects the association of these pollutants with traffic and the poor dispersion conditions during the evening hours, which are characterized by strong atmospheric stability and light winds [38,39]. During the weekend, this pattern is not observable, indicating the lower traffic influence in this period.

The monitoring stations under the influence of industry ("HOSO" and "Praias Sado") presented a different behavior of PM_{10} levels. PM emissions in industrial areas are a complex mixture of stationary and diffuse emissions associated with general site operations such as stocking and transportation of raw materials [40].

Traffic influence can also be confirmed regarding the daily pattern during weekdays of NO, NO_2 and NOx where a high level related to traffic peak hours could be found. This is mainly visible in vehicle peak hours at monitoring stations with traffic influence, such as "Quebedo" (urban traffic) and "Subestação" (suburban traffic), and also at monitoring stations considered as urban background ("Camarinha" and "Arcos"), and this is clearly related to engine combustion emissions [41].

NO concentrations were higher in "Quebedo", "Subestação", "Arcos" and "Camarinha" than in the other stations. "Quebedo" and "Arcos" presented a strong correlation with each other (0.82). These stations have two daily peaks: the first between 7:00 and 9:00 and the second between 17:00 and

19:00 reflecting the morning and evening rush hours. The trend in "Subestação" was characterized by only one NO peak between 7:00 and 9:00 and two NO_2 peaks in the morning and afternoon, which may indicate a larger influence of the traffic source and NO emissions during the morning that resulted in lower NO_2/NOx ratios. In the afternoon, the site was less affected by traffic, which increased the NO_2/NOx ratio due to the oxidation of NO.

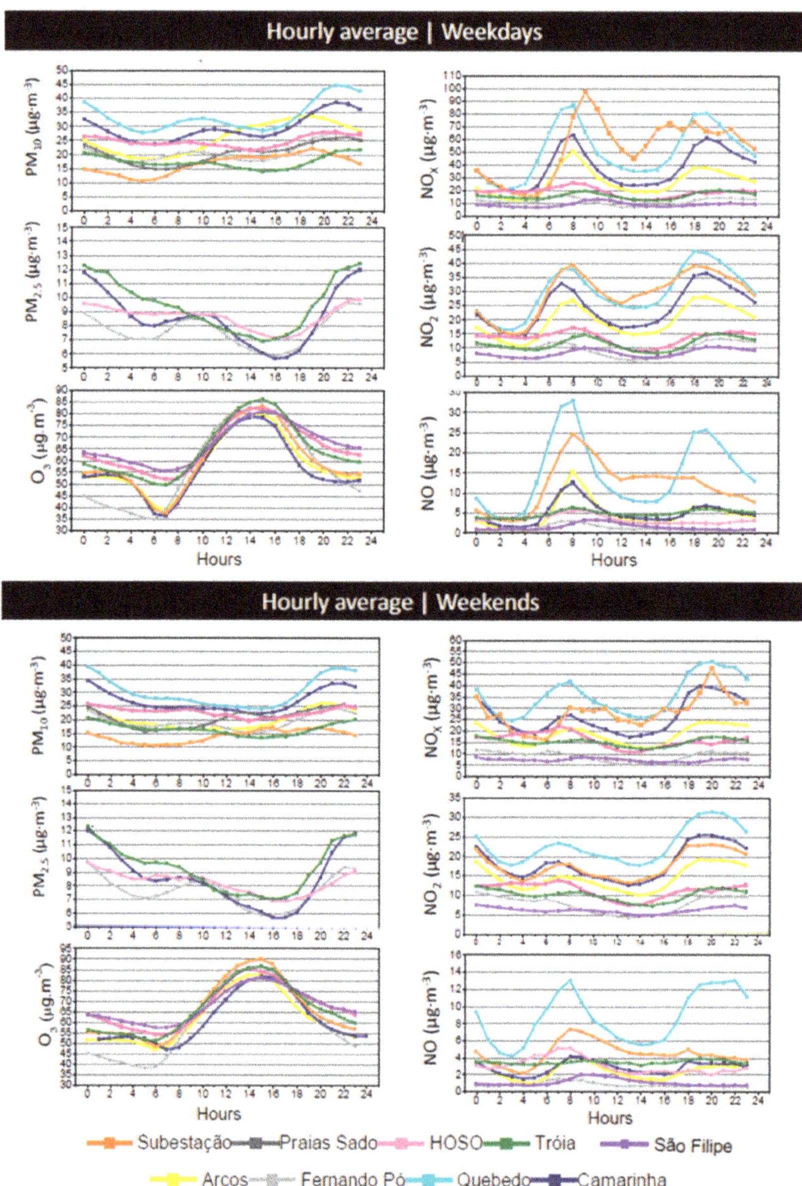

Figure 9. Hourly levels during weekdays (top) and weekends (bottom) of PM_{10}, $PM_{2.5}$, O_3, NO_2, NO_x and NO registered in the monitoring stations during the period 2003–2012 at Setúbal area.

A relevant difference between weekdays and weekend is possible to observe for the levels of NO, NO$_2$ and NO$_x$. During weekends the levels go down by half during the peak hours in the urban stations, when compared with weekdays, which highlights the traffic source of these pollutants.

Regarding ozone, the hourly trends indicate that this pollutant is directly related to the presence of solar radiation, showing lower values in the evening. Since O$_3$ is a secondary pollutant, which means that it is not emitted directly into the atmosphere, its production is achieved in the presence of sunlight by photochemical reactions between NO$_x$ and VOCs, explaining the observed pattern. Overall, O$_3$ levels did not differ significantly among the studied monitoring stations.

3.4. Characterisation of Particulate Matter

In order to understand the pollution sources of particulate matter affecting the study area, a sampling campaign of fine (PM$_{2.5}$) and coarse (PM$_{2.5-10}$) particles was conducted during two different seasons (winter and summer) in 2011 at two different study sites. This section presents the seasonal evaluation of PM levels, their chemical characterization and the source apportionment.

3.4.1. Mass Concentrations

PM levels at both studied monitoring stations ("Quebedo" and "Mitrena") during daytime and nighttime are shown in Figure 10 for winter and summer, along with their compliance with European legislation.

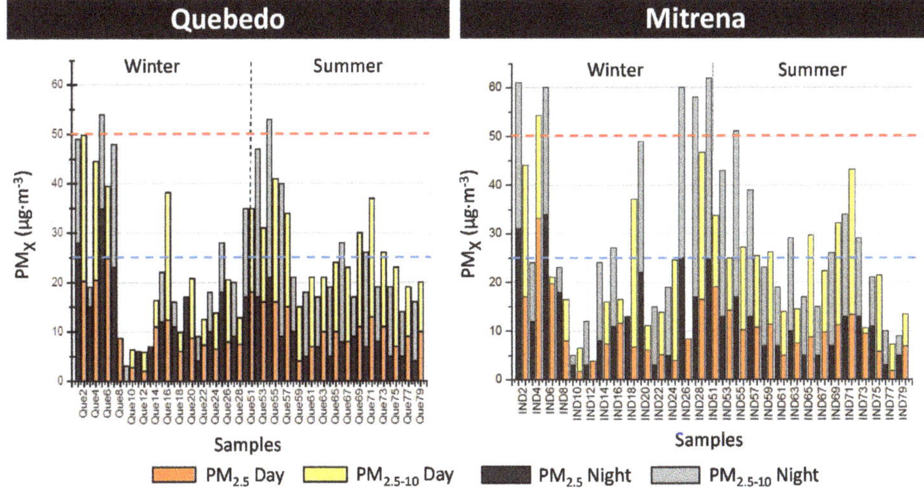

Figure 10. PM concentrations (fine and coarse fractions) sampled in "Quebedo" (**left**) and "Mitrena" (**right**) monitoring stations, during winter and summer. Red and blue dash lines stand for the daily limit value of PM$_{10}$ (50 µg·m^{-3}) and annual limit value of PM$_{2.5}$ (25 µg·m^{-3}), respectively, established by European legislation.

In the urban monitoring station "Quebedo", levels of fine particles ranged from 2 to 35 µg·m^{-3} with a mean value of (12.7 ± 8.1) µg·m^{-3} during the winter period, while during the summer period PM$_{2.5}$ levels ranged from 4 to 21 µg·m^{-3} with a mean value of (9.9 ± 4.5) µg·m^{-3}. Only two exceedances of the limit value of 25 µg·m^{-3} were registered in winter and both occurred during the night period. Regarding the coarse fraction, mean levels of (11.7 ± 8.2) µg·m^{-3} and (16.2 ± 6.7) µg·m^{-3} were registered during winter and summer, respectively. PM$_{10}$ levels ranged from 3 to 55 µg·m^{-3} during winter with a mean level of 22 µg·m^{-3} and from 14 to 53 µg·m^{-3} during summer with a mean level of 26 µg·m^{-3}.

Only two exceedances to the daily PM_{10} limit value of 50 $\mu g \cdot m^{-3}$ were registered, one in each sampling season, but both during the night period.

In the industrial monitoring station "Mitrena", $PM_{2.5}$ levels ranged from 2 to 36 $\mu g \cdot m^{-3}$ with a mean value of (13.0 ± 9.6) $\mu g \cdot m^{-3}$ during winter, while during summer levels ranged from 2 to 19 $\mu g \cdot m^{-3}$ with a mean value of (9.4 ± 4.1) $\mu g \cdot m^{-3}$. These levels were very similar to the ones registered in the "Quebedo" monitoring station. Regarding the coarse fraction, the mean levels registered in Mitrena (16.4 ± 11.7) $\mu g \cdot m^{-3}$ during winter were higher than in Quebedo, principally during the night. This can be explained by the fact that in the industrial site the PM emissions from industry (mainly fugitive emissions) occur during 24 h, while the influence of non-exhaust traffic emissions in the urban area occurs mainly during the day. PM_{10} levels ranged from 4 to 62 $\mu g \cdot m^{-3}$ during winter with a mean value of 29 $\mu g \cdot m^{-3}$, while in summer PM_{10} levels ranged from 7 to 52 $\mu g \cdot m^{-3}$ with a mean value of 25 $\mu g \cdot m^{-3}$. In winter, six exceedances of the daily PM_{10} limit value of 50 $\mu g \cdot m^{-3}$ were registered (five during the night period and one during daytime), while during summer only one exceedance was recorded (during the night period).

3.4.2. Chemical Characterisation

Tables 3 and 4 present the characterization of both fractions of particulate matter ($PM_{2.5}$ and $PM_{2.5-10}$) sampled in "Quebedo" and "Mitrena" monitoring stations, respectively, regarding the mass concentrations and the content of chemical elements and water soluble ions.

For both monitoring stations, the most abundant ions in the fine fraction ($PM_{2.5}$) were SO_4^{2-}, NO_3^- and NH_4^+, which are associated with secondary aerosols [42], resulting from emissions of industry activities and traffic [43]. In the coarse fraction ($PM_{2.5-10}$), the main components were Cl^- and Na^+, which are typically associated with a sea salt source [26], and Ca^{2+}, which is associated with a crustal origin [42].

The Mann-Whitney test showed that in "Quebedo" only Ca^{2+}, in both fractions, and NH_4^+, in the coarse fraction, presented significant differences between day and night. The higher concentrations of Ca^{2+} during the day are probably due to dust re-suspension associated with traffic [37]. NH_4^+ presented higher concentrations overnight. In the industrial monitoring station "Mitrena", no significant differences between day and night concentrations were found.

During summer in "Quebedo", considerably high concentrations were registered for: (1) SO_4^{2-}, due to the strong solar radiation that increases temperature and stimulates the formation of OH radicals, thus promoting the formation of secondary sulphates [44]; (2) La and Sm in the coarse fraction, which are associated with increased dust re-suspension in the dry period; and (3) NO_3^- in the coarse fraction, which can be partly attributed to the reaction of HNO_3 with mineral species, such as calcium carbonate and sea salt to form $Ca(NO_3)_2$ and $NaNO_3$, respectively. These reactions are prevalent in the warm season, while in winter NO_3^- preferentially reacts with NH_3 to form NH_4NO_3 [45]. During winter, high concentrations of NO_3^- were observed in the fine fraction and a strong contribution from the wood burning used in dwellings for house heating was also observed corroborated by high concentrations of K and Sb observed in both fractions and as in the fine fraction when compared with summer [26,46].

In the "Mitrena" monitoring station, an increase of NO_3^- levels in the fine fraction during winter and in the coarse faction during the summer was observed. SO_4^{2-} only presented high concentrations in summer for the coarse fraction. As, Sb, Zn and K showed significantly high concentrations in winter for the fine fraction.

Table 3. Characterization of $PM_{2.5}$ and $PM_{2.5-10}$ sampled at "Quebedo" urban traffic station, regarding mass concentration and contents of chemical elements and water soluble ions. Ion balance stands for the ratio of cations to anions and LoD stands for Limit of Detection.

Parameter	Unit	Annually		Winter						Summer					
		$PM_{2.5}$	$PM_{2.5-10}$	$PM_{2.5}$			$PM_{2.5-10}$			$PM_{2.5}$			$PM_{2.5-10}$		
				24 h	Day	Night	24 h	Day	Night	24 h	Day	Night	24 h	Day	Night
PM_X	$\mu g \cdot m^{-3}$	11.3 ± 6.6	14.1 ± 7.7	12.7 ± 8.1	10.4 ± 6.8	15.0 ± 8.9	11.7 ± 8.2	11.8 ± 9.2	11.7 ± 7.4	9.87 ± 4.47	11.0 ± 3.9	8.68 ± 4.85	16.2 ± 6.7	15.6 ± 4.6	16.8 ± 8.5
Cl^-		195 ± 181	971 ± 974	195 ± 200	154 ± 142	234 ± 241	444 ± 504	472 ± 565	416 ± 453	195 ± 162	220 ± 176	168 ± 149	1480 ± 1050	1100 ± 700	1890 ± 1230
NO_3^-		1070 ± 1130	1070 ± 790	1620 ± 1330	1160 ± 910	2060 ± 1540	858 ± 715	869 ± 834	847 ± 614	487 ± 333	582 ± 393	392 ± 237	1290 ± 820	1240 ± 840	1330 ± 820
SO_4^{2-}		1230 ± 990	392 ± 326	907 ± 699	847 ± 665	964 ± 748	320 ± 312	317 ± 310	322 ± 325	1550 ± 1140	1490 ± 800	1620 ± 1430	465 ± 328	372 ± 234	564 ± 390
Na^+	$ng \cdot m^{-3}$	239 ± 241	831 ± 728	74.6 ± 59.8	75.2 ± 60.9	74.1 ± 61.4	340 ± 337	340 ± 350	341 ± 336	387 ± 247	419 ± 263	353 ± 233	1310 ± 690	1060 ± 580	1570 ± 720
NH_4^+		519 ± 472	53.3 ± 45.9	619 ± 558	474 ± 446	753 ± 631	61.6 ± 47.0	50.9 ± 41.1	72.2 ± 51.6	419 ± 346	419 ± 314	418 ± 390	45.0 ± 44.0	28.8 ± 22.4	61.2 ± 54.4
K^+		126 ± 141	74 ± 102	187 ± 172	147 ± 107	224 ± 213	66.0 ± 60.6	68.0 ± 60.9	64.2 ± 62.3	64.4 ± 57.5	57.3 ± 31.0	71.9 ± 77.3	81 ± 132	48.6 ± 26.6	116 ± 186
Mg^{2+}		23.5 ± 21.2	85.9 ± 64.6	9.94 ± 5.90	10.2 ± 5.58	9.72 ± 6.37	43.8 ± 29.6	46.2 ± 31.6	41.4 ± 28.4	37.0 ± 22.4	41.1 ± 22.8	32.5 ± 21.9	128 ± 63	111 ± 57	146 ± 66
Ca^{2+}		210 ± 185	836 ± 829	192 ± 204	227 ± 228	150 ± 173	957 ± 1060	1210 ± 1250	708 ± 795	225 ± 169	266 ± 175	177 ± 156	724 ± 511	786 ± 329	658 ± 660
Ion balance	n/a	1.65 ± 2.22	2.25 ± 1.22	1.21 ± 0.23	1.31 ± 0.18	1.11 ± 0.24	2.71 ± 1.40	3.23 ± 1.73	2.35 ± 0.93	2.08 ± 3.10	2.71 ± 4.27	1.40 ± 0.27	1.80 ± 0.79	1.88 ± 0.47	1.71 ± 1.00
As		0.46 ± 0.44	0.17 ± 0.13	0.63 ± 0.41	0.67 ± 0.45	0.58 ± 0.39	0.16 ± 0.12	0.18 ± 0.14	0.14 ± 0.10	0.04 ± 0.04	0.04 ± 0.04	0.04 ± 0.06	0.22 ± 0.18	0.17 ± 0.22	0.32
Ce		0.35 ± 0.24	0.28 ± 0.14	0.35 ± 0.25	0.40 ± 0.29	0.30 ± 0.21	0.25 ± 0.13	0.21 ± 0.12	0.32 ± 0.16	0.35 ± 0.26	0.38 ± 0.36	0.31 ± 0.08	n.a.	n.a.	<LoD
Co		0.09 ± 0.12	5.19 ± 13.3	0.13 ± 0.17	0.13 ± 0.20	0.11 ± 0.11	0.05 ± 0.05	0.05 ± 0.06	0.05 ± 0.05	0.07 ± 0.05	0.07 ± 0.06	0.07 ± 0.05	18.6 ± 21.1	6.18 ± 8.12	37.1 ± 22.2
Cr		3.55 ± 2.72	7.93 ± 5.13	1.29 ± 0.99	1.10 ± 0.78	1.61 ± 1.41	5.08 ± 3.16	4.94 ± 2.80	5.22 ± 3.58	4.41 ± 2.68	4.86 ± 2.48	3.99 ± 2.90	13.2 ± 3.62	12.7 ± 2.50	13.6 ± 4.32
Cs		0.68 ± 2.61	0.06 ± 0.01	0.06 ± 0.04	0.07 ± 0.03	0.05 ± 0.06	0.06 ± 0.01	0.06 ± 0.01	<LoD	1.09 ± 3.36	0.07 ± 0.09	2.12 ± 4.72	<LoD	<LoD	<LoD
Fe	$ng \cdot m^{-3}$	138 ± 96	286 ± 236	153 ± 96	159 ± 88	147 ± 105	294 ± 262	348 ± 278	232 ± 238	121 ± 95	138 ± 113	99.9 ± 66.9	254 ± 95	293 ± 102	201 ± 66
K		137 ± 122	99.0 ± 62.8	184 ± 130	154 ± 117	212 ± 138	89.2 ± 66.5	100 ± 73	77.9 ± 59.7	61.7 ± 49.2	62.3 ± 46.5	60.9 ± 55.6	119 ± 51	106 ± 37	133 ± 65
La		0.06 ± 0.06	0.15 ± 0.13	0.06 ± 0.03	0.06 ± 0.04	0.05 ± 0.02	0.09 ± 0.09	0.10 ± 0.10	0.09 ± 0.09	0.07 ± 0.08	0.06 ± 0.09	0.07 ± 0.07	0.23 ± 0.14	0.23 ± 0.17	0.23 ± 0.12
Na		194 ± 181	697 ± 618	65.7 ± 47.0	61.1 ± 39.0	69.9 ± 54.5	302 ± 304	301 ± 294	303 ± 324	322 ± 175	348 ± 181	293 ± 170	1120 ± 590	852 ± 516	1410 ± 540
Sb		0.64 ± 0.46	0.71 ± 0.71	0.82 ± 0.51	0.82 ± 0.39	0.83 ± 9.61	0.95 ± 0.76	1.18 ± 0.76	0.73 ± 0.72	0.43 ± 0.26	0.41 ± 0.21	0.44 ± 0.32	0.23 ± 0.14	0.26 ± 0.17	0.18 ± 0.08
Sc		0.02 ± 0.02	0.05 ± 0.04	0.01 ± 0.01	0.01 ± 0.01	0.01 ± 0.01	0.02 ± 0.02	0.03 ± 0.02	0.02 ± 0.01	0.02 ± 0.03	0.02 ± 0.03	0.02 ± 0.02	0.06 ± 0.04	0.07 ± 0.03	0.06 ± 0.04
Se		0.29 ± 0.23	0.07 ± 0.05	0.30 ± 0.27	0.20 ± 0.21	0.37 ± 0.30	0.07 ± 0.05	0.02 ± 0.02	0.10 ± 0.04	0.29 ± 0.20	0.24 ± 0.17	0.34 ± 0.22	<LoD	<LoD	<LoD
Sm		0.01 ± 0.01	0.04 ± 0.06	0.01 ± 0.00	0.01 ± 0.01	0.01 ± 0.00	0.02 ± 0.02	0.02 ± 0.02	0.02 ± 0.01	0.02 ± 0.02	0.02 ± 0.03	0.01 ± 0.01	0.07 ± 0.10	0.03 ± 0.01	0.10 ± 0.13
Zn		11.0 ± 12.0	13.9 ± 14.3	13.4 ± 15.8	15.2 ± 14.0	11.8 ± 17.4	13.2 ± 14.1	12.7 ± 12.0	13.8 ± 17.3	8.44 ± 4.77	8.94 ± 5.38	7.93 ± 4.22	19.8 ± 17.9	19.8 ± 17.9	<LoD

Table 4. Characterization of $PM_{2.5}$ and $PM_{2.5-10}$ sampled at "Mitrena" suburban industrial station, regarding mass concentration and contents of chemical elements and water soluble ions. Ion balance stands for the ratio of cations to anions and LoD stands for Limit of Detection.

Parameter	Unit	Annually		Winter								Summer					
		$PM_{2.5}$	$PM_{2.5-10}$	$PM_{2.5}$			$PM_{2.5-10}$			$PM_{2.5}$			$PM_{2.5-10}$				
				24 h	Day	Night	24 h	Day	Night	24 h	Day	Night	24 h	Day	Night		
PM_X	$\mu g \cdot m^{-3}$	11.2 ± 7.5	15.7 ± 10.0	13.0 ± 9.6	10.6 ± 8.2	15.3 ± 10.3	16.4 ± 11.7	13.8 ± 9.7	18.8 ± 12.9	9.44 ± 4.10	9.69 ± 4.03	9.18 ± 4.16	15.1 ± 8.1	13.8 ± 9.7	16.9 ± 8.6		
Cl^-		456 ± 685	1050 ± 900	437 ± 692	265 ± 243	597 ± 917	479 ± 552	402 ± 471	550 ± 613	476 ± 689	308 ± 218	656 ± 949	1620 ± 820	1320 ± 740	1940 ± 760		
NO_3^-		917 ± 920	996 ± 735	1320 ± 1120	907 ± 752	1700 ± 1280	670 ± 495	574 ± 471	760 ± 500	519 ± 369	556 ± 415	480 ± 303	1320 ± 800	1220 ± 740	1430 ± 840		
SO_4^{2-}	$ng \cdot m^{-3}$	1570 ± 1120	726 ± 473	1310 ± 1010	1310 ± 1110	1310 ± 910	613 ± 494	556 ± 392	666 ± 573	1820 ± 1170	1790 ± 1100	1860 ± 1250	840 ± 430	751 ± 469	936 ± 355		
Na^+		294 ± 371	751 ± 662	88.2 ± 66.4	98.4 ± 72.0	78.8 ± 58.6	314 ± 343	281 ± 298	344 ± 379	499 ± 434	504 ± 267	494 ± 569	1190 ± 610	1040 ± 610	1350 ± 570		
NH_4^+		534 ± 483	136 ± 224	552 ± 568	452 ± 524	646 ± 593	124 ± 152	83.0 ± 64.2	161 ± 198	516 ± 388	451 ± 341	585 ± 425	149 ± 280	98 ± 139	203 ± 374		
K^+		146 ± 151	146 ± 117	201 ± 196	170 ± 101	230 ± 254	138 ± 83	123 ± 65	152 ± 95	90.8 ± 40.7	101 ± 46	80.0 ± 29.2	155 ± 144	147 ± 90	163 ± 188		
Mg^{2+}		35.0 ± 63.5	93 ± 103	29.5 ± 88.1	16.8 ± 9.8	41 ± 123	70 ± 130	48.4 ± 39.7	90 ± 176	39.9 ± 19.5	44.0 ± 21.3	35.5 ± 15.8	117 ± 61	98.0 ± 56.5	137 ± 60		
Ca^{2+}		280 ± 261	1080 ± 1180	331 ± 313	334 ± 294	329 ± 332	1470 ± 1480	1150 ± 920	1780 ± 1820	229 ± 187	288 ± 208	166 ± 130	684 ± 565	694 ± 559	674 ± 572		
Ion balance	n/a	1.29 ± 0.57	2.33 ± 1.61	1.28 ± 0.75	1.40 ± 0.64	1.19 ± 0.84	3.35 ± 1.74	3.52 ± 1.80	3.20 ± 1.73	1.29 ± 0.32	1.38 ± 0.36	1.20 ± 0.26	1.35 ± 0.49	1.41 ± 0.62	1.29 ± 0.31		
As		0.87 ± 3.14	0.66 ± 1.64	1.49 ± 4.38	2.17 ± 6.06	0.85 ± 0.85	0.30 ± 0.37	0.33 ± 0.45	0.28 ± 0.26	0.25 ± 0.26	0.26 ± 0.27	0.24 ± 0.24	1.01 ± 2.25	0.57 ± 1.01	1.49 ± 3.04		
Ce		0.11 ± 0.12	0.33 ± 0.31	0.13 ± 0.13	0.12 ± 0.10	0.14 ± 0.16	0.39 ± 0.39	0.30 ± 0.26	0.47 ± 0.47	0.09 ± 0.11	0.10 ± 0.12	0.08 ± 0.10	0.28 ± 0.22	0.24 ± 0.18	0.32 ± 0.24		
Co		0.04 ± 0.05	0.09 ± 0.07	0.04 ± 0.05	0.04 ± 0.07	0.03 ± 0.03	0.09 ± 0.08	0.10 ± 0.09	0.08 ± 0.06	0.04 ± 0.04	0.04 ± 0.03	0.04 ± 0.04	0.08 ± 0.06	0.07 ± 0.06	0.10 ± 0.06		
Cr		2.31 ± 1.60	4.27 ± 5.27	2.56 ± 1.49	2.74 ± 1.66	2.40 ± 1.28	6.62 ± 6.25	4.46 ± 3.32	8.64 ± 7.63	2.06 ± 1.70	2.20 ± 1.68	1.90 ± 1.70	1.92 ± 2.45	2.08 ± 2.41	1.75 ± 2.49		
Fe		91 ± 101	296 ± 484	61.3 ± 38.2	64.9 ± 37.0	58.0 ± 38.9	177 ± 114	185 ± 106	170 ± 120	121 ± 132	111 ± 121	132 ± 144	416 ± 660	253 ± 248	589 ± 897		
K	$ng \cdot m^{-3}$	123 ± 119	146 ± 92	187 ± 137	168 ± 130	205 ± 142	146 ± 97	154 ± 93	139 ± 100	59.4 ± 36.7	62.6 ± 28.1	56.0 ± 44.3	146 ± 89	144 ± 84	147 ± 94		
La		0.11 ± 0.10	0.43 ± 0.56	0.14 ± 0.12	0.14 ± 0.12	0.13 ± 0.12	0.63 ± 0.71	0.41 ± 0.41	0.83 ± 0.87	0.08 ± 0.06	0.06 ± 0.04	0.09 ± 0.07	0.23 ± 0.21	0.18 ± 0.16	0.29 ± 0.24		
Na		271 ± 293	786 ± 642	98.7 ± 68.8	105 ± 73	93.0 ± 63.2	381 ± 389	354 ± 354	406 ± 419	443 ± 329	434 ± 257	452 ± 396	1190 ± 590	1050 ± 570	1340 ± 570		
Sb		0.47 ± 0.42	0.27 ± 0.20	0.67 ± 0.50	0.49 ± 0.25	0.84 ± 0.62	0.31 ± 0.23	0.28 ± 0.20	0.33 ± 0.25	0.28 ± 0.18	0.26 ± 0.13	0.30 ± 0.23	0.22 ± 0.15	0.22 ± 0.11	0.23 ± 0.18		
Sc		0.01 ± 0.01	0.06 ± 0.07	0.01 ± 0.01	0.01 ± 0.01	0.01 ± 0.01	0.08 ± 0.09	0.05 ± 0.05	0.10 ± 0.10	0.01 ± 0.01	0.01 ± 0.01	0.01 ± 0.01	0.04 ± 0.03	0.04 ± 0.03	0.04 ± 0.04		
Se		0.37 ± 0.23	0.14 ± 0.13	0.33 ± 0.16	0.29 ± 0.15	0.37 ± 0.16	0.06 ± 0.08	0.05 ± 0.08	0.07 ± 0.07	0.40 ± 0.28	0.38 ± 0.28	0.43 ± 0.27	0.21 ± 0.12	0.20 ± 0.13	0.22 ± 0.11		
Sm		0.01 ± 0.01	0.06 ± 0.09	0.01 ± 0.02	0.01 ± 0.01	0.02 ± 0.02	0.09 ± 0.12	0.05 ± 0.06	0.13 ± 0.15	0.01 ± 0.01	0.01 ± 0.01	0.01 ± 0.01	0.03 ± 0.02	0.02 ± 0.02	0.03 ± 0.03		
Zn		12.6 ± 14.7	11.2 ± 9.2	17.9 ± 19.0	20.4 ± 19.6	15.6 ± 18.0	13.0 ± 11.2	12.6 ± 9.6	13.4 ± 12.7	7.28 ± 4.89	7.39 ± 5.32	7.16 ± 4.34	9.44 ± 6.25	8.22 ± 4.87	10.8 ± 7.3		

Figure S1 (supplementary material) provides the comparison between the PM levels and its components found in both monitoring stations, which allows us to understand the existence of local or non-local sources for PM. High correlations between both monitoring stations, along with similar levels in both, were found regarding SO_4^{2-}, NO_3^- and NH_4^+, which are from secondary aerosols, along with the ions Na^+ and Cl^-, which are associated with sea salt spray [42]. These correlations suggest that these species and elements are from non-local sources. Low correlations were found for Zn, Sb, As, Cl^-, Ca^{2+}, Fe, Sm, K^+ and Cr between both monitoring stations, revealing that there were local sources contributing to the air concentrations of these species. Cr was not associated with a preferential sampling station, indicating the existence of multiple sources for this element. As, Zn, Cl^-, Ca^{2+} and K^+ presented higher concentrations in "Mitrena", indicating the existence of local sources for these species. Sb, Fe and Sm had high levels in "Quebedo" urban traffic monitoring station, probably due to the contribution of vehicles traffic, namely due to tire and break wear and road dust re-suspension [42].

3.4.3. Identification of Emission Sources

In order to identify and assess the contribution of emission sources to the sampled PM levels, a source apportionment study was conducted using the PMF model. Figure 11 presents the contribution of the different sources for the PM levels, where six main chemical sources were identified in both $PM_{2.5}$ and PM_{10}. Figure S2 (supplementary material) provides the mass contribution of the assessed sources to the PM levels of each sampling period.

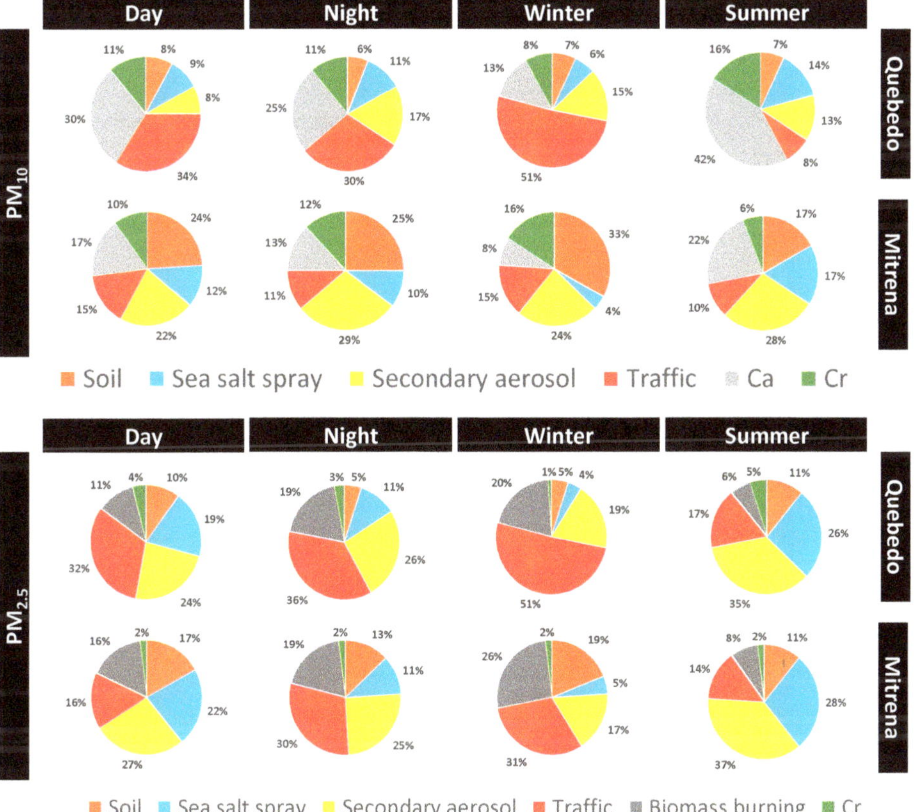

Figure 11. Source contributions (in %) to PM_{10} (**top**) and $PM_{2.5}$ (**bottom**) mass sampled in the monitoring stations "Quebedo" and "Mitrena".

Regarding $PM_{2.5}$, the six main chemical profiles/factors assessed were the following:

(1) Factor 1 was associated with soil since a high association with typical soil elements was found, namely Ca, Sc, Sm and La [42]. This crustal source contributed, on average, to 8% of the total $PM_{2.5}$ mass in "Quebedo" and 15% in "Mitrena". The contribution of this source was higher during daytime probably due to the resuspension related to traffic and due to an increase of the activities dealing with materials handling in the industrial area ("Mitrena").

(2) Factor 2 presented high associations between the water-soluble ions NH_4^+ and SO_4^{2-} and the elements Co and Se, which are associated with secondary aerosols [47]. This source made an average contribution of 26% to the total $PM_{2.5}$ mass in both monitoring stations. In both sites, a higher contribution was found in summer than in winter and contribution during day and night was similar.

(3) Factor 3 was associated with sea salt spray since a high association was found between the water soluble ions Na^+, Mg^{2+} and Cl^-, which are associated with this source [42]. This source contributed around 15% to the total $PM_{2.5}$ mass in both monitoring stations, with a higher contribution during summer probably due to the more intense sea breeze and the dominant S/SW winds registered during that season (Figure 3).

(4) Factor 4 was characterized by water soluble ions K^+ and Cl^- and by As, which are associated with wood burning [48]. This source made a high contribution during the winter in both monitoring stations due to the use of wood burning for house heating. Overall, this source contributed, on average, 15% and 17% to the total $PM_{2.5}$ in "Quebedo" and "Mitrena", respectively.

(5) Factor 5 was associated with a traffic source due to the high association between Sb, NO_3^-, As and Fe. NO_3^- is associated with car emissions and As and Sb are typically from mechanical abrasion of brakes and tires [37,42,49]. Fe may be associated with dust resuspension since this element is typically associated with crustal sources [42]. The traffic contribution was higher in "Quebedo" (34%) than in "Mitrena" monitoring station. In both monitoring stations, the contribution of this traffic source was higher in winter than in summer.

(6) Factor 6 was characterized by Cr, which represents the industrial contribution [50] that was, on average, 3% and 2% for the monitoring stations "Quebedo" and "Mitrena", respectively.

For PM_{10}, the same sources were identified, with the exception of the contribution of wood burning and with the identification of a new source of Ca^{2+}. This calcium source is probably from the cement industry [41] that exists in the area of Setúbal. Overall, the contribution of this source to the total PM_{10} was 27% and 15% in monitoring stations "Quebedo" and "Mitrena", respectively. The traffic contributed to 32% of PM_{10} load in "Quebedo" and 13% in "Mitrena". The industrial emissions characterized by Cr and sea salt spray contributed, on average, to around 11% to the total PM_{10} in both sites. The secondary aerosols contributed to, on average, 13% and 26% of the total PM_{10} in "Quebedo" and "Mitrena", respectively.

4. Conclusions

This study allowed an understanding of the temporal evolution of the air quality during a period of ten years (2003–2012) in Setúbal, an urban area with a high influence of industrial activities. Overall, the air quality index has been improving during the studied period. Setúbal has a set of climate variables, which favors good dispersion of pollutants and, ultimately, confer good air quality in this region despite strong industrial emissions.

With the exception of ozone, all pollutants have demonstrated a decreasing trend, probably due to the implementation of cleaner technologies in the industries, the development of less polluting vehicles; during the study period, the global economic crisis situation also had an impact on the region, which promoted a decrease in production and the closure of some industrial units. However, despite this trend, some pollutants still presented exceedances of the European and WHO guidelines, namely, particulate matter ($PM_{2.5}$ and PM_{10}), NOx and ozone.

Characterization of $PM_{2.5}$ and PM_{10} levels in the area allowed the identification of the main sources contributing to local PM levels, namely, traffic, industry and wood burning, which should be addressed by specific mitigation measures in order to minimize their impact on local air quality.

Supplementary Materials: The following are available online at http://www.mdpi.com/1660-4601/17/15/5447/s1, Figure S1: Spearman correlations between PM and PM components sampled in the studied monitoring stations: "Quebedo" (urban traffic type) and "Mitrena" (industrial type). All elements/ions are displayed in ng·m^{-3} and PM mass concentration is displayed in mg·m^{-3}, Figure S2: Contribution of each source to total PM_{10} mass (top) and total $PM_{2.5}$ mass (bottom) sampled in the monitoring stations Quebedo and Mitrena, Table S1: Statistical data (yearly mean [minimum-maximum]) of meteorological variables (relative humidity and temperature, with the number of measurements, n) registered in the two meteorological stations during the monitoring period of 2004–2012, Table S2: Air quality limit and target values established by the European Commission's Directive 2008/50/EC [51] and air quality guidelines (AQG) defined by the World Health Organization (WHO) [29].

Author Contributions: Conceptualization, A.V.S., S.M.A. and A.I.M.; methodology, A.V.S., S.M.A., A.I.M., C.M.O.; data research and data analysis, A.V.S., C.M.O. and S.M.A.; original draft preparation of the manuscript, A.V.S. and S.M.A.; review of the manuscript, A.V.S., A.I.M., C.M.O., N.C. and S.M.A. All authors have read and agreed to the published version of the manuscript.

Funding: This research was funded by Fundação para a Ciência e Tecnologia, I.P. (FCT) by the grant project PTDC/AAC-AMB/098825/2008 ("PMfugitive—Mitigating the Environmental and Health Impacts of Particles from Fugitive Emissions") and the PhD fellowship SFRH/BD/78698/2011. The FCT support is also acknowledged by C^{2}TN/IST authors (UIDB/04349/2020+UIDP/04349/2020), by CESAM authors (UIDB/50017/2020+UIDP/50017/2020) and by CQE author (UIDB/00100/2020).

Conflicts of Interest: The authors declare no conflict of interest.

References

1. Schraufnagel, D.E.; Balmes, J.R.; Cowl, C.T.; De Matteis, S.; Jung, S.H.; Mortimer, K.; Perez-Padilla, R.; Rice, M.B.; Riojas-Rodriguez, H.; Sood, A.; et al. Air Pollution and Noncommunicable Diseases: A Review by the Forum of International Respiratory Societies' Environmental Committee, Part 1: The Damaging Effects of Air Pollution. *Chest* **2019**, *155*, 409–416. [CrossRef] [PubMed]

2. WHO. *Ambient Air Pollution: A Global Assessment of Exposure and Burden of Disease*; World Health Organization: Geneva, Switzerland, 2016; ISBN 9789241511353.

3. Kuklinska, K.; Wolska, L.; Namiesnik, J. Air quality policy in the U.S. and the EU—A review. *Atmos. Pollut. Res.* **2015**, *6*, 129–137. [CrossRef]

4. European Environment Agency. *Air Quality in Europe—2019 Report—EEA Report No 10/2019*; Publications Office of the European Union: Brussels, Belgium, 2019.

5. European Environment Agency. *Air Quality in Europe—2017 Report—EEA Report No 13/2017*; Publications Office of the European Union: Brussels, Belgium, 2017.

6. Freitas, M.C.; Reis, M.A.; Marques, A.P.; Almeida, S.M.; Farinha, M.M.; De Oliveira, O.; Ventura, M.G.; Pacheco, A.M.G.; Barros, L.I.C. Monitoring of environmental contaminants: 10 Years of application of k 0-INAA. *J. Radioanal. Nucl. Chem.* **2003**, *257*, 621–625. [CrossRef]

7. Farinha, M.M.; Almeida, S.M.; Freitas, M.C.; Verburg, T.G.; Wolterbeek, H.T. Influence of meteorological conditions on PM2.5 and PM 2.5–10 elemental concentrations on Sado estuary area, Portugal. *J. Radioanal. Nucl. Chem.* **2009**, *282*, 815–819. [CrossRef]

8. Almeida, S.M.; Ramos, C.A.; Marques, A.M.; Silva, A.V.; Freitas, M.C.; Farinha, M.M.; Reis, M.; Marques, A.P. Use of INAA and PIXE for multipollutant air quality assessment and management. *J. Radioanal. Nucl. Chem.* **2012**, *294*, 343–347. [CrossRef]

9. Almeida, S.M.; Silva, A.I.V.; Freitas, M.C.; Marques, A.M.; Ramos, C.A.; Silva, A.I.V.; Pinheiro, T. Characterization of dust material emitted during harbour activities by k0-INAA and PIXE. *J. Radioanal. Nucl. Chem.* **2012**, *291*, 77–82. [CrossRef]

10. Silva, A.I.V.; Almeida, S.M.; Freitas, M.C.; Marques, A.M.; Silva, A.I.V.; Ramos, C.A.; Pinheiro, T. INAA and PIXE characterization of heavy metals and rare earth elements emissions from phosphorite handling in harbours. *J. Radioanal. Nucl. Chem.* **2012**, *294*, 277–281. [CrossRef]

11. Garcia, S.M.; Domingues, G.; Gomes, C.; Silva, A.V.; Marta Almeida, S. Impact of road traffic emissions on ambient air quality in an industrialized area. *J. Toxicol. Environ. Heal. Part A Curr. Issues* **2013**, *76*, 429–439. [CrossRef]

12. INE—Instituto Nacional de Estatística. *Censos 2011 Resultados Definitivos*; Instituto Nacional de Estatística I.P.: Lisboa, Portugal, 2011.

13. Agência Portuguesa do Ambiente (APA). *Manual de Métodos e de procedimentos operativos das redes de monitorização da Qualidade do Ar—Amostragem e análise*; Agência Portuguesa do Ambiente: Lisboa, Portugal, 2010.

14. Maenhaut, W.; Francois, F.; Cafmeyer, J. *The "Gent" Stacked Filter Unit (SFU) Sampler for the Collection of Atmospheric Aerosols in Two Size Fractions: Description and Instructions for Installation and Use*; International Atomic Energy Agency: Vienna, Austria, 1994; pp. 249–263.

15. Almeida, S.M.; Freitas, M.C.; Reis, M.; Pinheiro, T.; Felix, P.M.; Pio, C.A. Fifteen years of nuclear techniques application to suspended particulate matter studies. *J. Radioanal. Nucl. Chem.* **2013**, *297*, 347–356. [CrossRef]

16. Dung, H.M.; Freitas, M.C.; Blaauw, M.; Almeida, S.M.; Dionisio, I.; Canha, N.H. Quality control and performance evaluation of k0-based neutron activation analysis at the Portuguese research reactor. *Nuclear Instrum. Meth. Phys. Res. Sect. A Accel. Spectrometers Spectrome Assoc. Equip.* **2010**, *622*, 392–398. [CrossRef]

17. Almeida, S.M.; Almeida-Silva, M.; Galinha, C.; Ramos, C.A.; Lage, J.; Canha, N.; Silva, A.V.; Bode, P. Assessment of the Portuguese k 0-INAA laboratory performance by evaluating internal quality control data. *J. Radioanal. Nucl. Chem.* **2014**, *300*, 581–587. [CrossRef]

18. EMEP. *EMEP Manual for Sampling and Chemical Analysis*; EMEP: Kjeller, Norway, 2001.

19. Carslaw, D.C.; Ropkins, K. Openair—An r package for air quality data analysis. *Environ. Model. Softw.* **2012**, *27*, 52–61. [CrossRef]

20. European Environment Agency. European Air Quality Index. Available online: https://airindex.eea.europa.eu/ (accessed on 16 September 2019).

21. United States Environmental Protection Agency. AirNow. Available online: https://airnow.gov/ (accessed on 16 September 2019).

22. Bao, J.; Yang, X.; Zhao, Z.; Wang, Z.; Yu, C.; Li, X. The Spatial-Temporal Characteristics of Air Pollution in China from 2001–2014. *Int. J. Environ. Res. Public Health* **2015**, *12*, 15875–15887. [CrossRef] [PubMed]

23. Monteiro, A.; Vieira, M.; Gama, C.; Miranda, A.I. Towards an improved air quality index. *Air Qual. Atmos. Health* **2017**, *10*, 447–455. [CrossRef]

24. APA. QualAR—Qualidade do AR. Available online: https://qualar.apambiente.pt/node/indice-qualar (accessed on 16 September 2019).

25. United States Environmental Protection Agency. *Positive Matrix Factorization Model for Environmental data Analyses*; U.S. Environmental Protection Agency: Washington, DC, USA, 2014.

26. Canha, N.; Almeida, S.M.; do Carmo Freitas, M.; Wolterbeek, H.T.; Cardoso, J.; Pio, C.; Caseiro, A. Impact of wood burning on indoor PM2.5in a primary school in rural Portugal. *Atmos. Environ.* **2014**, *94*, 663–670. [CrossRef]

27. Monteiro, A.; Russo, M.; Gama, C.; Lopes, M.; Borrego, C. How economic crisis influence air quality over Portugal (Lisbon and Porto)? *Atmos. Pollut. Res.* **2018**, *9*, 439–445. [CrossRef]

28. Almeida, S.M.; Silva, A.I.; Freitas, M.C.; Dzung, H.M.; Caseiro, A.; Pio, C.A. Impact of maritime air mass trajectories on the western european coast urban aerosol. *J. Toxicol. Environ. Heal. Part A Curr. Issues* **2013**, *76*, 252–262. [CrossRef]

29. WHO. *Air Quality Guidelines for Particulate Matter, Ozone, Nitrogen Dioxide and Sulfur Dioxide: Global Update 2005: Summary of Risk Assessment*; World Health Organization: Geneva, Switzerland, 2006.

30. Guerreiro, C.B.B.; Foltescu, V.; de Leeuw, F. Air quality status and trends in Europe. *Atmos. Environ.* **2014**, *98*, 376–384. [CrossRef]

31. Dayan, U.; Ricaud, P.; Zbinden, R.; Dulac, F. Atmospheric pollution over the eastern Mediterranean during summer—A review. *Atmos. Chem. Phys.* **2017**, *17*, 13233–13263. [CrossRef]

32. Adame, J.A.; Sole, J.G. Surface ozone variations at a rural area in the northeast of the Iberian Peninsula. *Atmos. Pollut. Res.* **2013**, *4*, 130–141. [CrossRef]

33. European Environment Agency. *Air Quality in Europe—2013 Report—European Environment Agency*; Publications Office of the European Union: Brussels, Belgium, 2013.

34. Zabalza, J.; Ogulei, D.; Elustondo, D.; Santamaría, J.M.; Alastuey, A.; Querol, X.; Hopke, P.K. Study of urban atmospheric pollution in Navarre (Northern Spain). *Environ. Monit. Assess.* **2007**, *134*, 137–151. [CrossRef] [PubMed]

35. Ta, W.; Wang, T.; Xiao, H.; Zhu, X.; Xiao, Z. Gaseous and particulate air pollution in the Lanzhou Valley, China. *Sci. Total Environ.* **2004**, *320*, 163–176. [CrossRef] [PubMed]

36. Yan, Y.; Lin, J.; Pozzer, A.; Kong, S.; Lelieveld, J. Trend reversal from high-to-low and from rural-to-urban ozone concentrations over Europe. *Atmos. Environ.* **2019**, *213*, 25–36. [CrossRef]

37. Almeida-Silva, M.; Canha, N.; Freitas, M.C.; Dung, H.M.; Dionísio, I. Air pollution at an urban traffic tunnel in Lisbon, Portugal-an INAA study. *Appl. Radiat. Isot.* **2011**, *69*. [CrossRef]

38. Vecchi, R.; Marcazzan, G.; Valli, G. A study on nighttime-daytime PM10 concentration and elemental composition in relation to atmospheric dispersion in the urban area of Milan (Italy). *Atmos. Environ.* **2007**, *41*, 2136–2144. [CrossRef]

39. Latif, M.T.; Dominick, D.; Ahamad, F.; Khan, M.F.; Juneng, L.; Hamzah, F.M.; Nadzir, M.S.M. Long term assessment of air quality from a background station on the Malaysian Peninsula. *Sci. Total Environ.* **2014**, *482*, 336–348. [CrossRef]

40. Almeida, S.M.; Lage, J.; Fernández, B.; Garcia, S.; Reis, M.A.; Chaves, P.C. Chemical characterization of atmospheric particles and source apportionment in the vicinity of a steelmaking industry. *Sci. Total Environ.* **2015**, *521–522*, 411–420. [CrossRef]

41. Casquero-Vera, J.A.; Lyamani, H.; Titos, G.; Borrás, E.; Olmo, F.J.; Alados-Arboledas, L. Impact of primary NO$_2$ emissions at different urban sites exceeding the European NO$_2$ standard limit. *Sci. Total Environ.* **2019**, *646*, 1117–1125. [CrossRef]

42. Calvo, A.I.; Alves, C.; Castro, A.; Pont, V.; Vicente, A.M.; Fraile, R. Research on aerosol sources and chemical composition: Past, current and emerging issues. *Atmos. Res.* **2013**, *120*, 1–28. [CrossRef]

43. Karagulian, F.; Belis, C.A.; Dora, C.F.C.; Prüss-Ustün, A.M.; Bonjour, S.; Adair-Rohani, H.; Amann, M. Contributions to cities' ambient particulate matter (PM): A systematic review of local source contributions at global level. *Atmos. Environ.* **2015**, *120*, 475–483. [CrossRef]

44. Almeida, S.M.; Freitas, M.C.; Repolho, C.; Dionísio, I.; Dung, H.M.; Pio, C.A.; Alves, C.; Caseiro, A.; Pacheco, A.M.G. Evaluating children exposure to air pollutants for an epidemiological study. *J. Radioanal. Nucl. Chem.* **2009**, *280*, 405–409. [CrossRef]

45. Almeida, S.M.; Pio, C.A.; Freitas, M.C.; Reis, M.A.; Trancoso, M.A. Source apportionment of fine and coarse particulate matter in a sub-urban area at the Western European Coast. *Atmos. Environ.* **2005**, *39*, 3127–3138. [CrossRef]

46. Vicente, A.; Alves, C.; Calvo, A.I.; Fernandes, A.P.; Nunes, T.; Monteiro, C.; Almeida, S.M.; Pio, C. Emission factors and detailed chemical composition of smoke particles from the 2010 wildfire season. *Atmos. Environ.* **2013**, *71*, 295–303. [CrossRef]

47. Canha, N.; Almeida, S.M.; Freitas, M.D.C.; Trancoso, M.; Sousa, A.; Mouro, F.; Wolterbeek, H.T. Particulate matter analysis in indoor environments of urban and rural primary schools using passive sampling methodology. *Atmos. Environ.* **2014**, *83*, 21–34. [CrossRef]

48. Canha, N.; Freitas, M.C.; Almeida-Silva, M.; Almeida, S.M.; Dung, H.M.; Dionísio, I.; Cardoso, J.; Pio, C.A.; Caseiro, A.; Verburg, T.G.; et al. Burn wood influence on outdoor air quality in a small village: Foros de Arrão, Portugal. *J. Radioanal. Nucl. Chem.* **2012**, *291*, 83–88. [CrossRef]

49. Grigoratos, T.; Martini, G. Brake wear particle emissions: A review. *Environ. Sci. Pollut. Res.* **2015**, *22*, 2491–2504. [CrossRef]

50. Cesari, D.; Genga, A.; Ielpo, P.; Siciliano, M.; Mascolo, G.; Grasso, F.M.; Contini, D. Source apportionment of PM2.5 in the harbour-industrial area of Brindisi (Italy): Identification and estimation of the contribution of in-port ship emissions. *Sci. Total Environ.* **2014**, *497*, 392–400. [CrossRef]

51. European Commission. Directive 2008/50/EC of the European Parliament and of the Council of 21 May 2008 on ambient air quality and cleaner air for Europe. *Off. J. Eur. Union* **2008**, *151*, 1–44.

International Journal of
*Environmental Research
and Public Health*

Article

Geochemical, Mineralogical and Morphological Characterisation of Road Dust and Associated Health Risks

Carla Candeias [1], Estela Vicente [2], Mário Tomé [3], Fernando Rocha [1], Paula Ávila [4] and Alves Célia [2,*]

[1] Geobiosciences, Geotechnologies and Geoengineering Research Centre (GeoBioTec),
   Department of Geosciences, University of Aveiro, 3810-193 Aveiro, Portugal; candeias@ua.pt (C.C.);
   tavares.rocha@ua.pt (F.R.)
[2] Centre for Environmental and Marine Studies (CESAM), Department of Environment, University of Aveiro,
   3810-193 Aveiro, Portugal; estelaavicente@ua.pt
[3] School of Technology and Management (ESTG), Polytechnic Institute of Viana do Castelo,
   Avenida do Atlântico, nº 644, 4900-348 Viana do Castelo, Portugal; mariotome@estg.ipvc.pt
[4] LNEG, National Laboratory of Energy and Geology, Rua da Amieira,
   4466-901 São Mamede de Infesta, Portugal; paula.avila@lneg.pt
* Correspondence: celia.alves@ua.pt

Received: 20 January 2020; Accepted: 26 February 2020; Published: 28 February 2020

Abstract: Road dust resuspension, especially the particulate matter fraction below 10 μm (PM_{10}), is one of the main air quality management challenges in Europe. Road dust samples were collected from representative streets (suburban and urban) of the city of Viana do Castelo, Portugal. PM_{10} emission factors (mg veh^{-1} km^{-1}) ranging from 49 (asphalt) to 330 (cobble stone) were estimated by means of the United Stated Environmental Protection Agency method. Two road dust fractions (<0.074 mm and from 0.0074 to 1 mm) were characterised for their geochemical, mineralogical and morphological properties. In urban streets, road dusts reveal the contribution from traffic emissions, with higher concentrations of, for example, Cu, Zn and Pb. In the suburban area, agriculture practices likely contributed to As concentrations of 180 mg kg^{-1} in the finest road dust fraction. Samples are primarily composed of quartz, but also of muscovite, albite, kaolinite, microcline, Fe-enstatite, graphite and amorphous content. Particle morphology clearly shows the link with natural and traffic related materials, with well-formed minerals and irregular aggregates. The hazard quotient suggests a probability to induce non-carcinogenic adverse health effects in children by ingestion of Zr. Arsenic in the suburban street represents a human health risk of 1.58×10^{-4}.

Keywords: road dust; traffic; PM_{10} emission factors; enrichment index; human health risk

---

1. Introduction

Particulates that are deposited on a road, usually called "road sediments", "street dust" or "road dust", are significant pollutants in the urban environment because they contain high levels of toxic metals and organic contaminants, such as polycyclic aromatic hydrocarbons [1,2]. These materials can be pulverised by the passing traffic and become aerolisable, making up a significant fraction of atmospheric particles. Another process contributing to the atmospheric particle loads is the resuspension of road dust, which is due to traffic induced turbulence, tyre friction or the action of the wind. Amato et al. [3] reported that local dust accounted for 7%–12% of the particulate matter <10 μm (PM_{10}) concentrations at suburban and urban background sites in southern European cities and 19% at a traffic site, revealing the contribution from road dust resuspension. In the case of particulate matter <2.5 μm ($PM_{2.5}$), the percentages decreased to 2%–7% at suburban and urban background sites and to 15% at the traffic site. Results for southern Spain indicated that road dust increased PM_{10} levels on average by 21%–35% at traffic sites, 29%–34% at urban background sites heavily affected by road traffic emissions, 17%–22% at urban-industrial sites and 9%–22% at rural sites [4]. In the urban area of Lanzhou, China, fugitive road dust was found to contribute to 24.6% of total $PM_{2.5}$ emissions [5]. In Harbin, located at the north region of China, road dust represented 7% to 26% of PM_{10} [6]. In Delhi, road dust is the largest contributor to PM_{10}, accounting for 35.6%–65.9% [7]. As motor exhaust emissions decay as a result of increasingly strict limits, the relative importance of emissions from resuspension and wear (brakes, tyres and road pavement) will grow. These emissions are recognised as non-exhaust sources. It has been estimated that in 2020 non-exhaust sources will represent about 90% of total road traffic emissions [8].

Long-term exposure to traffic-generated dust was estimated to cause every year 1.5 to 2 million premature deaths (mostly women and children) in low-income countries [9]. There is increasing awareness and concern of the potential adverse impacts of dust on health in high- and middle-income countries. A systematic literature review of articles on road dust and its effects on health was recently carried out by Khan and Strand [2]. The components of road dust particles have been associated with multiple health effects, in particular on the respiratory and cardiovascular systems. The list of health effects reported in the reviewed articles included chronic obstructive pulmonary disease, asthma, allergy, carcinoma, emergency cardiovascular disease issues, increased mortality due to cardiovascular disease, low birth weight and non-specific carcinoma. Numerous studies on the costs and efficacy of various products (e.g., suppressants) aimed at reducing road dust have been undertaken in the USA, Canada, Australia, New Zealand, South Africa and several European countries [9–12].

The amount of dust that is generated and then re-settles on the road surface depends on various factors including traffic speed, vehicle weight, local road conditions and rainfall. In regions with scarce precipitation, such as the Mediterranean countries, road dust resuspension is one of the major sources triggering PM_{10} exceedances. The composition of dust presents huge geographical and seasonal variations, so it is difficult to create a universal chemical fingerprint for this source [13]. To more accurately apportion the contribution of road dust to the atmospheric particulate matter levels, to better assess health risks, and to establish cause-effect relationships by seeking potentially toxic constituents, specific road dust chemical profiles should be obtained for each region or location. Moreover, to account for this source in pollutant inventories, appropriate emission factors are required. In Europe, some information on the chemical composition, mainly elemental, of resuspendable road dust samples from Barcelona, Girona and Zurich [14], four of the most polluted Polish cities [15], Oporto [1], and Paris [16], have been documented. This paper aims to present not only emission factors and the chemical composition, but also the mineralogical and morphological characteristics of road dust, as well as an assessment of health risks, in a medium-sized city with low-pollution levels for which no previous study has been conducted.

2. Methodologies

2.1. Sampling

Road dust samples were collected in September and October 2018 on representative streets of Viana do Castelo (latitude: 41°41′35.63″ N; longitude: −8°49′ 58.33″ W), the most northern Atlantic city in Portugal with a population of about 40,000 in the most central urbanised area, although the municipality as a whole reaches 90,000 inhabitants. The city is located between Santa Luzia hill and the mouth of the river Lima. Its major industries are related to naval construction and repair, with one of the few large shipyards still in operation. It is also home to a large cluster of wind green electricity and car-parts industries. The old downtown streets are closed to traffic.

Three streets were selected for road dust sampling (Figure S1):

(S1) Suburban context—Rua Alto Xisto is a street with cobbled pavement made of granite cubes in a residential area on the outskirts of the city. One side of the street consists of terraced houses. The other side is flanked by an agricultural farm with some animals, and vineyard, corn and potato cultivation;

(S2) Urban context—Local road within the campus of the Higher School of Technology and Management. It is composed of stone mastic asphalt pavement and located a few meters from the beach and the shipyards;

(S3) Urban context—Avenida Combatentes da Grande Guerra is a central artery connecting to the train station. It is an avenue with shops, services and an elementary school. Its cobbled pavement is made of granite cubes.

The selection of streets was carried out in collaboration with the city council. It was considered that the urban area could be subdivided into three sectors: (i) a residential area with a lower population density and some agricultural influence, (ii) a central area with a higher volume of traffic where the main commercial activities and public services take place, (iii) and another area on the coast line, also with busy streets, but with industrial influence. For each sector, a street representing the dominant pavements and traffic was chosen. The various samplings took place on 3 weekends in September and October 2018 and implied the traffic cut by the police authorities. Road dust was collected on delimited lane sections using a vacuum cleaner, following the procedure described by the United Stated Environmental Protection Agency (USEPA) in its AP-42 document [17]. In each street, several subsamples were obtained by vacuuming segments of 20 m in length and width corresponding to the lanes. The collection of the first subsample started at a distance of 50 m from the intersection with another street. Distances of approximately 50 m were maintained between sampled segments. The weight of subsamples ranged from about 200 to 600 g. For each road, a composite sample from a minimum of 3 incremental subsamples was created.

2.2. Geochemical, Mineralogical and Morphological Characterisation

After sampling, the vacuum cleaner bags were stored in polyethylene bags and brought to the laboratory, where the USEPA methodology for analysis of surface/bulk dust loading samples was followed [18]. Samples were wet sieved with addition of distilled water through standard Taylor screens, dried at ~40 °C, and weighed in a microbalance with 1 µg sensitivity. While particles >1 mm were rejected, the fraction <0.074 mm (F1) and that between 0.0074 and 1 mm (F2) were retained for further analyses.

The chemical composition of the road dust fractions was determined by X-ray fluorescence (XRF) using an Axios PW4400/40 X-ray (Marvel Panalytical, Almelo, The Netherlands) fluorescence wavelength dispersive spectrometer. The limit of detection (LOD) (i.e., the concentration at which it is possible to distinguish the signal from the background) is provided in Table 1. Mineralogy was determined by X-ray diffraction (XRD) using a X'Pert-Pro MPD diffractometer (Marvel Panalytical, Almelo, The Netherlands) with a Cu-Kα radiation at 45 kV, 40 mA, and with a step size of 0,02°2θ, in 2°–90° 2θ angle range. A Hitachi S-4100 scanning electron microscope (SEM) equipped with a Quantax 400 Energy Dispersive Spectrometer (EDS) (Bonsai Advanced, Madrid, Spain) was used for point and area chemical analyses. Particle size distributions of road dusts were determined by X-ray sedimentation technique with a Sedigraph III Plus grain size analyser (Micromeritics Instrument Corp., Norcross, GA, USA). This technique for determining the relative mass distribution of a sample by particle size is based on two physical principles: sedimentation theory (Stokes' law) and the absorption of X-radiation (Beer-Lambert law). Precision and accuracy of analyses and digestion procedures were monitored using internal standards, certified reference material and quality control blanks.

2.3. Estimation of Emission Factors

Road dust emission factors (*EFs*) were estimated using the equation outlined in the AP-42A document of USEPA [19] for paved roads:

$$EF = k\,(sL/2)^{0.65} \times (W/3)^{1.5} - C, \tag{1}$$

where:

EF = PM_{10} emission factor (g veh^{-1} km^{-1}),

k = particle size multiplier for particle size range and units of interest (0.46 g veh^{-1} km^{-1} for PM_{10}),

s = surface material silt content, defined as particles that pass through a 200-mesh screen, which corresponds to 74 µm (%),

L = surface material loading, defined as mass of particles per unit area of the travel surface (g m^{-2}),

W = average weight (tons) of the vehicles travelling the road (a value of 2 tons was assumed),

C = emission factor for 1980's vehicle fleet exhaust, brake wear and tyre wear (0.1317 g veh^{-1} km^{-1} for PM_{10}).

2.4. Enrichment Index

The enrichment index (*EI*) of an element is based on its concentration and the concentration of a reference element. The latter is chosen based on its low occurrence variability in order to identify geogenic and anthropogenic contributions. Due to the abundant natural occurrence on Earth's crust, Al was selected for this study. *EI* was calculated as follows:

$$EI_x = \left(X/C_{ref}\right)sample / \left(X/C_{ref}\right)crust, \tag{2}$$

with EI_x is the enrichment index of the element X, X the concentration of the element and C_{ref} the concentration of the reference element (Al). The Earth's crust individual elemental composition was taken from Riemann and Caritat [20]. $EI < 1$ indicates no enrichment (natural contribution), $1 \leq EI < 3$ minor anthropogenic enrichment, $3 \leq EI < 5$ moderate anthropogenic enrichment, $5 \leq EI < 10$ moderately severe anthropogenic enrichment, $10 \leq EI < 25$ severe anthropogenic enrichment, $25 \leq EI < 50$ very severe anthropogenic enrichment, and $EI \geq 50$ extremely severe anthropogenic enrichment [21].

2.5. Human Health Risk Assessment of Exposure to Potential Toxic Elements in Road Dust

A human health risk assessment assumes that residents, both children and adults, are directly exposed to potential toxic elements through three different routes: ingestion, dermal absorption and inhalation of particles [22–24]. For road dust, it was assumed that intake rates and particle emission are similar to those established for soils. Equations (3)–(5) were used to estimate the chronic daily intake (CDI_{route}, ingestion and dermal in mg $kg^{-1}\cdot d^{-1}$; inhalation in mg m^{-3} for non-carcinogenic effects, and µg m^{-3} for carcinogenic effects) of each exposure route and for separated elements:

$$CDI_{ing-nc} = \frac{C \times IngR \times EF \times ED}{BW \times AT_{nc}} \times 10^{-6}, \tag{3}$$

$$CDI_{drm-nc} = \frac{C \times SA \times SAF \times DA \times EF \times ED}{BW \times AT_{nc}} \times 10^{-6}, \tag{4}$$

$$CDI_{inh-nc} = \frac{C \times InhR \times EF \times ED}{PEF \times BW \times AT_{nc}}, \tag{5}$$

where, C (mg kg^{-1}) is the concentration of the element in road dust, $IngR$ is the ingestion rate (200 and 100 mg d^{-1} for children and adults, respectively), $InhR$ is the inhalation rate (7.6 and 20 m^3 d^{-1} for children and adults, respectively), EF is the exposure frequency (180 d yr^{-1}), ED is the exposure duration (6 and 24 years for non-carcinogenic effects in children and adults, respectively, and 70 years is the lifetime for carcinogenic effects), BW is the average body weight (15 and 70 kg for children and adults, respectively), AT_{nc} is the averaging time for non-carcinogenic effects (AT days = $ED \times 365$), SA is the exposed skin area (2800 and 5700 cm^2 for children and adults, respectively), SAF is the skin adherence factor (0.2 and 0.07 mg cm^{-2} for children and adults, respectively), DA is the dermal absorption factor (0.03 for As and 0.001 for other elements), and PEF is the particulate emission factor (1.36×10^9 m^3 kg^{-1}) [22–26].

For each element and route, the non-cancer toxic risk was estimated by computing the chronic hazard quotient (HQ, Equation (6)) for systemic toxicity [24]. A HQ less than or equal to 1 indicates that adverse effects are not likely to occur, and thus can be considered to have negligible hazard, while $HQ > 1$ suggests that there is a high probability of non-carcinogenic effects to occur. To estimate the overall developing hazard of non-carcinogenic effects, it is assumed that toxic risks have additive effects. Therefore, it is possible to calculate the cumulative non-carcinogenic hazard index (HI), which corresponds to the sum of HQ for each pathway (Equation (7)) [27].

$$HQ_{route} = \frac{CDI_{route}}{R_f D_{route}}, \tag{6}$$

$$HI = \sum HQ = HQ_{ing} + HQ_{drm} + HQ_{inh}, \tag{7}$$

with $R_f D$ being the chronic reference dose (values estimated as given in RAIS) [24].

The probability of an individual to develop any type of cancer over his lifetime (Risk), as a result of exposure to the carcinogenic hazards, was computed for each route according to Equation (8). The carcinogenic Total Risk is the sum of risk for each route (Equation (9)). A cancer risk below 1×10^{-6} is considered insignificant, being this value the carcinogenic target risk. A risk above 1×10^{-4} (a probability of 1 in 10,000 of an individual developing cancer) is considered unacceptable [22–24,27]:

$$Risk_{route} = CDI_{route} \times CSF_{route}, \tag{8}$$

$$\begin{aligned} Total\ Risk\ &=\ \sum Risk\ =\ Risk_{ing} + Risk_{drm} + Risk_{ihn} \\ &=\ CDI_{ing-ca} \times CFS_{ing} + CDI_{inh-ca} \times IUR + \frac{CDI_{drm-ca} \times CSF_{ing}}{ABS_{gi}}, \end{aligned} \tag{9}$$

$$CDI_{inh-ca} = \frac{c \times IngR_{adj} \times EF}{AT_{ca}} \times 10^{-6}, \tag{10}$$

$$IngR_{adj} = \frac{ED_{child} \times IngR_{child}}{BW_{child}} + \frac{(ED_{resident} - ED_{child}) \times IngR_{adult}}{BW_{adult}}, \tag{11}$$

$$CDI_{inh-ca} = \frac{C \times EF \times ET \times ED}{PEF \times 24 \times AT_{ca}}, \tag{12}$$

$$CDI_{drm-ca} = \frac{c \times DA_d \times EF \times DSF_{adj}}{AT_{ca}} \times 10^{-6}, \tag{13}$$

$$DSF_{adj} = \frac{ED_{child} \times SA_{child} \times SAF_{child}}{BW_{child}} + \frac{(ED_{resident} - ED_{child}) \times SA_{adult} \times SAF_{adult}}{BW_{adult}} \tag{14}$$

where CSF is the chronic slope factor (ingestion, (mg kg^{-1} d^{-1})$^{-1}$; dermal, CSF_{ing}/ABS_{gi}), ABS_{gi} is the gastrointestinal absorption factor, IUR is the chronic inhalation unit risk ((µg m^{-3})$^{-1}$), DFS_{adj} is the soil dermal contact factor-age-adjusted (362 mg yr kg^{-1} d^{-1}), AT_{ca} is the averaging time carcinogenic effects (AT days = $LT \times 365$) and ET is the exposure time (8 h d^{-1}) [27].

3. Results and Discussion

3.1. PM$_{10}$ Emission Factors

For granite cube paved streets, very close PM$_{10}$ emission factors were found: 277 mg veh^{-1} km^{-1} (central avenue, S3) and 330 mg veh^{-1} km^{-1} (street on the outskirts of the city, S1). On the asphalt paved street (S2), a much lower PM$_{10}$ emission factor was obtained (49 mg veh^{-1} km^{-1}). Using an in-situ resuspension chamber, PM$_{10}$ emission factors in the range 12.0–29.4 mg veh^{-1} km^{-1}, averaging 18.6 mg veh^{-1} km^{-1}, were determined for asphalt roads in the city of Oporto [1]. However, a much higher value of 1082 mg veh^{-1} km^{-1} was estimated for a cobbled pavement in the same city. According to the researchers, this great emission factor was due to the very high roughness of cobblestones, which may have promoted the build-up of road sediments. Moreover, the joints between granite cubes filled with soil may have constituted an additional source of resuspendable dust. Following the USEPA methodology, PM$_{10}$ emission factors for typical urban roads in Milan within 13–32 mg veh^{-1} km^{-1} were recently reported, which were found to be in the central range of the literature values in Europe [28]. Based on micro-scale vertical profiles of the deposited mass, PM$_{10}$ emission factors from 5.4 to 9.0 mg veh^{-1} km^{-1} were obtained at inner-roads of Paris, whilst a higher value was estimated for the ring road (17 mg^{-1} veh^{-1} km^{-1}) [16]. Based on a literature review, Boulter et al. [29] summarised the available information on emissions from road dust resuspension for the United Kingdom, USA, central and northern Europe. Differences in emission factors may be associated with local factors, such as road pavement type, regularity of street washing and precipitation events. On the other hand, while in Southern Europe high solar radiation is usually recorded, favouring the mobility of deposited particles, winter sanding or salting of roads and the use of studded tyres in Scandinavian and Alpine regions may promote the enhancement of road dust loadings. Road PM$_{10}$ emission factors documented for Europe are, in general, lower than those reported for the USA. However, it must not be forgotten that American databases are older than European ones.

3.2. Geochemical Characterization of Dust and Enrichment Index

Road dusts are known for their heterogeneous mix with diverse natural and anthropogenic origins. The composition can vary depending on geographical location, resuspended soil, atmospheric deposition and anthropogenic sources, which include traffic related particles such as metallic components, eroded road pavement, but also building construction and demolition, and power generation [14,30].

Cluster analysis of the XRF results (Figure S2) of the studied samples (S1 suburban environment influenced by agricultural activities; S2 and S3 urban streets) confirmed the chemical difference between the two analysed fractions (F1 with samples <0.074 mm; F2 with sizes >0.074 mm and <1 mm). Elemental concentrations (Table 1) revealed Si as the most representative constituent, followed by Al > K > Fe > Ca > Na. The highest Si content was found in F2 (fraction >0.074 mm and <1 mm). The enrichment index (Figure S3) suggests a low influence of anthropogenic activities in both size fractions (EI < 1.5). Crustal erosion and parent materials (e.g., Variscan granite parent rock) may also influence the concentrations of other elements, such as Fe, Al, Mg, Mn, Rb, Na, Ti, Mo, V and Zr [31]. Previous studies suggest that heavy metals such as Mn, V, Cu, Fe, Ni, Pb and Zn are linked to traffic emissions [32,33]. Low enrichment indices (<3) were also obtained for manganese, indicating a major natural source. However, this element is used in fuel additives, aluminium based alloys offering a high corrosion resistance and formability, vehicle applications, such as inner panels, and heater and radiator tubes.

Table 1. Summary descriptive statistics of elements analysed in two size fractions of road dusts from three selected locations. Concentrations in mg kg^{-1}.

	LOD	F1		F2	
		min–max	Med ± SD	min–max	Med ± SD
Al	0.50	64,013–95,874	75,836 ± 13,150	54,873–67,749	66,241 ± 5748
As	4.06	28–180	35 ± 70	12–21	18 ± 3.5
Ba	6.90	270–560	390 ± 119	250–490	290 ± 105
Br	0.78	37–62	49 ± 11	7.7–17	11 ± 4.0
Ca	0.50	17,146–58,412	19,669 ± 18,887	9,806–27,938	13,687 ± 7795
Cl	0.50	3090–11,610	6890 ± 3485	220–400	330 ± 74
Cr	1.96	48–230	210 ± 81	49–67	60 ± 7.5
Cu	2.84	95–870	260 ± 333	31–210	42 ± 82
F	0.60	920–1350	1170 ± 176	1160–1640	1220 ± 214
Fe	0.50	27,613–39,895	35,188 ± 5059	17,353–18,633	18,416 ± 559
Ga	0.94	14–18	15 ± 1.5	9.6–12	10 ± 0.8
K	0.60	31,969–48,614	43,318 ± 6943	38,146–47,095	43,102 ± 3661
Mg	0.50	5947–6887	6610 ± 395	2883–5030	3920 ± 877
Mn	5.62	472–488	480 ± 6.3	248–279	263 ± 13
Mo	0.78	1.3–2.9	2.1 ± 0.7	2.9–11	6.9 ± 3.3
Na	0.50	10,416–14,214	10,638 ± 1740	8583–16,454	10,349±3372
Ni	2.00	15–29	16 ± 6.2	4.4–8.7	5.7 ± 1.8
P	0.60	3561–5158	3 749 ± 713	1554–2453	1667 ± 400
Pb	1.72	81–310	86 ± 107	37–79	40 ± 19
Rb	0.64	230–350	240 ± 54	200–250	220 ± 21
S	0.50	2908–19,384	4870 ± 7348	1173–1970	1950 ± 371
Sb	4.18	nd	nd	5.1–35	5.2 ± 14
Si	0.50	175,454–239,759	229,176 ± 28,153	312,012–320,823	316,480 ± 3597
Sn	3.02	35–75	36 ± 19	19–25	23 ± 2.6
Sr	0.72	90–250	190 ± 66	55–350	76 ± 134
Ti	0.50	3717–5809	4310 ± 880	2668–2836	2668 ± 79
U	1.22	3.5–4.1	3.6 ± 0.3	8.5–13.2	11.2 ± 1.9
V	2.78	10–23	15 ± 5.2	28–47	38 ± 7.6
W	3.70	nd	nd	7.7–25	16 ± 8.7
Zn	1.28	680–1870	1180 ± 488	160–510	220 ± 153
Zr	0.80	270–550	360 ± 117	41–74	70 ± 15.0

F1—fraction <0.074 mm; F2—fraction >0.074 mm and <1 mm; min—minimum; max—maximum; med—median; SD—standard deviation; nd—not detected.

Iron, with higher concentrations in fraction F1 (max = 39,895 mg kg^{-1}), showed comparable results to those obtained for road dusts in Xi'an [34], Shanghai [35], Hong Kong [35], Beijing [35], Delhi [36] and Madrid [37], suggesting a significant geogenic contribution, which is confirmed by $EI < 3$. Elements such as Al, Fe, Ti, Zr and Na have a potential source in soil resuspended dusts and marine spray ($EI < 3$). The anthropogenic contribution may be linked to the application of these elements in the production of metal alloys commonly used in vehicles (e.g., Fe as a major component of steel and associated rust); Al alloys associated with other elements such as Mg, Mn and Cu to reduce vehicle weight; alloys with Mn to avoid corrosion and deformation; sulphides (such as Sb_2S_3, MoS_2 and SnS) and sulphates such as barite ($BaSO_4$) applied in brakes; titanium oxide (TiO_2) used as a pigment in white paints; and aluminium oxide and iron oxides also employed in brakes [33,38]. The detection of Zr may be, in part, related to vehicle exhaust emissions, since ceria/zirconia (CeO_2/ZrO_2) mixed oxides have become an essential component of three-way catalysts [39]. Asphalt materials are usually rich in Al, Si, K, and Ca, with smaller amounts of Fe, S, Mg, Zn, and Ti [40]. Elements such as P and K are commonly used in agriculture activities (e.g., phosphate, potassium nitrate). Higher concentrations of these elements were found in samples from the suburban location, indicating the contribution of the surrounding agricultural environment to road dust. The median concentrations of lead, chromium and copper in samples of urban streets, particularly in F1 fraction, were similar to those of other studies (e.g., Thessaloniki [41], Shiraz [42], Urumqi and Zhuzhou [43]). Lead, chromium and copper concentrations in the ranges 48–375, 2.0–498 and 47–995 mg kg^{-1}, respectively, have been reported for street dusts of different cities on various continents [44], and references therein.

The enrichment index of Zn in the finest fraction (F1) ranged from 10.6 to 43.5, with a moderate (suburban location) to very severe anthropogenic enrichment (urban streets). Zn has been reported to be present as a minor constituent of silicate minerals and linked to fly-ash particles and to traffic related materials. Tyre rubber (ZnO and Cu/Zn layers formed during vulcanisation) and break wear resuspended particles are major sources of Zn, together with Cu, in urban areas [45]. Zn is also used as engine oil additive. Sternbeck et al. [46] suggested that fuel combustion may be a significant source of Zn. Nickel is a ubiquitous natural metal, used in the production of stainless steel and other nickel alloys with high corrosion and temperature resistance. It is considered a tracer of oil combustion [47]. The high concentration of Ba is likely related to brake wear since BaSO$_4$ (barite) is used as filler for brake lining materials [48].

Antinomy concentrations revealed a very high enrichment index (F1 = 29.3; F2 = 158) in road dust from the central avenue (S3). This is a busy street with frequent braking, especially in front of the elementary school. Since there is no dedicated parking, parents stop at the lane to drop off or pick up their children. Antimony increases the hardness and mechanical strength of lead and is a significant metallic component in brake linings. It is also used in batteries and antifriction alloys, as additive in some types of oils, and applied in semiconductors and Sb$_2$O$_3$ in rubber vulcanisation flame retardants [40,49]. Cooper in the coarser fraction (F2) of samples from urban streets showed an *EI* of 73.6 and 18.6, while the corresponding value for the suburban location was 5.3. This element is commonly associated with traffic related activities, e.g., a component of brake pads wear. A mean Cu/Sb concentration ratio of 6.3 was obtained, in line with the 4.6 ± 2.3 diagnostic criteria proposed by Sternbeck et al. [46] for brake wear particles.

The maximum lead concentration in the coarse fraction did not exceed the minimum value recorded in the finest fraction. The highest concentration was found in the city centre (310 mg kg^{-1}), while the minimum was recorded in the suburban area with rural influence (81 mg kg^{-1}). According to Ferreira [50], the concentration of Pb in natural soils in this region is in the range 30–45 mg kg^{-1}. Thus, the higher values observed in the present study suggest anthropogenic sources. This element presented an *EI* of 22.1 in fraction F2 from urban locations, whilst a value of 4 was determined for the suburban area. Lead and lead compounds are used in batteries and as pigments in paints. It has been documented that lead weights, which are used to balance motor vehicle wheels, are lost and deposited in urban streets, that they accumulate along the outer curb, and that they are rapidly abraded and ground into tiny pieces by vehicle traffic [51]. Lead oxide is a component of brake friction materials [15]. Its elevated concentrations in urban dust could be a consequence of common use of PbO$_4$ as a gasoline additive in Portugal until the year 2000. Most of Pb is bound to stable fractions and only a negligible percentage is mobile, which contributes to its long persistence in the environment [15,45]. Tin revealed high *EI*, ranging from 11.6 in road dust samples from the suburban location to 36.3 in samples from urban streets. Tin was found to be one of the major elements in bulk brake pad samples, as well in airborne nano/micro-sized wear particles released from low-metallic automotive brakes [52]. Urban samples revealed a high W anthropogenic enrichment (*EI* up to 20.9). Tungsten is linked to break pads and tyre wear [53]. The abundant use of Br as flame retardant in several types of materials may explain the high *EI* of this element (23.3 and 47 in F1 of urban samples) [54].

An arsenic concentration of 180 mg kg^{-1} (*EI* = 72.7), six times higher than in urban samples, was observed in fraction F1 of the suburban sample. Arsenic rich particles in this road dust sample may originate from resuspended agricultural soil. It should be borne in mind that the street where the sampling took place is flanked by a farm. In addition to the abundant natural origin, there are many agricultural sources of arsenic to the soil, from pesticide application, livestock dips, organic manure to phosphate fertilisers [55]. Elemental As is used in the manufacture of alloys, particularly with lead (e.g., in lead acid batteries) and copper. Arsenic compounds are also extensively used in the semiconductor and electronics industries, in catalysts, pyrotechnics, antifouling agents in paints, among others. Fossil fuel combustion is considered a major contributor of anthropogenic As emissions to the air (mainly As^{III}). Arsenic is present in the air of suburban and urban areas mainly in the inorganic As^V form. Background concentrations in soil range from 1 to 40 mg kg^{-1}, with a mean value of 5 mg kg^{-1} [56], much lower than the levels found in this study.

3.3. Mineralogical Composition

The mineralogy of the two particle size fractions of the road dust samples was compared. X-ray diffraction results showed that road dust samples consist primarily of mineral matter accounting for a minimum of ~60%. The most abundant mineral was quartz [SiO_2], especially in the coarse fraction. Quartz has higher structural hardness than other minerals preventing physical weathering and abrasion of road surfaces. Other minerals also present were muscovite [$KAl_3Si_3O_{10}(OH)_2$], albite [$NaAlSi_3O_8$], kaolinite [$Al_2Si_2O_5(OH)_4$], microcline [$KAlSi_3O_8$], Fe-enstatite [$(Fe,Mg)_2Si_2O_6$] and graphite [C]. Minor proportions of calcite [$CaCO_3$] and rutile [TiO_2] were also observed, mostly in the coarse fraction of road dust collected in the suburban location. Clay forming minerals increase with the decrease of particle size. A significant proportion of amorphous content was detected in all samples, particularly in the finest fraction. This content may originate from weathered minerals, clays with low detection limit, organic matter and anthropogenic sources [57].

The bulk composition suggests that the SiO_2 content of all samples reflects the abundance of leucocratic phases (quartz and feldspar) in the coarser fractions of road dust, while Fe, Ti and Mg indicate the higher abundance of melanocratic phases (biotite, garnet and hornblende) in the finer fractions. The ternary diagram Al_2O_3/$CaO+Na_2O+K_2O$/$FeO+MgO$ (Figure 1) [58] of the studied samples defines three compositional trends, suggesting slightly different sources for each location. The composition of road dust from the suburban street (S1) reveals a feldspars/muscovite trend, S2 (urban) plots in the feldspars/biotite, and S3 (central avenue) in the feldspars/biotite with more compositional Ca. Feldspars are the most abundant mineral in all fractions. The ternary diagram Al_2O_3/$CaO+Na_2O$/K_2O suggests that all samples are closer to K-feldspar composition than plagioclase, and present also high content of micas (lever rule). This graphical analysis is in line with the XRD analysis, being quartz the most abundant mineral that is not plotted in these ternary diagrams.

The chemical index of alteration (CIA = $(Al_2O_3/(Al_2O_3+CaO^*+Na_2O+K_2O)) \times 100$, where CaO* is CaO if CaO < Na_2O, but if CaO > Na_2O, CaO = Na_2O) [58,59] indicates: (a) low weathering if 50–60; (b) intermediate weathering if 60–80; and (c) intense weathering when >80 [60]. The coarser fraction of road dust from urban streets (S2 and S3) reveal a low weathering, with CIA = 57.3 and 58.4, respectively. The same road dust fraction of the sample collected at the suburban location (F2 of S1) indicate a low intermediate weathering with CIA = 61.6. The finest fractions (F1) of all samples present intermediate weathering, ranging from 61.3 to 67.6.

To assess the alumina abundance compared to other major cations, the index of compositional variation (ICV = $((Fe_2O_3+K_2O+Na_2O+CaO+MgO+MnO+TiO_2)/Al_2O_3)$) was calculated [61,62]. Minerals such as kaolinite, illite and muscovite present ICV < 1, whereas minerals such as plagioclase, K-feldspar, biotite, amphiboles and pyroxenes have ICV > 1. The road dust sample from the suburban location influenced by agricultural activities, for both size fractions, presented an ICV < 1 (0.84 and 0.92), while the urban samples S2 and S3 revealed an ICV > 1 for both fractions (1.06 to 1.73), indicating an enrichment in rock forming minerals. The limitation of using the elemental composition to identify road dust sources should be noted, as not only the natural but also the anthropogenic contribution must be considered.

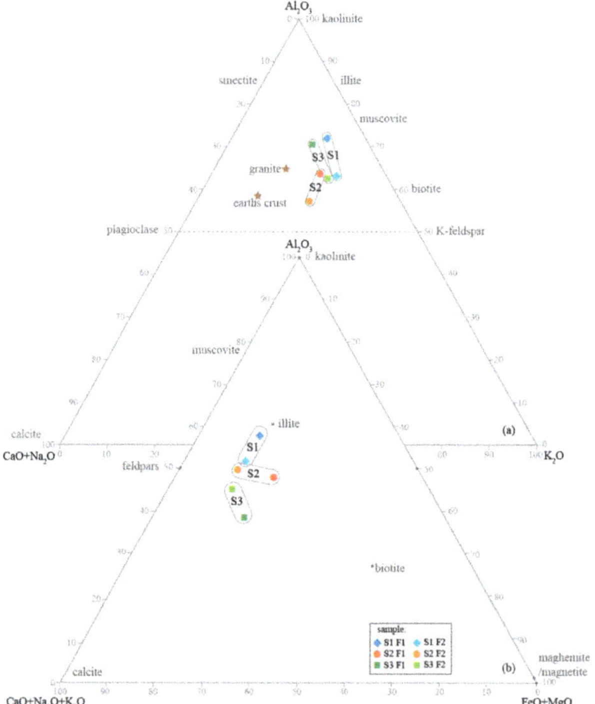

Figure 1. Ternary diagrams (**a**) Al_2O_3—$CaO+Na_2O+K_2O$—$FeO+MgO$, and (**b**) Al_2O_3—$CaO+Na_2O$ $+K_2O$—$FeO+MgO$ of the studied samples (adapted from [58]). Stars plotted indicate the average composition of the Earth's crust and granite [20]. S1—suburban street, cobbled pavement; S2—urban street, asphalt pavement; S3—central avenue, cobbled pavement.

3.4. Morphology

The SEM analysis revealed the presence of well-formed minerals and irregular aggregates, with abundant silicate minerals and an un-sorted mixture of geogenic and anthropogenic particles (Figure 2). Particles with irregular and subangular to angular shapes, including plate like morphology, with variable chemical composition that comprises Fe, Cu, Zn, S, Al, Ti and Sb, suggest abrasion processes on their formation, such as tyre, break and pavement wear and vehicle corrosion interfaces. Rounded, longish and plate-like particles were also found. Although some of the particles present a larger size, they are formed by several smaller particles, usually <10 μm, which are composed of a mixture of anthropogenic and natural substances. It is known that brake abrasion generates particles containing Zn, Cu, Ti, Fe, Cu, and Pb, and other specific compounds such as sulphate silicate and barium sulphate. Particles from tyre abrasion comprise Cd, Cu, Pb, and Zn [63,64]. Silicates and Fe oxides/hydroxides tied to Cl/S can also be associated with traffic or resuspension. Particles with spherical morphology are produced in high-temperature processes (e.g., asphalt and industry/metallurgy). Silicate and iron plerospheres and cenospheres in fly ashes were also found, with typical Si-Al-Fe-Cu-Ca and Fe-Si-Al-Na compositions. Although carbon could not be calculated due to the SEM analysis technique, its presence was confirmed by the XRD analyses. Asphalt paving, or bituminous, materials are mainly made of carbonaceous components (e.g., saturated hydrocarbons, aromatics, asphaltenes, etc.) that are mixed with mineral aggregates. In addition to metals, particles from brake-pad and brake-disc abrasion consists of carbon fibres and graphite, while particulate material from tyre wear comprises various carbonaceous constituents (organic compounds such as natural rubber copolymer, organotin compounds, and soot) [64]. Particle size indicates that road dust might be an important source of resuspended atmospheric particulate matter associated with non-exhaust emissions (e.g., brake, tyre, and road wear), as suggested in previous studies [65]. The size of most of the non-exhaust particles is commonly much larger than that of exhaust particles, since its formation comprises processes such as corrosion, crushing and mechanical abrasion.

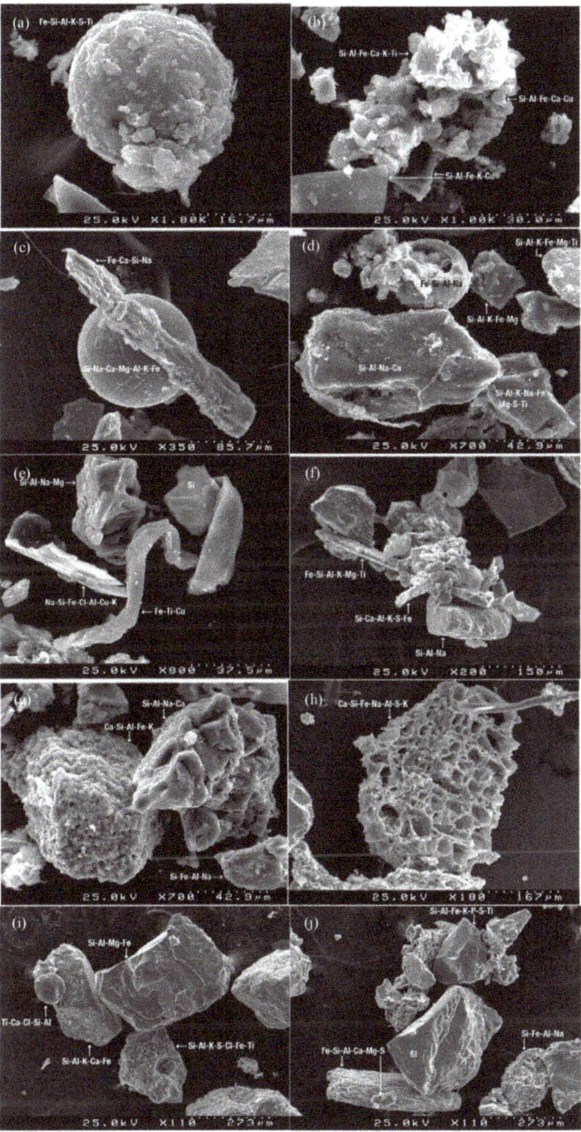

Figure 2. SEM images and composition of anthropogenic and geogenic road dust particles: (**a–f**) plerospheres, cenospheres, aggregates, plate like particles and fibrous steel with considerable plastic deformation and fragmentation; (**g**) irregular aggregates; (**h**) porous Ca-Si-Fe rich particle; (**i,j**) well-formed minerals.

3.5. Grain Size Distribution

The grain size distribution of road dusts is presented in Figure 3. Results show a similar pattern for both suburban (S1) and urban areas (S2 and S3) with a marked unimodal distribution. The mass volume of particles peaked in the range from 10 to 106 µm, although small modes, barely noticeable, were observed below 5 µm. Particles < 100 µm can easily be resuspended in the wake of passing traffic or by the blowing wind and might enter the mouth and nose while breathing. Particulate matter of 10 and 2.5 µm or less in diameter (PM_{10} and $PM_{2.5}$, respectively) can get deep into the lungs and some may even get into the bloodstream. Urban sample S3 presented a higher percentage of PM_{10}, with $S3_{PM10}$ 37.8% > $S1_{PM10}$ 27.2% > $S2_{PM10}$ 25.0%, while $PM_{2.5}$ represented a smaller fraction, with 4.5% (S3) > 3.0% (S1) > 2.7% (S2) of the total mass volume. A literature review by Grigoratos and Martini [66] documented unimodal mass size distributions of brake wear PM_{10}, with a mass weighed mean diameter of 2–6 µm. On the other hand, tyre wear PM_{10} often exhibits a bimodal distribution with one peak lying within the fine particle size range and the other one within the coarse range (5–9 µm). It is estimated that almost 40%–50% by mass of generated brake wear particles and 0.1%–10% by mass of tyre wear particles is emitted as PM_{10}. In terms of mass, more than 85% of diesel particulate exhaust emissions are below 1 µm. Gasoline vehicles emit an even higher proportion of smaller particles than diesel vehicles [67].

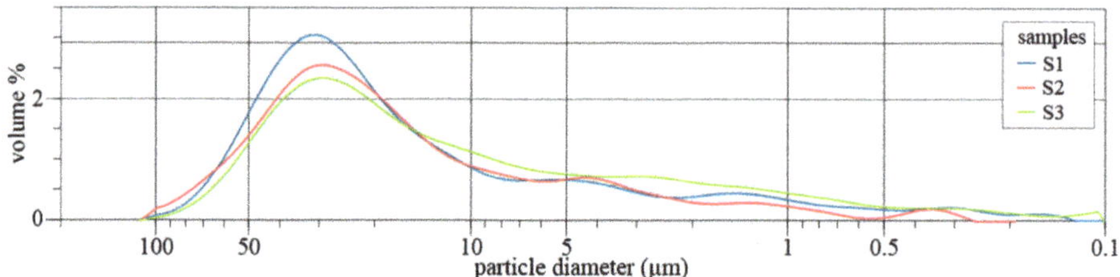

Figure 3. Size distribution (%) of road dust particles, Ø < 106 µm.

3.6. Human Health Exposure Assessment

The hazard quotient (*HQ*), or non-carcinogenic effects, suggest that ingestion is the major route of children's exposure to road dusts, with $HQ_{ing} \approx HI$ (Figure 4), both by hand-to-mouth common habits and by resuspended particles. Dermal and inhalation routes can be considered negligible. Fraction F1 revealed a higher probability to induce non-carcinogenic health effects in children. The *HI* values for adults were approximately an order of magnitude lower than those for children. For both children and adults, Zr is the element that most contributes to possible non-carcinogenic effects. Adverse health effects may occur mostly by ingestion of resuspended particles (children: F1 $HQ_{ing\text{-}Zr} \approx HI$ ranging from 27.51 to 553.10; F2 $HQ_{ing\text{-}Zr} \approx HI$ ranging from 5.66 to 8.71; adults: F1 $HQ_{ing\text{-}Zr} \approx HI$ ranging from 2.52 to 5.13; F2 $HQ_{ing\text{-}Zr} \approx HI$ ranging from 0.38 to 0.69). Studies suggest that Zr and its compounds represent a risk for pulmonary health effects (benign) potentially associated with short-term exposure [68].

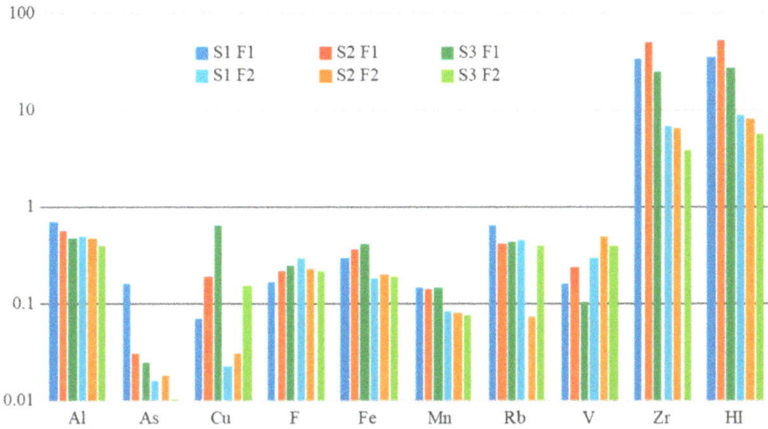

Figure 4. Non-carcinogenic chronic hazard quotient (*HQ*) of Al, As, Cu, F, Fe, Mn, Rb, V and Zr by ingestion in children and cumulative non-carcinogenic hazard index (*HI*). Logarithmic scale.

The probability of an individual to develop any type of cancer over lifetime (Risk) by As (Risk$_{As}$) content in fraction F1 of the suburban sample is 1.58×10^{-4} (Table 2), a Risk above the acceptable target of 1×10^{-4} proposed by USEPA [24], so the adoption of local measures is suggested. Fraction F2 of the same sample showed a Risk$_{As}$ = 1.58×10^{-5}. Risk$_{As}$ values of 3.07×10^{-5} and 2.46×10^{-5} were obtained for fraction F1 of road dust from urban streets, while the corresponding values for fraction F2 were 1.84×10^{-5} and 1.05×10^{-5}. These cancer risks are in the range for which management measures are required. The dermal risk for As in all samples and fractions was also within this range. Arsenic and its inorganic compounds are classified as carcinogenic to humans since 2012 [69]. The Pb content in fraction F1 of sample S3 is also indicative that, by the ingestion pathway, it may pose a risk (2.24×10^{-5}) to human health.

Table 2. Estimated human health risk for elements As and Pb.

ID		F1				F2			
		ing	inh	drm	Total	ing	inh	drm	Total
	As	1.38×10^{-4}	1.2038×10^{-7}	2.0238×10^{-5}	1.5838×10^{-4}	1.3838×10^{-5}	1.2038×10^{-8}	2.0238×10^{-6}	1.5838×10^{-5}
S1	Pb	5.8538×10^{-7}	1.5038×10^{-10}	–	5.8638×10^{-7}	2.6738×10^{-7}	6.8738×10^{-11}	–	2.6738×10^{-7}
	total	1.3838×10^{-4}	1.2038×10^{-7}	2.0238×10^{-5}	1.5938×10^{-4}	1.4038×10^{-5}	1.2038×10^{-8}	2.0238×10^{-6}	1.6138×10^{-5}
	As	2.6838×10^{-5}	2.3338×10^{-8}	3.9338×10^{-6}	3.0738×10^{-5}	1.6138×10^{-5}	1.4038×10^{-8}	2.3638×10^{-6}	1.8438×10^{-5}
S2	Pb	6.2238×10^{-7}	1.6038×10^{-10}	–	6.2238×10^{-7}	2.8938×10^{-7}	7.4338×10^{-11}	–	2.8938×10^{-7}
	total	2.7438×10^{-5}	2.3538×10^{-8}	3.9338×10^{-6}	3.1438×10^{-5}	1.6438×10^{-5}	1.4138×10^{-8}	2.3638×10^{-6}	1.8738×10^{-5}
	As	2.1438×10^{-5}	1.8638×10^{-8}	3.1438×10^{-6}	2.4638×10^{-5}	9.1838×10^{-6}	7.9938×10^{-9}	1.3538×10^{-6}	1.0538×10^{-5}
S3	Pb	2.2438×10^{-5}	5.76×10^{-10}	-	2.24×10^{-5}	5.71×10^{-7}	1.47×10^{-10}	–	5.71×10^{-7}
	total	2.3738×10^{-5}	1.92×10^{-8}	3.14×10^{-6}	2.68×10^{-5}	9.75×10^{-6}	8.13×10^{-9}	1.35×10^{-6}	1.11×10^{-5}

F1—fraction <0.074 mm; F2—fraction >0.074 mm and <1 mm; ing—ingestion; inh—inhalation; drm—dermal.

4. Conclusions

Road dust resuspension is one of the major sources causing PM$_{10}$ exceedances, with detrimental effects on climate and human health. It has been demonstrated that emission inventories must use locally determined emission factors as these vary not only with weather and road conditions, but also with vehicle fleet. This study offers the first experimental estimates for road dust emission factors in the region of Viana do Castelo. A PM$_{10}$ emission factor of 49 mg veh^{-1} km^{-1} was derived for an asphalt-paved road, which is in the range of values documented for the same type of pavement in a few other southern European cities, while much higher emissions (around 300 mg veh^{-1} km^{-1}) were found for cobble stone streets. Although sampling took place in only three streets, these were carefully selected to represent different sectors of the city. On the other hand, because composite samples were obtained from multiple road segments and the fact that the emission factors of the present study are similar to those determined in a previous work for several streets in Oporto, also an Atlantic city with comparable pavements and traffic fleet, the representativeness of the values now estimated is broadened.

The chemical composition and the enrichment indices suggest an anthropogenic contribution of traffic-related elements, such as Br, Cl, Cr, Cu, P, Pb, S, Sn, W and Zn, especially to the finest road dust fraction (F1, <0.074 mm) from urban streets. In these locations, the fingerprint of marine spray is also identified with higher concentrations of Cl, Na and Mg. Samples from the suburban area presented an extremely severe anthropogenic enrichment in the finest fraction for the element As, possibly linked to agricultural activities and fossil fuel combustion. Quartz was mostly present in the coarser fraction, in which it was the most abundant mineral. Other minerals derived from natural sources were also observed (muscovite, albite, kaolinite, Fe-enstatite and graphite), as well as a significant amount of anthropogenic-related materials (amorphous). SEM and EDS analyses indicated that the main constituents in the amorphous content originated from non-exhaust traffic sources (brake, tyre and road abrasion) and from fuel combustion.

The estimation of non-carcinogenic health risks due to exposure to heavy metals indicated that children may experience adverse effects due to ingestion of the finest size fraction of road dust. For the suburban location, the risk associated with the ingestion of this finest road dust was above the acceptable target proposed by USEPA. Other samples and size fractions presented a risk by ingestion and dermal contact within a range indicating that management measures are required.

Supplementary Materials: The following are available online at http://www.mdpi.com/1660-4601/17/5/1563/s1, Figure S1. Location of the sampling sites; parameters used in the calculations of Sections 2.3 and 2.4; Figure S2. Cluster analysis of the XRF results of road dust samples (S1 suburban environment influenced by agricultural activities; S2 and S3 urban streets); Figure S2. Enrichment index for elements detected in road dust fractions (F1—below 74 μm, F2—between 74 μm and 1 mm).

Author Contributions: Conceptualisation, C.A.; sampling, C.A., E.V. and M.T.; sieving, C.A., C.C. and E.V.; geochemical, mineralogical and morphological characterisation, C.C., P.Á. and F.R.; data treatment, C.A. and C.C.; writing of original draft, C.A. and C.C.; review of original draft, C.A. and C.C., in collaboration with all co-authors. All authors have read and agreed to the published version of the manuscript.

Funding: This work was financially supported by the project "Chemical and toxicological SOurce PROfiling of particulate matter in urban air (SOPRO)", POCI-01-0145-FEDER-029574, funded by FEDER, through Compete2020 - Programa Operacional Competitividade e Internacionalização (POCI), and by national funds (OE), through FCT/MCTES. An acknowledgement is also given to the Portuguese Foundation of Science and Technology (FCT) and to the POHP/FSE funding programme for the fellowship SFRH/BD/117993/2016. We are also grateful for the financial support to CESAM (UID/AMB/50017/2019), to GeoBioTec (UID/GEO/04035/2019; UIDB/04035/2020), and to FCT/MCTES through national funds, and co-funding by FEDER, within the PT2020 Partnership Agreement and Compete 2020.

Acknowledgments: Special thanks are given to the Viana do Castelo City Council, and especially to Ricardo Carvalhido, for all the logistic support.

Conflicts of Interest: The authors declare no competing interests.

References

1. Alves, C.A.; Evtyugina, M.; Vicente, A.M.P.; Vicente, E.D.; Nunes, T.V.; Silva, P.M.A.; Duarte, M.A.C.; Pio, C.A.; Amato, F.; Querol, X. Chemical profiling of PM_{10} from urban road dust. *Sci. Total Environ.* **2018**, *634*, 41–51. [CrossRef] [PubMed]

2. Khan, R.K.; Strand, M.A. Road dust and its effect on human health: A literature review. *Epidemiol. Health* **2018**, *40*, e2018013. [CrossRef] [PubMed]

3. Amato, F.; Alastuey, A.; Karanasiou, A.; Lucarelli, F.; Nava, S.; Calzolai, G.; Severi, M.; Becagli, S.; Gianelle, V.L.; Colombi, C.; et al. AIRUSE-LIFE+: A harmonized PM speciation and source apportionment in five southern European cities. *Atmos. Chem. Phys.* **2016**, *16*, 3289–3309. [CrossRef]

4. Amato, F.; Alastuey, A.; de la Rosa, J.; Gonzalez Castanedo, Y.; Sánchez de la Campa, A.M.; Pandolfi, M.; Lozano, A.; Contreras González, J.; Querol, X. Trends of road dust emissions contributions on ambient air particulate levels at rural, urban and industrial sites in southern Spain. *Atmos. Chem. Phys.* **2014**, *14*, 3533–3544. [CrossRef]

5. Chen, S.; Zhang, X.; Lin, J.; Huang, J.; Zhao, D.; Yuan, T.; Huang, K.; Luo, Y.; Jia, Z.; Zang, Z.; et al. Fugitive road dust $PM_{2.5}$ emissions and their potential health impacts. *Environ. Sci. Technol.* **2019**, *53*, 8455–8465. [CrossRef]

6. Huang, L.; Wang, K.; Yuan, C.-S.; Wang, G. Study on the seasonal variation and source apportionment of PM_{10} in Harbin, China. *Aerosol Air Qual. Res.* **2010**, *10*, 86–93. [CrossRef]

7. Jalan, I.; Dholakia, H.H. *Understanding Uncertainties in Emissions Inventories*; What is Polluting Delhi's Air; Council on Energy, Environment and Water: New Delhi, India, 2019.

8. Denier van der Gon, H.A.C.; Gerlofs-Nijland, M.E.; Gehrig, R.; Gustafsson, M.; Janssen, N.; Harrison, R.M.; Hulskotte, J.; Johansson, C.; Jozwicka, M.; Keuken, M.; et al. The policy relevance of wear emissions from road transport, now and in the future—An international workshop report and consensus statement. *J. Air Waste Manag. Assoc.* **2013**, *63*, 136–149. [CrossRef]

9. Greening, T. *Quantifying the Impacts of Vehicle-Generated Dust: A Comprehensive Approach*; The International Bank for Reconstruction and Development/The World Bank: Washington, DC, USA, 2011.

10. AIRUSE. *The Efficacy of Dust Suppressants to Control Road Dust Re-Suspension in Northern and Central Europe*; Report 14, LIFE 11/ENV/ES/584; AIRUSE: Barcelona, Spain, 2016.

11. AIRUSE. *Methods Used in Barcelona to Evaluate the Effectiveness of CMA and MgCl2 in Reducing Road Dust Emissions*; Action B7, LIFE11 ENV/ES/584; AIRUSE: Barcelona, Spain, 2017.

12. Gulia, S.; Goyal, P.; Goyal, S.K.; Kumar, R. Re-suspension of road dust: Contribution, assessment and control through dust suppressants—A review. *Int. J. Environ. Sci. Technol.* **2019**, *16*, 1717–1728. [CrossRef]

13. Amato, F.; Cassee, F.R.; Denier van der Gon, H.A.C.; Gehrig, R.; Gustafsson, M.; Hafner, W.; Harrison, R.M.; Jozwicka, M.; Kelly, F.J.; Moreno, T.; et al. Urban air quality: The challenge of traffic non-exhaust emissions. *J. Hazard. Mater.* **2014**, *275*, 31–36. [CrossRef]

14. Amato, F.; Pandolfi, M.; Moreno, T.; Furger, M.; Pey, J.; Alastuey, A.; Bukowiecki, N.; Prevot, A.S.H.; Baltensperger, U.; Querol, X. Sources and variability of inhalable road dust particles in three European cities. *Atmos. Environ.* **2011**, *45*, 6777–6787. [CrossRef]

15. Adamiec, E. Chemical fractionation and mobility of traffic-related elements in road environments. *Environ. Geochem. Health* **2017**, *39*, 1457–1468. [CrossRef] [PubMed]

16. Amato, F.; Favez, O.; Pandolfi, M.; Alastuey, A.; Querol, X.; Moukhtar, S.; Bruge, B.; Verlhac, S.; Orza, J.A.G.; Bonnaire, N.; et al. Traffic induced particle resuspension in Paris: Emission factors and source contributions. *Atmos. Environ.* **2016**, *129*, 114–124. [CrossRef]

17. USEPA. *AP-42, Appendix C.1: Procedures For Sampling Surface/Bulk Dust Loading*; U.S. Environmental Protection Agency: Washington, DC, USA, 2003.

18. USEPA. *AP-42, Appendix C.2: Procedures For Laboratory Analysis Of Surface/Bulk Dust Loading Samples*; U.S. Environmental Protection Agency: Washington, DC, USA, 2003.

19. USEPA. *Emission Factor Documentation for AP-42, Section 13.2.1, Paved Roads—January 2011*; U.S. Environmental Protection Agency: Washington, DC, USA, 2011.

20. Reimann, C.; de Caritat, P. *Chemical Elements in the Environment*; Springer: Berlin, Germany, 1998.

21. Chen, C.W.; Kao, C.M.; Chen, C.F.; Dong, C. Distribution and accumulation of heavy metals in the sediments of Kaohsiung Harbor, Taiwan. *Chemosphere* **2007**, *66*, 1431–1440. [CrossRef]

22. USEPA. *Soil Screening Guidance: Technical Background Document*; EPA 540-R-95-128; U.S. Environmental Protection Agency: Washington, DC, USA, 1996.

23. USEPA. *Screening Levels (RSL) for Chemical Contaminants at Superfund Sites*; U.S. Environmental Protection Agency: Washington, DC, USA, 2013.

24. USEPA. The Risk Assessment Information System. Available online: https://rais.ornl.gov/ (accessed on 1 November 2019).

25. INE. Portal do Instituto Nacional de Estatística—Statistics Portugal. Available online: http://www.ine.pt/ (accessed on 1 November 2019).

26. Berg, R. *Human Exposure to Soil Contamination: A Qualitative and Quantitative Analysis towards Proposals for Human Toxicological Intervention Values (Partly Revised Edition)*; Report N°. 725201011; National Institute for Public Health and the Environment: Bilthofen, The Netherlands, 1994.

27. USEPA. *Risk Assessment Guidance for Superfund, Volume I: Human Health Evaluation Manual*; EPA 540-1-89-002; U.S. Environmental Protection Agency: Washington, DC, USA, 1989.

28. Amato, F.; Bedogni, M.; Padoan, E.; Querol, X.; Ealo, M.; Rivas, I. Characterization of road dust emissions in milan: Impact of vehicle fleet speed. *Aerosol Air Qual. Res.* **2017**, *17*, 2438–2449. [CrossRef]

29. Boulter, P.G. *A Review of Emission Factors and Models for Road Vehicle Non-exhaust Particulate Matter*; Project Report PPR065; TRL: Berks, UK, 2005.

30. Li, F.; Zhang, J.; Huang, J.; Huang, D.; Yang, J.; Song, Y.; Zeng, G. Heavy metals in road dust from Xiandao District, Changsha City, China: Characteristics, health risk assessment, and integrated source identification. *Environ. Sci. Pollut. Res.* **2016**, *23*, 13100–13113. [CrossRef] [PubMed]

31. Pamplona, J.; Ribeiro, A. Evolução geodinâmica de Viana do Castelo (Zona Centro-Ibérica, NW de Portugal. In *Geologia de Portugal*; Dias, R., Araújo, A., Terrinha, P., Kullberg, J., Eds.; Escolar Editora: Lisboa, Portugal, 2013; Volume I, pp. 149–204. ISBN 978-972-592-364-1.

32. Straffelini, G.; Ciudin, R.; Ciotti, A.; Gialanella, S. Present knowledge and perspectives on the role of copper in brake materials and related environmental issues: A critical assessment. *Environ. Pollut.* **2015**, *207*, 211–219. [CrossRef]

33. Fujiwara, F.; Rebagliati, R.J.; Dawidowski, L.; Gómez, D.; Polla, G.; Pereyra, V.; Smichowski, P. Spatial and chemical patterns of size fractionated road dust collected in a megacitiy. *Atmos. Environ.* **2011**, *45*, 1497–1505. [CrossRef]

34. Li, X.; Liu, B.; Zhang, Y.; Wang, J.; Ullah, H.; Zhou, M.; Peng, L.; He, A.; Zhang, X.; Yan, X.; et al. Spatial distributions, sources, potential risks of multi-trace metal/metalloids in street dusts from barbican downtown embracing by Xi'an ancient city wall (NW, China). *Int. J. Environ. Res. Public Health* **2019**, *16*, 2992. [CrossRef]

35. Tanner, P.A.; Ma, H.-L.; Yu, P.K.N. Fingerprinting metals in urban street dust of Beijing, Shanghai, and Hong Kong. *Environ. Sci. Technol.* **2008**, *42*, 7111–7117. [CrossRef]

36. Rajaram, B.S.; Suryawanshi, P.V.; Bhanarkar, A.D.; Rao, C.V.C. Heavy metals contamination in road dust in Delhi city, India. *Environ. Earth Sci.* **2014**, *72*, 3929–3938. [CrossRef]

37. De Miguel, E.; Llamas, J.F.; Chacón, E.; Berg, T.; Larssen, S.; Røyset, O.; Vadset, M. Origin and patterns of distribution of trace elements in street dust: Unleaded petrol and urban lead. *Atmos. Environ.* **1997**, *31*, 2733–2740. [CrossRef]

38. Adachi, K.; Tainosho, Y. Characterization of heavy metal particles embedded in tire dust. *Environ. Int.* **2004**, *30*, 1009–1017. [CrossRef] [PubMed]

39. Chen, H.-Y.; Chang, H.-L.R. Development of low temperature three-way catalysts for future fuel efficient vehicles. *Johns. Matthey Technol. Rev.* **2015**, *59*, 64–67. [CrossRef]

40. Gietl, J.K.; Lawrence, R.; Thorpe, A.J.; Harrison, R.M. Identification of brake wear particles and derivation of a quantitative tracer for brake dust at a major road. *Atmos. Environ.* **2010**, *44*, 141–146. [CrossRef]

41. Bourliva, A.; Christophoridis, C.; Papadopoulou, L.; Giouri, K.; Papadopoulos, A.; Mitsika, E.; Fytianos, K. Characterization, heavy metal content and health risk assessment of urban road dusts from the historic center of the city of Thessaloniki, Greece. *Environ. Geochem. Health* **2017**, *39*, 611–634. [CrossRef] [PubMed]

42. Keshavarzi, B.; Tazarvi, Z.; Rajabzadeh, M.A.; Najmeddin, A. Chemical speciation, human health risk assessment and pollution level of selected heavy metals in urban street dust of Shiraz, Iran. *Atmos. Environ.* **2015**, *119*, 1–10. [CrossRef]

43. Zhang, C.; Yang, Y.; Li, W.; Zhang, C.; Zhang, R.; Mei, Y.; Liao, X.; Liu, Y. Spatial distribution and ecological risk assessment of trace metals in urban soils in Wuhan, central China. *Environ. Monit. Assess.* **2015**, *187*. [CrossRef]

44. Cai, K.; Li, C. Street dust heavy metal pollution source apportionment and sustainable management in a typical city—Shijiazhuang, China. *Int. J. Environ. Res. Public Health* **2019**, *16*, 2625. [CrossRef]

45. Valotto, G.; Rampazzo, G.; Visin, F.; Gonella, F.; Cattaruzza, E.; Glisenti, A.; Formenton, G.; Tieppo, P. Environmental and traffic-related parameters affecting road dust composition: A multi-technique approach applied to Venice area (Italy). *Atmos. Environ.* **2015**, *122*, 596–608. [CrossRef]

46. Sternbeck, J.; Sjödin, Å.; Andréasson, K. Metal emissions from road traffic and the influence of resuspension—Results from two tunnel studies. *Atmos. Environ.* **2002**, *36*, 4735–4744. [CrossRef]

47. USNLM. ToxNet—Toxicology Data Network. U.S. National Library of Medicine. Available online: https://toxnet.nlm.nih.gov/ (accessed on 1 November 2019).

48. Sugözü, B.; Dağhan, B. Effect of $BaSO_4$ on tribological properties of brake friction materials. *Int. J. Innov. Res. Sci. Eng. Technol.* **2016**, *5*, 30–35.

49. Bi, X.; Li, Z.; Zhuang, X.; Han, Z.; Yang, W. High levels of antimony in dust from e-waste recycling in southeastern China. *Sci. Total Environ.* **2011**, *409*, 5126–5128. [CrossRef] [PubMed]

50. Ferreira, I.M. Dados Geoquímicos de Base de Solos de Portugal Continental, Utilizando Amostragem de Baixa Densidade. Ph.D. Thesis, University of Aveiro, Aveiro, Portugal, 2004.

51. Root, R.A. Lead loading of urban streets by motor vehicle wheel weights. *Environ. Health Perspect.* **2000**, *108*, 937–940. [CrossRef] [PubMed]

52. Kukutschová, J.; Moravec, P.; Tomášek, V.; Matějka, V.; Smolík, J.; Schwarz, J.; Seidlerová, J.; Šafářová, K.; Filip, P. On airborne nano/micro-sized wear particles released from low-metallic automotive brakes. *Environ. Pollut.* **2011**, *159*, 998–1006. [CrossRef] [PubMed]

53. Apeagyei, E.; Bank, M.S.; Spengler, J.D. Distribution of heavy metals in road dust along an urban-rural gradient in Massachusetts. *Atmos. Environ.* **2011**, *45*, 2310–2323. [CrossRef]

54. Abbasi, G.; Saini, A.; Goosey, E.; Diamond, M.L. Product screening for sources of halogenated flame retardants in Canadian house and office dust. *Sci. Total Environ.* **2016**, *545–546*, 299–307. [CrossRef]

55. Punshon, T.; Jackson, B.P.; Meharg, A.A.; Warczack, T.; Scheckel, K.; Guerinot, M. Lou Understanding arsenic dynamics in agronomic systems to predict and prevent uptake by crop plants. *Sci. Total Environ.* **2017**, *581–582*, 209–220. [CrossRef]

56. WHO. *Arsenic and Arsenic Compounds (Environmental Health Criteria 224)*, 2nd ed.; International Programme on Chemical Safety; World Health Organization: Geneva, Switzerland, 2001.

57. Gunawardana, C.; Goonetilleke, A.; Egodawatta, P.; Dawes, L.; Kokot, S. Source characterisation of road dust based on chemical and mineralogical composition. *Chemosphere* **2012**, *87*, 163–170. [CrossRef]

58. Nesbitt, H.W.; Young, G.M. Petrogenesis of sediments in the absence of chemical weathering: Effects of abrasion and sorting on bulk composition and mineralogy. *Sedimentology* **1996**, *43*, 341–358. [CrossRef]

59. Nesbitt, H.W.; Young, G.M. Formation and diagenesis of weathering profiles. *J. Geol.* **1989**, *97*, 129–147. [CrossRef]

60. Selvaraj, K.; Chen, C.T.A. Moderate chemical weathering of subtropical Taiwan: Constraints from solid-phase geochemistry of sediments and sedimentary rocks. *J. Geol.* **2006**, *114*, 101–116. [CrossRef]

61. Cox, R.; Lowe, D.R.; Cullers, R.L. The influence of sediment recycling and basement composition on evolution of mudrock chemistry in the southwestern United States. *Geochim. Cosmochim. Acta* **1995**, *59*, 2919–2940. [CrossRef]

62. Dehghani, S.; Moore, F.; Vasiluk, L.; Hale, B.A. The geochemical fingerprinting of geogenic particles in road deposited dust from Tehran metropolis, Iran: Implications for provenance tracking. *J. Geochem. Explor.* **2018**, *190*, 411–423. [CrossRef]

63. Bukowiecki, N.; Lienemann, P.; Hill, M.; Furger, M.; Richard, A.; Amato, F.; Prévôt, A.S.H.; Baltensperger, U.; Buchmann, B.; Gehrig, R. PM_{10} emission factors for non-exhaust particles generated by road traffic in an urban street canyon and along a freeway in Switzerland. *Atmos. Environ.* **2010**, *44*, 2330–2340. [CrossRef]

64. Penkała, M.; Ogrodnik, P.; Rogula-Kozłowska, W. Particulate matter from the road surface abrasion as a problem of non-exhaust emission control. *Environments* **2018**, *5*, 9. [CrossRef]

65. Weinbruch, S.; Worringen, A.; Ebert, M.; Scheuvens, D.; Kandler, K.; Pfeffer, U.; Bruckmann, P. A quantitative estimation of the exhaust, abrasion and resuspension components of particulate traffic emissions using electron microscopy. *Atmos. Environ.* **2014**, *99*, 175–182. [CrossRef]

66. Grigoratos, T.; Martini, G. *Non-exhaust Traffic Related Emissions. Brake and Tyre Wear PM*; Publications Office of the European Union: Luxembourg, 2014.

67. CONCAWE. *A Study of the Number, Size & Mass of Exhaust Particles Emitted from European Diesel and Gasoline Vehicles under Steady-state and European Driving Cycle Conditions*; Report N°. 98/51; CONCAWE: Brussels, Belgium, 1998.

68. NIOSH. National Institute for Occupational Safety and Health. Available online: https://www.cdc.gov/niosh (accessed on 1 November 2019).
69. IARC. Monographs on the Identification of Carcinogenic Hazards to Humans. International Association for Research on Cancer. Available online: https://monographs.iarc.fr/monographs-available/ (accessed on 1 November 2019).

International Journal of
*Environmental Research
and Public Health*

Article

How Has the Hazard to Humans of Microorganisms Found in Atmospheric Aerosol in the South of Western Siberia Changed over 10 Years?

Alexandr Safatov [1,*], Irina Andreeva [1], Galina Buryak [1], Olesia Ohlopkova [1], Sergei Olkin [1], Larisa Puchkova [1], Irina Reznikova [1], Nadezda Solovyanova [1], Boris Belan [2], Mikhail Panchenko [2] and Denis Simonenkov [2]

[1] Department of Biophysics and Ecological Researches, FBRI SRC VB "Vector" of Rospotrebnadzor, Koltsovo, 630559 Novosibirsk rgn., Russia; andreeva_is@vector.nsc.ru (I.A.); buryak@vector.nsc.ru (G.B.); ohlopkova_ov@vector.nsc.ru (O.O.); olkin@vector.nsc.ru (S.O.); puchkova@vector.nsc.ru (L.P.); reznikova@vector.nsc.ru (I.R.); solovyanova_na@vector.nsc.ru (N.S.)

[2] Laboratory of Atmosphere Composition Climatology, V.E. Zuev Institute Of Atmospheric Optics SB RAS, 634055 Tomsk, Russia; bbd@iao.ru (B.B.); pmv@iao.ru (M.P.); simon@iao.ru (D.S.)

* Correspondence: safatov@mail.ru; Tel.: +7-913-927-2690

Received: 27 December 2019; Accepted: 28 February 2020; Published: 3 March 2020

Abstract: One of the most important components of atmospheric aerosols are microorganisms. Therefore, it is necessary to assess the hazard to humans, both from individual microorganisms which are present in atmospheric bioaerosols as well as from their pool. An approach for determining the hazard of bacteria and yeasts found in atmospheric bioaerosols for humans has previously been proposed. The purpose of this paper is to compare our results for 2006–2008 with the results of studies obtained in 2012–2016 to identify changes in the characteristics of bioaerosols occurring over a decade in the south of Western Siberia. Experimental data on the growth, morphological and biochemical properties of bacteria and yeasts were determined for each isolate found in bioaerosol samples. The integral indices of the hazards of bacteria and yeast for humans were constructed for each isolate based on experimentally determined isolate characteristics according to the approach developed by authors in 2008. Data analysis of two datasets showed that hazard to humans of culturable microorganisms in the atmospheric aerosol in the south of Western Siberia has not changed significantly for 10 years (trends are undistinguishable from zero with a confidence level of more than 95%) despite a noticeable decrease in the average annual number of culturable microorganisms per cubic meter (6–10 times for 10 years).

Keywords: atmospheric aerosols; bioaerosols; culturable bacteria; long-term trends; hazard for human

---

1. Introduction

Human exposure to air pollution is among major health problems. Some pollutants can be easily monitored, doses obtained by individuals may be calculated and the results of such exposure may be predicted if the "dose–effect" dependencies are known. These dependencies are not known for most bioaerosols. An atmospheric bioaerosol is omnipresent part of atmospheric aerosols, accounting for up to 95% of their total number or up to 80% of their mass [1–4]. The peculiarity of atmospheric bioaerosols is, besides the usual effects of aerosols on atmospheric processes and climate [5–7], the ability to cause or provoke various infectious or non-infectious diseases in humans [5,8–12]. The most important component of atmospheric bioaerosols is microorganisms. Since microorganisms are usually hazardous to humans, it is important to be able to assess what this hazard is in each air sample. It is also important to be able to track the change in this hazard in the air of controlled points over time. Modern molecular biological methods can quite quickly reveal the biological diversity of microorganisms present in a sample [13–27], but for the vast majority of known microorganisms, their danger to humans has not been studied. Therefore, an approach that is based on growth, morphological and biochemical properties of culturable microorganisms (even if a microorganism has never been isolated from the natural environment early or most of its properties have not been studied) may be very useful.

We have previously proposed such an approach for determining the hazard of bacteria and yeasts found in atmospheric bioaerosols for humans and demonstrated its capabilities using the available experimental data as an example [28]. In the framework of the developed approach, a dimensionless integral index of the hazard to humans of microorganisms isolated from atmospheric bioaerosols was constructed, which is the product of four lower-level dimensionless integral indices characterizing the complex properties of the microorganism. One of them is an integral index of the concentration of culturable bacteria in an aerosol sample. Other components of the integral index of the hazard of microorganisms to humans are [28]:

- an index that evaluates the pathogenicity or hazard of individual isolates of microorganisms to humans;
- an index that evaluates the resistance of microorganisms to adverse environmental factors;
- an index that evaluates the resistance of microorganisms to antibiotics or other drugs.

Obviously, the more culturable microorganisms (identical or different) are in the air, the more dangerous they are to humans. The value normalized to the maximum number of culturable microorganisms in all samples studied (or in one sample; in this case, of course, this index is equal to 1) represents the first index [28]. The second index assesses the hazard (pathogenicity or conditional pathogenicity) of microorganisms to humans. Its normalized value reaches 1 for pathogens and is strictly equal to zero for completely non-pathogenic microorganisms. This index is based on the morphological and biochemical properties of microorganisms and allows us to predict their individual or collective pathogenicity in the range of 0 to 1. The third normalized index assesses the resistance of microbial cells to adverse environmental factors. It considers their growth, morphological and biochemical properties. The more resistant the microorganism is in the external environment, the greater the likelihood of it entering the human body and maintaining the ability to trigger negative reactions. Finally, the fourth index is determined by the resistance of microorganisms to the action of antibiotics or other drugs. The higher this resistance, the more difficult it is to overcome the negative effects of microorganisms in the body and, consequently, the higher the hazard of such a microorganism. The approach proposed in [28] was illustrated using experimental data obtained in 2006–2008.

Changes occurring in nature (climate change, atmospheric processes changes, changes of habitats areas of animals, insects and vegetation, changes in water systems [29–34]) should also be manifested in changes in the abundance and biodiversity of bioaerosols in the atmosphere.

Western Siberia is a region in which global climatic changes are clearly manifested: permafrost thawing, decrease in snow cover time, temperature increase, powerful greenhouse gas emissions [35–45]. At the same time, changes in the state of health of the population of the region caused by these climatic changes are poorly studied (see, for example, [37,46,47]). The method we developed allows us to assess the change in the hazard to humans of cultivated microorganisms located in atmospheric aerosols, and as a result, the influence of this factor, which undoubtedly affects the health of the population of the region. Thus, the aim of this work is to obtain new data on the hazard to humans in the context of ongoing climate change, both for individual microorganisms (only bacteria and yeasts in this paper), which are present in atmospheric bioaerosol, as well as their pool, and compare the results with the results obtained previously. This study includes two on-ground sampling sites where a person breathes directly and where the presence of microorganisms is largely determined by local sources of bioaerosols. In addition, microorganisms in the atmosphere at altitudes of up to 7 km were studied and its bioaerosol composition was largely determined by bioaerosols remote sources.

2. Materials and Methods

To ensure the comparability of the results obtained in 2006–2008 and in 2012–2016, materials and methods used were mainly the same as described in [28]. These materials and methods which were used in 2006–2008 and in 2012–2016 are described in details below, and their differences are highlighted in special Section 2.4.

2.1. Atmospheric Air Sampling

Sampling of atmospheric air was carried out at three points in the studied region: on the site of the FBRI SRC VB "Vector" of Rospotrebnadzor (Vector), 4 times a day (the sampling starts at 10:00, 16:00, 22:00 and 4:00 on the next day) in the middle of the month; in the Klyuchi village, once a season for 7 consecutive days (usually the sampling starts at 8:00–9:00); about 50 km south of Novosibirsk using the "Optik-E" laboratory (Aircraft, Figures 1 and 2) on one of the last days of each month. A map showing the position of two sampling points and a typical flight trajectory of a laboratory airplane is shown in Figure 3. The "Optic-E" laboratory mounted on an Antonov-30 aircraft (Figure 2, see also [48–51]) for 2006–2008 session includes device for bioaerosol sampling and additional devices for registration of physical and chemical aerosol characteristics, meteorological conditions, etc. [48,52]. Isokinetic air sampling was performed through the special inlet outside the cabin (inserts in Figures 2 and 3) [48,53] at cruising speed of aircraft approximately 360 km/hour. The operation of the air intake with reduced dynamic pressure is based on the principle of Venturi. Further, outboard air with a pressure equal to the air pressure inside the aircraft cabin was supplied to a stand with impingers (Figure 4), into which it was sampled for analysis of the presence and concentration of culturable microorganisms. It is obvious that both the device for outboard air intake and the tubes leading to the impinger are characterized by a certain percentage of losses of aerosol particles during their transportation. However, since both the intake air device and the length of the tubes are unchanged in all samples, these losses are the same for all samples. The samples were taken during daytime (usually the time of flight was in interval 12:00–15:00) over the Karakan pine forest (sees Figure 3) successively at eight altitudes: 7000, 5500, 4000, 3000, 2000, 1500, 1000, and 500 m [28]. To create sterile conditions, before each sampling or each flight, all incoming tubes were rinsed with ethanol.

Figure 1. Photo of Antonov-30 aircraft. Insert is the samplers' inlets.

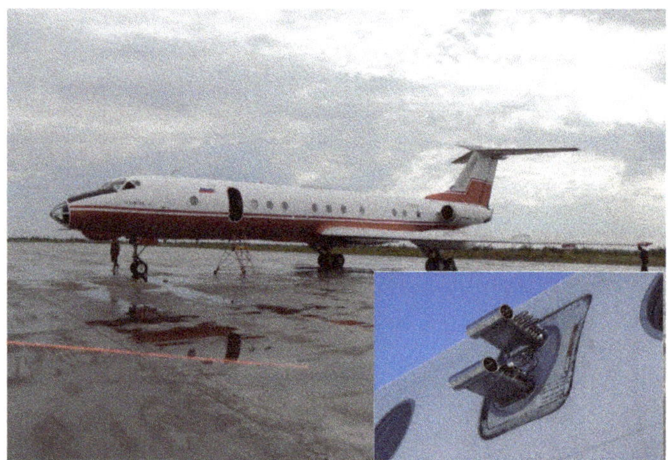

Figure 2. Photo of Tupolev-134 aircraft. Insert is the samplers' Inlets.

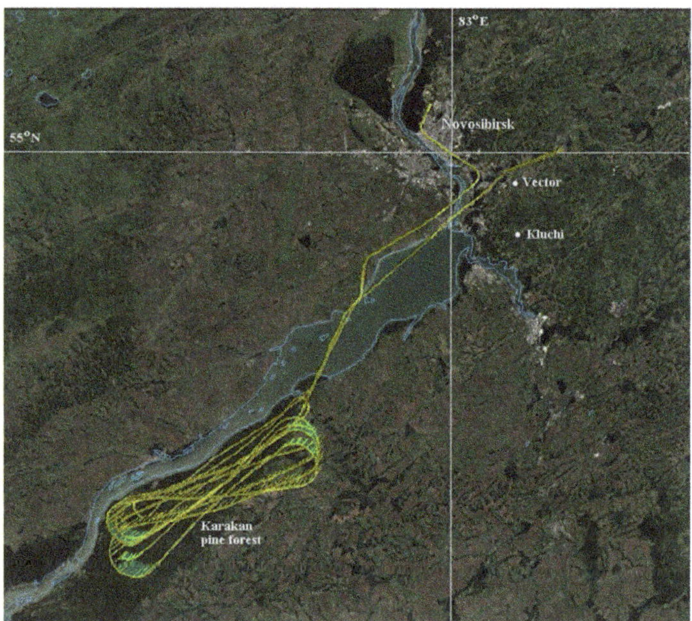

Figure 3. The map of sampling sites and aircraft's flight typical trajectory.

Stainless steel impingers with a critical nozzle [28] (its analogue is described in [54]) were used for air sample collection. These devices (manufactured by JSC "Experimental-design bureau of biological precision engineering", Kirishi, Russia) maintain a constant flow rate at a pressure differential of more than 4×10^4 Pa of air through the device. An A-D1-04 pump (JSC "Kot", St. Petersburg, Russia) was used to pump air samples through the impinger. Fifty milliliters of noncolored Hanks' solution (SIGMA) was used as a sorbing fluid. Above-ground samples were taken at the flow rate of 50 ± 5 L/min for 30 minutes and altitude samples for 5 or 10 minutes (10-minute sampling was used during 2006–2008 and 5-minute sampling was used during 2012–2016) at each altitude. The retention efficiency of this impinger for aerosols of more than 0.3 μm exceeds 80% making up a constant value of 90% ± 15% for particles with a diameter of more than 2 μm.

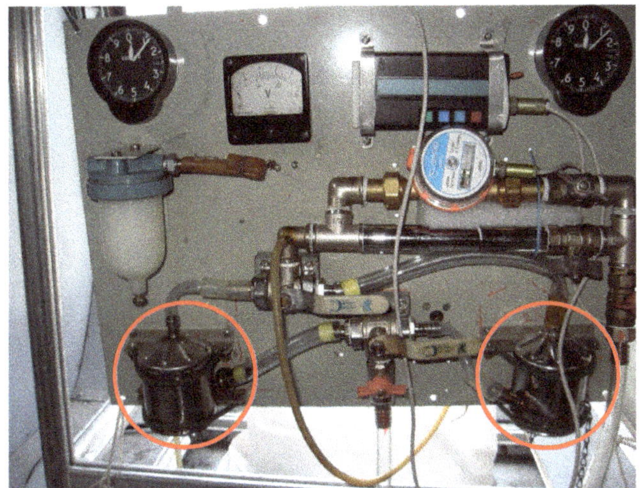

Figure 4. Photo of two impingers (red circles) at a sampling stand of microorganisms located in atmospheric aerosol.

2.2. Culturable Microorganisms' Concentration

The concentrations of culturable microorganisms were determined by standard microbiological methods. Samples were seeded onto Petri dishes containing agarized media. LB [55] was used to detect saprophyte bacteria; depleted LB medium (diluted 1:10) was used to isolate microorganisms inhibited by the excess of organic substances, starch–ammoniac medium [56] was used to detect actinomyces; soil agar was used for soil microorganisms, and Sabouraud medium [56] was used for lower fungi and yeast. Successive sample dilutions were prepared when necessary. The seedings were incubated in a thermostat at a temperature of 28–30 °C for 3–14 days. Some isolates were additionally incubated at a temperature of 6–10 °C in 2012–2016. Phase contrast light microscopy was used for the study of morphological characteristics of bacteria (live cells and fixed Gram-stained ones too) and its colonies. Taxonomic groups the detected microorganisms referred to were determined according to [57–59]. Nucleotide sequences of PCR products corresponding to the fragments of 16S rDNA gene was performed for some bacteria [60,61]. The numbers of culturable microorganisms in samples were calculated according to standard methods [62]; the number of microorganisms was averaged over 3–4 parallels of samples 4–5 seeded on different media.

2.3. Microorganisms' Biochemical and Morphological Characteristics

2.3.1. Pathogenic Properties of the Isolates

Isolated microorganisms were tested for the presence of the following signs of pathogenicity:

- plasma-coagulase activity was determined by placing a loop of the test culture in a test tube with 0.5 mL of rabbit citrate plasma diluted with 0.9% sterile sodium chloride solution and incubating the suspension at 37 °C for up to 24 hours; a positive reaction was determined by plasma coagulation [28];
- hemolytic activity was determined by seeding cultures on nutrient agar containing 5% defibrinated blood and the seeding was incubated at 37 °C for 24–48 hours; a positive result was the presence of hemolysis zones around the grown colonies [59];

- fibrinolytic activity was determined by sowing 0.25 mL of an 18–20-hour broth culture of the test strain in test tubes containing 0.1 mL of citrate plasma, 0.4 mL of 0.9% sodium chloride, 0.25 mL of 0.25% $CaCl_2$ solution; tubes with suspension were kept in an incubator at a temperature of 37 °C for 15–20 minutes; a clot formed in a test tube with a positive reaction (the presence of fibrinolysin) liquefies after the next two hours of incubation in a thermostat; the test culture does not have fibrinolytic properties if the clot persists, as in the control tube, where the culture was not added;
- gelatinase activity was determined by sowing microorganisms by injection in test tubes with meat–peptone broth containing 12% gelatin, kept in an incubator for up to 20 days; a liquefaction of the nutrient medium is noted in the presence of the gelatinase enzyme [57].

2.3.2. Growth Characteristics of Bacteria at Increased Salt Concentration

The resistance of the studied microorganisms to high salt concentrations was determined when they were sown on a complete nutrient medium with the addition of NaCl at concentrations of 1%, 5% or 10%. After incubation at the optimum temperature, the range of resistance of the tested microorganism to the concentration of salts in the nutrient medium was established by the nature of growth or its absence.

2.3.3. The Determination of Enzymatic Activity of Isolated Bacteria

Isolated microorganisms were tested for the presence of the following signs of enzymatic activities:

- proteolytic activity was determined by plating the studied microorganisms on an agar medium containing milk casein (milk agar); to prepare agar with casein, two components were prepared: 3% "hungry" agar (distilled water + 3% agar), sterilized at 1 atm for 30 minutes, and 12% milk sterilized at 1 atm for 20 minutes; then, agar was cooled to 50–55 °C, mixed with milk heated to the same temperature under aseptic conditions in a 1: 1 ratio, and poured into Petri dishes; after solidification of the medium, cultures were streaked and incubated under optimal conditions; the formation of transparent hydrolysis zones around crops in casein-containing agar indicated protease production [57];
- the amylolytic activity of cultures of microorganisms was determined when they were streaked on starch–ammonia agar; after incubation in a thermostat for 24–48 hours, the grown cultures in the dishes were poured into 5 mL of Lugol's solution; the appearance of bright areas around crops within 3–5 minutes is a positive result;
- determination of lecithinase and lipase activity was carried out by two methods:

    1. seeding cultures with a stroke on yolk nutrient agar, for the preparation of which, under aseptic conditions, in yolk of molten and cooled to 50–55 °C fish-peptone agar (FPA) medium, yolk from a chicken egg is introduced; then medium is thoroughly mixed and poured into Petri dishes; the studied culture is streaked, incubated at the required temperature for 24–48 hours and the result is taken into account; when observed in an oblique light, lipase production was judged by the formation of a pearly shiny hydrolysis zone on agar around grown colonies; lecithinase (phospholipase) hydrolyzes the yolk lecithin; as a result of the precipitation reaction, a turbid whitish zone forms around the lecithin-fermenting colonies [57];

    2. plating cultures on a complete LB or RPA medium with 1% Tween-20 or Tween-40 and 0.01% $CaCl_2$ as a substrate; sown cultures were incubated in a thermostat for 3–4 days and the result of the presence or absence of hydrolysis zones was determined [57];

- testing of cultures for the production of alkaline phosphatase was carried out using a reaction mixture of the composition: 0.3 mL of 0.85% NaCl added to 0.3 mL of substrate solution containing 0.04 M glycine buffer pH 10.5 and 0.01 M disodium-n-nitrophenyl phosphate (Sigma); the reaction mixture was incubated at 37 °C for 3 hours; Positive reaction manifested itself as yellow staining of the reaction mixture [59]; enzyme activity was determined within 3 hours of incubation by absorption on Uniplan apparatus (Russia) with a color filter at the wavelength of 450 nm.

- nuclease activity was determined by streaking cultures on RPA medium with the addition of an aqueous DNA solution to a final concentration of 2 mg/mL; before filling the medium, a sterile solution of $CaCl_2$ (0.8 mg/mL) and 0.01% toluidine blue were added to the Petri dishes; in the case of the formation of DNase, a pink colored zone arose around the bacterial culture [59];

- the concentration of plasmid DNA in the strains was determined with screening method using a standard procedure; cells from a solid medium were suspended with a loop in 100 µl of buffer (50 mM Tris pH 8.0, 50 mM Na2-EDTA, 15% sucrose), 200 µl of alkaline solution (0.2 N NaOH, 1% SDS) and 150 µl of 3 M sodium acetate pH 5.0 were added, and centrifugation was performed for 5 minutes on a desktop centrifuge, then 1 mL of 96% ethanol was added to the sediment. The obtained DNA was analyzed in 0.8 % agarose in Tris-borate buffer pH 8.0 [48];

- when screening the strains for the presence of restriction endonucleases, individual colonies collected from a solid culture were suspended in 100–200 µl of TEN-buffer (0.1 M Tris, pH 7.5, 0.01 M EDTA, 0.05 M NaCl), lysocime and triton X-100 were used to destroy the cell wall of bacteria; the obtained cell extract was used for analysis for the presence of restriction endonucleases. DNAs of phages λcI857 and T7 were used as substrates for hydrolysis; electrophoresis of DNA after restriction was performed in 1% agarose (Sigma) [63]; the presence of restriction endonucleases in microorganism strains was revealed by the appearance of discrete fragments of substrate DNA in electrophoregram in UV light.

2.3.4. Microorganisms' Antibiotic Resistance

The sensitivity of microorganisms to antibiotics was determined by diffusion using the discs method [57]; the concentration of antibiotics in the discs used was: ampicillin (10 µg/disk), neomycin (30 µg/disk), benzyl-penicillin (100 U/disk), levomycetin (30 µg/disk), carbenicillin (100 µg/disk), canamycin (30 µg/disk), oleandomycin (15 µg/disk), rifampicin (5 µg/disk), streptomycin (30 µg/disk), polymxin (300 U), erythromycin (15 µg/disk), lincomycin (15 µg/disk), oxacillin (10 µg/disk), gentamycin (10 µg/disk), tetracycline (30 µg/disk), vancomycin (30 µg/disk), amikacin (30 µg/disk), netilmycin (30 µg/disk), monomycin (30 µg/disk). Note that lists of antibiotics used were not the same for 2006–2008 and for 2012–2016.

2.4. Changes in the Methods Used in the 2006–2008 and 2012–2016 Studies

1. A Tupolev-134 aircraft (Figure 3) was used for sounding of the atmosphere since 2011 [64]. The speed of this aircraft was maintained at the level of the Antonov-30 aircraft to maintain stable operation of the air intake. For sampling air containing microorganisms, the same impingers were used on both airplanes and in on-ground sampling; Figure 4. As noted above, the sampling time for airborne atmospheric sounding was 10 minutes for 2006–2008 and 5 minutes for 2012–2016.
2. When analyzing samples by cultural methods in 2012–2016, in addition to the cultivation temperature of 28–30 °C, some isolates were additionally incubated at the temperature of 6–10 °C.
3. And the last difference between the methods of samples analysis in the 2006–2008 and 2012–2016 investigations-the lists of antibiotics the sensitivity of isolates to which was determined are slightly different. For 2012–2016, the sensitivity of isolates was determined for a larger number of antibiotics, but this was done for not all identified isolates.

2.5. Data Analysis and Statistics

The initial data for all measured values for which lower-level integral indices were calculated using the formulas published in [28] and these integral indices themselves are given in Supplementary Materials (Tables S1–S12 and List of description). The average values of the integral indices and their standard deviations from the mean are given in these tables for each year of observation. Additionally, the maximum values of the corresponding integral for each year are given.

In Table S13, the integral index of hazard of bacteria and yeasts found in atmospheric bioaerosols for humans are calculated from the lower level integral indices for each of the isolates according to the procedure described in [28]. Furthermore, using the method of regression and ANOVA analysis (built-in Microsoft Excel software), a statistical analysis of the data was carried out. All analysis results were obtained at a reliability level of 95%.

3. Results

Based on characteristics of isolated microorganisms using the methods described above, four indices were calculated, which form the integral index of the hazard of microorganisms found out in atmospheric aerosols for humans. These indices were calculated both for each isolate and averaged for each year of research.

Let us compare the data presented in [28] with the results obtained in this research using the same methods. It should be noted that due to resource constraints for the company 2012–2016, not all the characteristics of the identified isolates were investigated, as it was done during the company 2006–2008. However, according to the authors' opinion, the results obtained on a truncated set of characteristics quite clearly demonstrate the changes occurring in the pool of cultured microorganisms isolated from atmospheric aerosols in the south of Western Siberia.

3.1. Long-Term Trends of Average Annual Concentrations of Culturable Microorganisms in the Atmospheric Aerosol of the South of Western Siberia

Figure 5 shows the long-term trends of average annual concentrations of culturable microorganisms in the atmospheric aerosol of southern Western Siberia. In constructing this graph, we used the results published in [65] and our own data obtained after the publication of this paper.

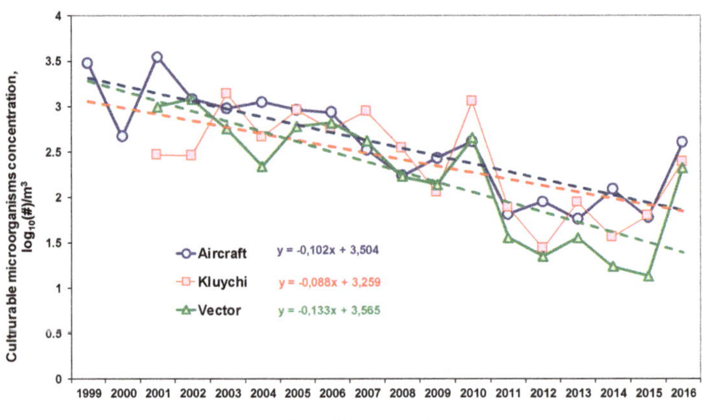

Figure 5. Trends of average annual concentrations of culturable microorganisms in the atmospheric aerosol in the south of Western Siberia. The average annual values with root mean square deviations (standard deviations) from the mean are given.

Long-term observations of the concentrations of culturable microorganisms (bacteria, fungi and yeasts) in the atmosphere indicate that the average annual values of these concentrations have a pronounced tendency to decrease. In particular, from 2008 to 2016, this decrease is at least three times.

It should be noted that, for the concentration of cultuarable microorganisms in atmospheric aerosols in the south of Western Siberia, the intra-annual variability was revealed [65]. Figure 6 presents the data for 2001–2016 values averaged for each month of years 2001–2016, normalized to the corresponding annual average. As can be seen from this graph, the differences between the maximum and minimum concentrations are on average up to three times. Naturally, for each sampling, the concentration of microorganisms depends not only on the revealed dependence, but also on the specific meteorological conditions, the effect of which on this concentration can be very significant.

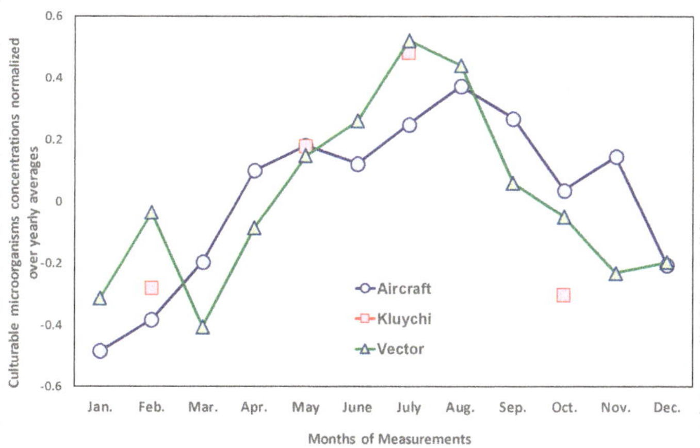

Figure 6. Intra-annual variability of concentrations of culturable microorganisms in the atmospheric aerosol of the south of Western Siberia. The average values for a given month for the entire observation period and the standard deviations from the average are given.

It should be noted that the data used to construct the integral index characterizing the concentration of isolates in the air make up only part of the raw data used to construct Figures 5 and 6. This is due to the fact that, to construct the integral index characterizing the concentration of specific isolates in the air, only data on the concentrations of bacteria and yeast, but not all culturable microorganisms, are used. To calculate the total concentration of culturable microorganisms in the air, data are used on all grown colonies in the seeded sample, but not all colonies succeed in obtaining isolates suitable for further studies, namely, isolate concentrations are used to construct the corresponding integral indices, and finally, Figures 5 and 6 are constructed for data for 2001–2016, and data obtained for a smaller number of years were used to construct integral indices.

3.2. Sustainability of Culturable Microorganisms in the Environment

The results of determining the integral index of resistance of microorganisms to adverse environmental factors are summarized in Table 1 for annual mean values and standard deviations from them.

Table 1. The annual average values and standard deviations from them of the integral index of the microorganisms' resistance to adverse environmental factors values.

Sampling Site	Values	2006–2007	2007–2008	2012	2013	2014	2015	2016
Vector	Average	0.359	0.372	0.324	0.247	0.244	0.135	0.252
	Standard deviation	0.223	0.206	0.256	0.243	0.229	0.132	0.209
Klyuchi	Average	0.315	0.335	0.452	0.399	0.381	0.354	0.354
	Standard deviation	0.216	0.182	0.246	0.222	0.215	0.108	0.149
Aircraft	Average	0.307	0.399	0.567	0.380	0.415	0.405	0.431
	Standard deviation	0.212	0.222	0.333	0.191	0.137	0.125	0.176

As follows from the analysis of the data presented in Table 1 and in Supplementary Materials (see Tables S1–S3, S13) for the sustainability index of culturable microorganisms in the atmospheric aerosol of the south of Western Siberia, differently directed trends are observed for different sites of observation. Changes in the sustainability index of culturable microorganisms in an atmospheric aerosol in the south of Western Siberia do not exceed 1.1–1.75 times for 10 years of observations. However, given the fairly large scatter of data for each year of measurement relative to average values, these changes cannot be considered non-zero.

There are those isolates among the isolates of bacteria and yeasts that have values of the microorganism sustainability index in the environment from 0 to 1 and the average annual values of the indexes are at the level of 0.2–0.45. Note that the maximum index values close to or equal to 1 exist for a small number of bacteria and yeasts, not in every sample, and not even in every year (Supplementary Materials, Tables S1–S3).

Thus, the obtained experimental data do not reliably reveal differences at a reliability level of 95% between the sustainability indexes of culturable microorganisms in the atmospheric aerosol of the south of Western Siberia determined for 2006–2008 and for 2012–2016. At a 95% reliability level, all three trends are significantly indistinguishable from zero.

3.3. Potential Pathogenicity of Culturable Microorganisms in the Atmospheric Aerosol of the South of Western Siberia

The results of determining the integral index of potential pathogenicity of pathogenicity of culturable bacteria and yeast for humans in an atmospheric aerosol in the south of Western Siberia are summarized in Table 2 for annual mean values and standard deviations from them.

Table 2. Annual average values and standard deviations from them of the integral index of potential pathogenicity of culturable bacteria and yeast in an atmospheric aerosol in the south of Western Siberia.

Sampling Site	Values	2006–2007	2007–2008	2012	2013	2014	2015	2016
Vector	Average	0.239	0.239	0.315	0.202	0.213	0.374	0.292
	Standard deviation	0.148	0.121	0.177	0.117	0.146	0.177	0.175
Klyuchi	Average	0.209	0.275	0.232	0.205	0.210	0.371	0.175
	Standard deviation	0.132	0.139	0.187	0.138	0.134	0.197	0.175
Aircraft	Average	0.235	0.270	0.306	0.247	0.203	0.409	0.303
	Standard deviation	0.139	0.150	0.196	0.130	0.116	0.182	0.177

As follows from the analysis of the data shown in Table 2 and in Supplementary Materials (see Tables S4–S6, S13), there is no noticeable positive trend in the index of the potential pathogenicity of cultivated microorganisms in the atmospheric aerosol for air samples taken at the site of Klyuchi only. This increase from 2006 to 2016 is 1.5 times. The obtained experimental data do not reliably reveal differences at a reliability level of 95% in indices of potential pathogenicity of culturable microorganisms in the atmospheric aerosol of the south of Western Siberia determined in 2006–2008 and in 2012–2016 for the Klyuchi sampling site. At a 95% reliability level, all three trends are significantly indistinguishable from zero.

The average annual level of this index for all isolates was 0.2–0.4, while only a few of the microorganisms detected over the entire period had indices of the potential pathogenicity of culturable microorganisms in the atmospheric aerosol, approaching but not reaching 1.

3.4. Culturable Microorganisms' Resistance to Antibiotics

The results of determining the integral index of antibiotic resistance of culturable bacteria and yeast in an atmospheric aerosol in the south of Western Siberia are summarized in Table 3 for annual mean values and standard deviations from them.

Table 3. Annual average values and standard deviations from them of the integral index of culturable microorganisms 'resistance to antibiotics bacteria and yeast in an atmospheric aerosol in the south of Western Siberia.

Sampling Site	Values	2006–2007	2007–2008	2012	2013	2014
Vector	Average	0.207	0.206	0.170	0.195	0.152
	Standard deviation	0.181	0.129	0.173	0.208	0.107
Klyuchi	Average	0.262	0.245	0.265	0.239	0.144
	Standard deviation	0.211	0.146	0.106	0.231	0.107
Aircraft	Average	0.232	0.206	0.200	0.314	0.278
	Standard deviation	0.184	0.149	0.188	0.319	0.207

The situation for this index is similar to the situation with the sustainability index of culturable microorganisms in the atmospheric aerosol of the south of Western Siberia. The trends of this index are multidirectional and over 8 years of observations of changes in the culturable microorganisms in the atmospheric aerosol in the south of Western Siberia resistance to the action of antibiotics do not exceed 1.5 times (Table 3 and Tables S4–S6, S13 in Supplementary Materials). Note that in 2015 and 2016, experimental studies on resistance to antibiotics of isolates from atmospheric aerosol have not been conducted. Given the fairly large scatter of data relative to the average values for each year of measurement, these changes can be considered indistinguishable from zero. Thus, the obtained experimental data do not reliably reveal differences at a reliability level of 95% between the indices of culturable microorganisms in the atmospheric aerosol in the south of Western Siberia resistance to antibiotics determined in 2006–2008 and in 2012–2014. At a 95% reliability level, all three trends are significantly indistinguishable from zero.

Among the identified isolates, there are both those that are susceptible and resistant to all studied antibiotics. It was found that the proportion of isolates sensitive to the action of all antibiotics or resistant to the action of no more than one antibiotic did not change significantly during the period studied. In 2006–2008, there were from 33% to 42% of such isolates, and in 2012–2014 there were from 28% to 42%. Two isolates, which are resistant to all studied antibiotics (11 out of 11) or 0.21%, were detected at all observation sites in 2006–2008 and in 2012–2014 there were three such isolates (8 out of 8, 9 out of 9, and 15 out of 15), or 0.95%, Supplementary Materials (Tables S10–S12). Consequently, the number of microorganisms that are most resistant to the action of antibiotics over 8 years has increased more than four times, while the indices of culturable microorganisms in the atmospheric aerosol in the south of Western Siberia resistance to the action of antibiotics have not changed.

The sources of atmospheric microorganisms, besides vegetation, animals, and human activity, are soil and water; most likely, bioaerosols should contain a similar percentage of bacteria resistant to the action of antibiotics. Indeed, about the same percentage of bacteria resistant to the action of at least one of the antibiotics studied was found for the water system [66–71] and in the soil [69,72].

3.5. Integral Index

The results of the determination of the integral index of hazard for humans of culturable bacteria and yeast in an atmospheric aerosol in the south of Western Siberia are summarized in Table 4 for annual mean values and standard deviations from them.

As already noted, this integral index is the product of the 4 indices described above (see Tables S1–S12 in Supplementary Materials). However, for some isolates of microorganisms, some indices were not determined, therefore, the average annual indices of this index for the corresponding year were substituted for the calculation. These values in the relevant cells of Table S13 in Supplementary Materials are highlighted in light blue. In general, the average annual values of this index are not large and are in the range of 0.001–0.007. Recall that for individual pathogenic microbial isolates that are sustainable in the environment and do not possess antibiotic resistance, this index equals 1. Consequently, there are very few such pathogenic microorganisms in the atmospheric aerosol of the south of Western Siberia (see Table S13 in Supplementary Materials). Moreover, the maximum value of this integral index over all the time of research for isolates from all observation sites is only 0.1074. At the same time, a fairly large number of isolates was found in the samples, for which the integral index is strictly equal to zero (see Table S13 in Supplementary Materials). Here is interesting situation: when the concentration of one of the isolates is very high (normalized index about 1) other indices are low; and when, for example, pathogenicity index is high, other indices are low. Hence, the integral index is not large for all studied isolates. The dependence of integral index for monthly averages is pronounced for 2006–2008 when the number of isolates is large for many months [28]. The isolates number for 2012–2016 is low, they are found not in every sample. So, integral index for monthly averages is not representative in 2012–2016 (Supplementary Materials, Tables S1–S13 for 2012–2016).

Table 4. The annual average values and standard deviations from them of the integral index of hazard of culturable bacteria and yeast in an atmospheric aerosol in the south of Western Siberia.

Sampling Site	Values	2006–2007	2007–2008	2012	2013	2014	2015	2016
Vector	Average	0.0035	0.0007	0.0001	0.0011	0.0003	0.0001	0.0018
	Standard deviation	0.0110	0.0017	0.0002	0.0026	0.0008	0.00015	0.0061
Klyuchi	Average	0.0010	0.0047	0.0002	0.0005	0.0001	0.0007	0.0032
	Standard deviation	0.0045	0.0109	0.0003	0.0009	0.0002	0.0012	0.0057
Aircraft	Average	0.0016	0.0019	0.0000	0.0000	0.0037	0.0004	0.0018
	Standard deviation	0.0034	0.0058	0.0001	0.0001	0.0059	0.0004	0.0024

As follows from the data presented in Table 4, all long-term trends of these integral indices are unidirectional, showing a tendency to decrease; however, the magnitude of this decrease is almost indistinguishable from zero at a reliability level of 95%. Based on this, one can conclude that in 10 years of research, the hazard to humans of microorganisms isolated from atmospheric aerosols of the south of Western Siberia remains almost unchanged.

As it was mentioned above, the large scatter of the experimentally determined values included in this index (see Table S13 in Supplementary Materials) makes it possible to speak only of the tendency of a small decrease in the integral index of the hazard of microorganisms isolated from atmospheric aerosols in the south of Western Siberia for humans. This decrease is primarily due to a decrease in the observed average annual number of culturable microorganisms in atmospheric aerosols in the south of Western Siberia.

4. Discussion

As it was mentioned above, a long-term series of bacteria and yeast concentration observations have not been presented in the literature for other regions of the world; therefore, the results presented in Figure 5 refer only to the south of Western Siberia. A similar result is described in [73] for 2014–2018 for fungi yearly averaged concentrations in the air in France: the decrease is about 1.5 times. At the same time, an increase in the average annual concentrations of fungi of the genera *Aspergillus* and *Penicillium* in Derby, UK, in 1970–2003 was recorded in [74]. As for the intra-annual dynamics of changes in the concentrations of bacteria and yeast, there is information in the literature, although such long-term observations that are presented in this paper are also absent in the literature. It should be noted that for different microorganisms (strains, genera, kingdoms), these dynamics can vary significantly. In particular, in many cases, the concentration of culturable bacteria in a warm season is higher than in a cold one [75–86], but there are also opposite examples [80,87–91].

The noted decrease in the average annual concentrations of culturable microorganisms in the south of Western Siberia is probably associated with the ongoing climate changes, which are manifested in a change in the location and power of bioaerosol sources, pathways of predominant transport of bioaerosols in the atmosphere, etc.

There is also no information in the literature about the change over time of microorganism characteristics that are included in the index of potential pathogenicity of culturable microorganisms in the atmospheric aerosol and in the index of their sustainability in the atmospheric aerosol to adverse environmental factors. Similar comprehensive studies have not been conducted previously.

With regard to antibiotic resistance, the literature presents data indicating that there is a tendency to an increase in the number of bacteria resistant to the action of antibiotics over time, see, for example, [92–96]. In the collections of bacteria isolates collected over a long period of time, the latest trend can be traced reliably [97,98]. Moreover, more and more isolates with multiple antibiotic resistances are detected in nature [99–102].

It should be noted that while for ground-based observation points, the obtained data directly provide an assessment of the hazard to humans of microorganisms in the air. Let us remind the reader that a large portion of aerosol particles in the on-ground atmospheric layer originated from local sources. It is difficult to relate the results to any particular place on the Earth's surface for high-altitude observations. Bioaerosols found in high-altitude samples of atmospheric air originated mainly from long-distance sources and can, due to the stochastic nature of their movement in the atmosphere, reach the surface at completely different points. In addition, during the transfer process, some microorganisms can simply be inactivated, or at least lose their ability to grow under cultivation conditions.

It should be noted that it is not easy to build the dynamics of the monthly change of all the above-mentioned indices because of their very high variability and a small number of identified microorganisms for some months. Very weak dependencies can be observed only.

All the results presented in Section 3 have very high variability for both the average annual values of integral indices and the values of these indices for individual isolates. This is due to the properties of an atmospheric aerosol, which is a complex mixture of particles of various compositions. These particles enter the atmosphere from various terrestrial sources of inorganic, organic or biological nature, and are also formed in the atmosphere during nucleation processes. While in the aerosol, particles are continuously transformed under the influence of changing temperature and humidity, due to the presence of volatile compounds in it, capable of entering into chemical interactions with particles under the influence of solar radiation, etc. These factors can lead to inactivation of microorganisms in the particles. It is also necessary to consider the processes of particle deposition (sedimentation, washout by precipitations). Because of this, the concentration and composition of an aerosol in the atmosphere can vary greatly in samples that are close in time or in distance. For example, in [65], it was mentioned that two air samples taken during aircraft sounding of the atmosphere (a difference of 1000 m in height and 15 minutes in time) contain radically different compositions of microorganisms. This also leads to the fact that, if the average annual dependencies of the total concentration of microorganisms shown in Figure 5 have a relatively small dispersion, the attempt to construct similar dependencies for individual phylums and lower taxonomic groups was unsuccessful due to the extremely large concentration dispersion [103]. All of the above is particularly pronounced in Siberia with its sharply continental climate, many local sources of bioaerosols (forests, meadows, swamps, industrial production) and the prevailing direction of the winds, which bring a lot of dust (and soil microorganisms to it) from north-western Kazakhstan to Western Siberia. Moreover, the authors are not sure that an increase in the number of samples analyzed for each month (season) will noticeably affect a decrease in the variance of the result and, therefore, an increase in the reliability of the dependencies identified in this article.

The approach proposed in [28] made it possible, for the first time, to quantitatively compare the potential danger to humans of the totality of microorganisms in various samples of atmospheric air. Using this unique approach, it was possible to assess the change in potential danger to humans over 10 years of observation at different points in Western Siberia. Currently, the authors are not aware of any other approaches that allow such estimates.

At the same time, the authors do not pretend to the universality of the method used in this article for assessing the hazard to humans of culturable bacteria and fungi located in atmospheric aerosol. The algorithm for estimating integral indices proposed in [28] is not exhaustive and, probably, there are still some characteristics of cultivated microorganisms (morphological, cultural, biochemical, and biophysical) that refine the constructed integral indices. In addition, the weight values of the various characteristics of microorganisms by which integral indicators are calculated may also need to be clarified, and finally, an approach similar to that proposed for bacteria and yeast was not developed for micromycetes. Therefore, the results obtained in this article reflect the hazard to humans of only bacteria and yeast present in the aerosol, but not for all culturable microorganisms. In addition, not all microorganisms are easily cultivated under standard conditions. The method used simply cannot be implemented for unculturable microorganisms.

5. Conclusions

Summing up the conducted research, one must conclude that, despite a noticeable decrease in the average annual number of culturable microorganisms in the atmospheric aerosol in the south of Western Siberia, the hazard of these bacteria and yeasts to humans has not practically changed for 10 years. Thus, neither passing climatic changes in the world, nor changes in the antibiotic resistance and sustainability of culturable microorganisms in the environment have affected the hazard to humans of bacteria and yeasts found in the atmospheric aerosol in the south of Western Siberia.

The authors published in 2012 [28] an approach on how to estimate the hazard of bacteria and yeasts in atmospheric aerosols to humans, but no similar investigation has been conducted elsewhere in the world. Usually, the data of experiments are given in the literature, in which the effect of a specific bioaerosol that has penetrated the respiratory tract is studied, see, for example [104]. It is clear that such an approach is difficult to use to characterize a large number of bioaerosol samples. The authors hope that very important studies on risks of human exposure to microorganisms in the air will be estimated using our approach [28] and will be published, not only for Western Siberia; the authors also hope that a longer study of the properties of culturable microorganisms in atmospheric aerosols will reliably confirm the identified weak dependencies in the future in the south of Western Siberia.

Supplementary Materials: The following are available online at http://www.mdpi.com/1660-4601/17/5/1651/s1, Table S1: Sustainability of culturable microorganisms in the environment. Site of the FBRI SRC VB "Vector", Table S2: Sustainability of culturable microorganisms in the environment. Site Klyuchi, Table S3: Sustainability of culturable microorganisms in the environment. Aircraft sounding of the atmosphere, Table S4: Hazard of culturable microorganisms in the atmospheric aerosol. Site of the FBRI SRC VB "Vector", Table S5: Hazard of culturable microorganisms in the atmospheric aerosol. Site Klyuchi, Table S6: Hazard y of culturable microorganisms in the atmospheric aerosol. Aircraft sounding of the atmosphere, Table S7: Indices of culturable bacteria and yeast resistance to antibiotics. Site of the FBRI SRC VB "Vector", Table S8: Indices of culturable bacteria and yeast resistance to antibiotics. Site Klyuchi, Table S9: Indices of culturable bacteria and yeast resistance to antibiotics. Aircraft sounding of the atmosphere, Table S10: Number of isolates in the samples. Site of the FBRI SRC VB "Vector", Table S11: Number of isolates in the samples. Site Klyuchi, Table S12: Number of isolates in the samples. Aircraft sounding of the atmosphere. Table S13: Integral index and data analysis.

Author Contributions: Conceptualization and methodology, A.S., M.P.; Investigation, B.B., O.O., D.S., S.O., I.R., G.B., I.A., L.P., N.S.; data analysis, A.S., G.B.; writing—original draft preparation, A.S.; writing—review and editing, A.S., G.B.; project administration, A.S. All authors have read and agreed to the published version of the manuscript.

Funding: This research was partially funded by Russian Federation Rospotrebnadzor, grant number 141-00069-18-01.

Conflicts of Interest: The authors declare no conflict of interest.

References

1. Artaxo, P.; Storms, H.; Bruynseels, F.; Van Grieken, R.; Maenhaut, W. Composition and sources of aerosols from the Amazon basin. *J. Geophys. Res.* **1988**, *93*, 1605–1615. [CrossRef]
2. Artaxo, P.; Maenhaut, W.; Storms, H.; Van Grieken, R. Aerosol characteristics and sources for the Amazon basin during the wet season. *Geophys. Res.* **1990**, *95*, 16971–16985. [CrossRef]
3. Matthias-Maser, S.; Jaenicke, R. Examination of atmospheric bioaerosol particles with radii >0.2 μm. *J. Aerosol Sci.* **1994**, *25*, 1605–1613. [CrossRef]
4. Zhang, Q.; Jimenez, J.L.; Canagaratna, M.R.; Allan, J.D.; Coe, H.; Ulbrich, I.; Alfarra, M.R.; Takami, A.; Middlebrook, A.M.; Sun, Y.L.; et al. Ubiquity and dominance of oxygenated species in organic aerosols

in anthropogenically-influenced Northern Hemisphere midlatitudes. *Geophys. Res. Lett.* **2007**, *34*, L13801:1–L13801:6. [CrossRef]

5. Fröhlich-Nowoisky, J.; Kampf, C.J.; Weber, B.; Huffman, J.A.; Pöhlker, C.; Andreae, M.O.; Lang-Yona, N.; Burrows, S.M.; Gunthe, S.S.; Elbert, W.; et al. Bioaerosols in the Earth system: Climate, health, and ecosystem interactions. *Atmos. Res.* **2016**, *182*, 346–376. [CrossRef]

6. Morris, C.E.; Sands, D.C.; Bardin, M.; Jaenicke, R.; Vogel, B.; Leyronas, C.; Ariya, P.A.; Psenner, R. Microbiology and atmospheric processes: Research challenges concerning the impact of airborne micro-organisms on the atmosphere and climate. *Biogeosciences* **2011**, *8*, 17–25. [CrossRef]

7. Sahyoun, M.; Wex, H.; Gosewinkel, U.; Santl-Temkiv, T.; Nielsen, N.W.; Finster, K.; Sørensen, J.H.; Stratmann, F.; Korsholm, U.S. On the usage of classical nucleation theory in quantification of the impact of bacterial INP on weather and climate. *Atmos. Environ.* **2016**, *139*, 230–240. [CrossRef]

8. Douwes, J.; Thorne, P.; Pearce, N.; Heederik, D. Bioaerosols health effects and exposure assessment: Progress and prospects. *Ann. Occup. Hyg.* **2003**, *47*, 187–200. [CrossRef]

9. Kim, K.-H.; Kabir, E.; Jahan, S.A. Airborne bioaerosols and their impact on human health. *J. Environ. Sci.* **2018**, *67*, 23–35. [CrossRef]

10. Walser, S.M.; Gerstner, D.G.; Brenner, B.; Bünger, J.; Eikmann, T.; Janssen, B.; Kolb, S.; Kolk, A.; Nowak, D.; Raulf, M.; et al. Evaluation of exposure–response relationships for health effects of microbial bio-aerosols—A systematic review. *Int. J. Hyg. Environ. Health* **2015**, *218*, 577–589. [CrossRef]

11. Fernstrom, A.; Goldblatt, M. Aerobiology and Its Role in the Transmission of Infectious Diseases. *J. Pathogens* **2013**, *2013*, 493960:1–493960:14. [CrossRef] [PubMed]

12. Ypma, R.J.F.; Jonges, M.; Bataille, A.; Stegeman, A.; Koch, G.; van Boven, M.; Koopmans, M.; van Ballegooijen, W.M.; Wallinga, J. Genetic Data Provide Evidence for Wind-Mediated Transmission of Highly Pathogenic Avian Influenza. *J. Infect. Dis.* **2013**, *207*, 730–735. [CrossRef] [PubMed]

13. An, H.R.; Mainelis, G.; White, L. Development and calibration of real-time PCR for quantification of airborne microorganisms in air samples. *Atmos. Environ.* **2006**, *40*, 7924–7939. [CrossRef]

14. Belgrader, P.; Elkin, C.J.; Brown, S.B.; Nasarabadi, S.N.; Langlois, R.G.; Milanovich, F.P.; Colston, B.W., Jr.; Marshall, G.D. A reusable flow-through polymerase chain reaction instrument for the continuous monitoring of infectious biological agents. *Anal. Chem.* **2003**, *75*, 3446–3450. [CrossRef]

15. Jiang, W.; Liang, P.; Wang, B.; Fang, J.; Lang, J.; Tian, G.; Jiang, J.; Zhu, T.F. Optimized DNA extraction and metagenomics sequencing of airborne microbial communities. *Nature Protocols* **2015**, *10*, 768–779. [CrossRef]

16. Maron, P.A.; Lejon, D.P.H.; Carvalho, E.; Bizet, K.; Lemanceau, P.; Ranjard, L.; Mougel, C. Assessing genetic structure and diversity of airborne bacterial communities by DNA fingerprinting and 16S rDNA clone libraries. *Atmos. Environ.* **2005**, *39*, 3687–3695. [CrossRef]

17. Mbareche, H.; Brisebois, E.; Veillette, M.; Duchaine, C. Bioaerosol sampling and detection methods based on molecular approaches: No pain no gain. *Sci. Total Environ.* **2017**, *599–600*, 2095–2104. [CrossRef]

18. Rinsoz, T.; Duquenne, P.; Greff-Mirguet, G.; Oppliger, A. Application of real-time PCR for total airborne bacterial assessment: Comparison with epifluorescence microscopy and culture-dependent methods. *Atmos. Environ.* **2008**, *42*, 6767–6774. [CrossRef]

19. Xu, S.; Yao, M. NanoPCR detection of bacterial aerosols. *J. Aerosol Sci.* **2013**, *65*, 1–9. [CrossRef]

20. Zhang, K.; Martiny, A.C.; Reppas, N.B.; Barry, K.W.; Malek, J.; Chisholm, S.W.; Church, G.M. Sequencing genomes from single cells by polymerase cloning. *Nature Biotechnol.* **2006**, *24*, 680–686. [CrossRef]

21. Després, V.R.; Nowoisky, J.F.; Klose, M.; Conrad, R.; Andreae, M.O.; Pöschl, U. Characterisation of primary biogenic aerosol particles in urban, rural, and high-alpine air by DNA sequence and restriction fragment analysis of ribosomal RNA genes. *Biogeosciences* **2007**, *4*, 1127–1141. [CrossRef]

22. Caporaso, J.G.; Lauber, C.L.; Walters, W.A.; Berg-Lyons, D.; Huntley, J.; Fierer, N.; Owens, S.M.; Betley, J.; Fraser, L.; Bauer, M.; et al. Ultra-high-throughput microbial community analysis on the Illumina HiSeq and MiSeq platforms. *ISME J.* **2012**, *6*, 1621–1624. [CrossRef] [PubMed]

23. Cha, S.; Srinivasan, S.; Jang, J.H.; Lee, D.; Lim, S.; Kim, K.S.; Jheong, W.; Lee, D.-W.; Park, E.-R.; Chung, H.-M.; et al. Metagenomic Analysis of Airborne Bacterial Community and Diversity in Seoul, Korea, during December 2014, Asian Dust Event. *PLoS ONE* **2017**, *12*, e0170693:1–e0170693:12. [CrossRef] [PubMed]

24. Duquenne, P. On the Identification of Culturable Microorganisms for the Assessment of Biodiversity in Bioaerosols. *Ann. Work Expo. Health.* **2018**, *62*, 139–146. [CrossRef]

25. Gao, J.-F.; Fan, X.-Y.; Li, H.-Y.; Pan, K.-L. Airborne Bacterial Communities of $PM_{2.5}$ in Beijing-Tianjin-Hebei Megalopolis, China as Revealed by Illumina MiSeq Sequencing: A Case Study. *Aerosol Air Qual. Res.* **2017**, *17*, 788–798. [CrossRef]

26. Serrano-Silva, N.; Calderón-Ezquerro, M.C. Metagenomic survey of bacterial diversity in the atmosphere of Mexico City using different sampling methods. *Environ. Pollut.* **2018**, *235*, 20–29. [CrossRef]

27. Yoo, K.; Lee, T.K.; Choi, E.J.; Yang, J.; Shukla, S.K.; Hwang, S.-I.; Park, J. Molecular approaches for the detection and monitoring of microbial communities in bio-aerosols: A review. *J. Environ. Sci.* **2017**, *51*, 234–247. [CrossRef]

28. Safatov, A.S.; Andreeva, I.S.; Belan, B.D.; Buryak, G.A.; Emel'yanova, E.K.; Jaenicke, R.; Panchenko, M.V.; Pechurkina, N.I.; Puchkova, L.I.; Repin, V.E.; et al. To what extent can viable bacteria in atmospheric aerosols be dangerous for humans? *Clean* **2008**, *36*, 564–571. [CrossRef]

29. Peñuelas, J.; Fernández-Martínez, M.; Ciais, P.; Jou, D.; Piao, S.; Obersteiner, M.; Vicca, S.; Janssens, I.A.; Sardans, J. The bioelements, the elementome, and the biogeochemical niche. *Ecology* **2019**, *100*, e02652:1–e02652:15. [CrossRef]

30. Jenhani, A.B.R.; Fathalli, A.; Djemali, I.; Changeux, T.; Romdhane, M.S. Tunisian reservoirs: Diagnosis and biological potentialities. *Aquat. Living Resour.* **2019**, *32*, 17:1–17:17. [CrossRef]

31. Huang, Y.; Guenet, B.; Ciais, P.; Janssens, I.A.; Soong, J.L.; Wang, Y.; Goll, D.; Blagodatskaya, E.; Huang, Y. ORCHIMIC (v1.0), a microbe-mediated model for soil organic matter decomposition. *Geosci. Model Dev.* **2018**, *11*, 2111–2138. [CrossRef]

32. Zouidi, M.; Borsali, A.H.; Allam, A.; Gros, R. Characterization of coniferous forest soils in the arid zone. *Environ. Sci.* **2018**, *68*, 64–74. [CrossRef]

33. Boyd, P.W.; Collins, S.; Dupont, S.; Fabricius, K.; Gattuso, J.-P.; Havenhand, J.; Hutchins, D.A.; Riebesell, U.; Rintoul, M.S.; Vichi, M.; et al. Experimental strategies to assess the biological ramifications of multiple drivers of global ocean change—A review. *Glob Change Biol.* **2018**, *24*, 2239–2261. [CrossRef]

34. Błażejczyk, K.; Baranowski, J.; Błażejczyk, A. Climate related diseases. Current regional variability and projections to the year 2100. *Quaestiones Geographicae* **2018**, *37*, 23–36. [CrossRef]

35. Masyagina, O.V.; Menyailo, O.V. The impact of permafrost on carbon dioxide and methane fluxes in Siberia: A meta-analysis. *Environ. Res.* **2020**, *182*, 109096. [CrossRef] [PubMed]

36. Zhang-Turpeinen, H.; Kivimäenpää, M.; Aaltonen, H.; Berninger, F.; Köster, E.; Köster, K.; Menyailo, O.; Prokushkin, A.; Pumpanen, J. Wildfire effects on BVOC emissions from boreal forest floor on permafrost soil in Siberia. *Sci. Total Environ.* **2020**, *711*, 134851. [CrossRef]

37. Revich, B.A.; Shaposhnikov, D.A. Extreme temperature episodes and mortality in Yakutsk, East Siberia. *Rural Remote Health.* **2010**, *10*, 1338.

38. Konovalov, I.B.; Lvova, D.A.; Beekmann, M.; Jethva, H.; Mikhailov, E.F.; Paris, J.-D.; Belan, B.D.; Kozlov, V.S.; Ciais, P.; Andreae, M.O. Estimation of black carbon emissions from Siberian fires using satellite observations of absorption and extinction optical depths. *Atmos. Chem. Phys.* **2018**, *18*, 14889–14924. [CrossRef]

39. Antokhina, O.Y.; Antokhin, P.N.; Martynova, Y.V. Methane emissions from wildfires in Siberia caused by the atmospheric blocking in the summertime. In Proceedings of the 25th International Symposium on Atmospheric and Ocean Optics: Atmospheric Physics, Novosibirsk, Russia, 1–5 June 2019.

40. Korotkova, E.M.; Zuev, V.V.; Pavlinsky, A.V. Trend and correlation analysis of air temperature and NDVI in Western Siberia over the period 1982-2015. In Proceedings of the 25th International Symposium on Atmospheric and Ocean Optics: Atmospheric Physics, Novosibirsk, Russia, 1–5 June 2019.

41. Babushkina, E.A.; Zhirnova, D.F.; Belokopytova, L.V.; Tychkov, I.I.; Vaganov, E.A.; Krutovsky, K.V. Response of Four Tree Species to Changing Climate in a Moisture-Limited Area of South Siberia. *Forests* **2019**, *10*, 999. [CrossRef]

42. Koenigk, T.; Fuentes-Franco, R. Towards normal Siberian winter temperatures? *Int. J. Climatol.* **2019**, *39*, 4567–4574. [CrossRef]

43. Song, L.; Wu, R. Intraseasonal snow cover variations over western Siberia and associated atmospheric processes. *J. Geophys. Res. Atmospheres.* **2019**, *124*, 8994–9010. [CrossRef]

44. Gorbatenko, V.P.; Sevastyanov, V.V.; Konstantinova, D.A.; Nosyreva, O.V. Characteristic of the snow cover for the Western Siberia territory. In *IOP Conference Series: Earth and Environmental Science*; IOP Publishing: Bristol, UK, 2019; Volume 232. [CrossRef]

45. Podnebesnykh, N.V.; Loginov, S.V.; Kharyutkina, E.V.; Usova, E.I. Vortex circulation and anomalous meteorological phenomena over the Asian territory of Russia in the context of climate change. In Proceedings of the 25th International Symposium on Atmospheric and Ocean Optics: Atmospheric Physics, Novosibirsk, Russia, 1–5 June 2019.
46. Costello, A.; Abbas, M.; Allen, A.; Ball, S.; Bell, S.; Bellamy, R.; Friel, S.; Groce, N.; Johnson, A.; Kett, M.; et al. Managing the health effects of climate change: Lancet and University College London Institute for Global Health Commission. *Lancet.* **2009**, *373*, 1693–1733. [CrossRef]
47. Ryti, N.R.I.; Guo, Y.; Jaakkola, J.J.K. Global association of cold spells and adverse health effects: A systematic review and meta-analysis. *Environ Health Perspect.* **2016**, *124*, 12–22. [CrossRef] [PubMed]
48. Zuev, V.E.; Belan, B.D.; Kabanov, D.M.; Kovalevskii, V.K.; Luk'yanov, O.Y.; Meleshkin, V.E.; Mikushev, M.K.; Panchenko, M.V.; Penner, I.E.; Pokrovskii, E.D.; et al. The "OPTIK – E,'" AN – 30 aircraft – laboratory for ecological investigations. *Atmos. Oceanic Optics* **1992**, *5*, 658–663.
49. Belan, B.D.; Zuev, V.E.; Panchenko, M.V. Main results of airborne sounding of aerosol conducted at the Institute of Atmospheric Optics from 1981 till 1991. *Atmos. Oceanic Optics* **1995**, *8*, 131–156.
50. Belan, B.D.; Ligotskii, A.V.; Luk'yanov, O.Y.; Mikushev, M.K.; Plokhikh, I.N.; Podanev, A.V.; Tolmachev, G.N. Database on the results of ecological survey of air basins. *Atmos. Oceanic Optics* **1994**, *7*, 585–590.
51. Belan, B.D. Airborne ecological sounding of the atmosphere. *Atmos. Oceanic Optics* **1993**, *6*, 205–222.
52. *Equipment for Remote Probing of Atmospheric Parameters*; V.E. Zuev, TP SB USSR AS, USSR: Tomsk, Russia, 1987; p. 156.
53. Nazarov, L.E. Isokinetic atmospheric aerosol sampling from an airplane. *Tr. Inst. Exp. Meteor.* **1985**, *9*, 76–81. (In Russian)
54. Griffiths, W.D.; DeCosemo, G.A.L. The assessment of bioaerosols: A critical review. *J. Aerosol Sci.* **1994**, *25*, 1425–1458. [CrossRef]
55. Miller, J.H. *Experiments in Molecular Genetics*; Cold Spring Harbor Laboratory Press: New York, NY, USA, 1972; p. 468.
56. Saggie, J. *The Methods of Soil Microbiology*; Kolos Publishers, USSR: Moscow, Russia, 1983; p. 295.
57. *Methods of General Bacteriology*, 2nd ed.; Gerhardt, F.; Murray, R.G.E.; Wood, W.A.; Krieg, N.R. (Eds.) Publisher American Society for Microbiology: Washington, DC, USA, 1994; p. 791.
58. *The Prokaryotes. A Handbook on Habitats, Isolation, and Identification of Bacteria*; Starr, M.P.; Stolp, H.; Truper, H.G.; Balows, A.; Schlegel, H.G. (Eds.) Springer: Berlin/Heidelberg, Germany, 1981; p. 2596.
59. Lebedeva, M.N. *A Guide for Practical Studies in Medical Microbiology*; Medicine, USSR: Moscow, Russia, 1973; p. 312.
60. Maniatis, T.; Fritsch, E.E.; Sambrook, J. *Molecular Cloning. A laboratory Manual*; Cold Spring Harbor Laboratory: New York, NY, USA, 1982; p. 545.
61. Weisburg, W.G.; Barns, S.M.; Pelletier, D.A.; Lane, D.J. 16S ribosomal DNA amplification for phylogenetic study. *J. Bacteriol.* **1991**, *173*, 697–703. [CrossRef]
62. *Statistical Methods in Microbiological Studies*; Ashmarin, I.P.; Vorobyov, A.A. (Eds.) Medgiz, USSR: Leningrad, Russia, 1962; p. 180.
63. Repin, V.E.; Lebedev, L.R.; Andreeva, I.S.; Puchkova, L.I.; Zernov, Y.P.; Serov, G.D.; Tereshchenko, T.A.; Afinogenova, G.N.; Pustoshilova, N.M. The producers of restriction endonucleases from natural microbe isolates and the development on this basis of enzymes production technologies. *Biotechnology* **1998**, *2*, 18–27. (In Russian)
64. Anokhin, G.G.; Antokhin, P.N.; Arshinov, M.Y.; Barsuk, V.E.; Belan, B.D.; Belan, S.B.; Davydov, D.K.; Ivlev, G.A.; Kozlov, A.V.; Kozlov, V.S.; et al. OPTIK Tu–134 Aircraft Laboratory. *Atmos. Oceanic Optics* **2011**, *24*, 805–816. [CrossRef]
65. Safatov, A.S.; Buryak, G.A.; Andreeva, I.S.; Olkin, S.E.; Reznikova, I.K.; Sergeev, A.N.; Belan, B.D.; Panchenko, M.V. Atmospheric bioaerosols. In *Aerosols – Science and Technology*; Agranovski, I., Ed.; Wiley – VCH Verlag GmbH & Co. KGaA: Wienheim, Germany, 2010; pp. 407–454.
66. Armstrong, J.L.; Shigeno, D.S.; Calomiris, J.J.; Seigler, R.J. Antibiotic-resistant bacteria in drinking water. *Appl. Environ. Microbiol.* **1981**, *42*, 277–283. [CrossRef] [PubMed]
67. Baquero, F.; Martinez, J.-L.; Cantón, R. Antibiotics and antibiotic resistance in water environment. *Curr. Opin. Biotechnol.* **2008**, *19*, 260–265. [CrossRef]

68. Boczek, L.A.; Rice, E.W.; Johnston, B.; Johnston, J.R. Occurrence of antibiotic-resistant uropathogenic *Escherichia coli* clonal group A in a wastewater effluent. *Appl. Environ. Microbiol.* **2007**, *73*, 4180–4184. [CrossRef]

69. Esiobu, N.; Armenta, L.; Ike, J. Antibiotic resistance in soil and water environments. *Int. J. Environ. Health Res.* **2002**, *12*, 133–144. [CrossRef]

70. Hermansson, M.; Jones, G.W.; Kjelleberg, S. Frequency of antibiotic and heavy metal resistance, pigmentation, and plasmids in bacteria of the marine air-water interface. *Appl. Environ. Microbiol.* **1987**, *53*, 2338–2342. [CrossRef]

71. Schwartz, T.; Kohnen, W.; Jansen, B.; Obst, U. Detection of antibiotic-resistant bacteria and their resistance genes in wastewater, surface water, and drinking water biofilms. *FEMS Microbiol. Ecol.* **2003**, *43*, 325–335. [CrossRef]

72. Sengeløv, G.; Agersø, Y.; Halling-Sørensen, B.; Baloda, S.B.; Andersen, J.S.; Jensen, L.B. Bacterial antibiotic resistance levels in Danish farmland as a result of treatment with pig manure slurry. *Environ. Int.* **2003**, *28*, 587–595. [CrossRef]

73. Sarda-Estève, R.; Baisnée, D.; Guinot, B.; Sodeau, J.; O'Connor, D.; Belmonte, J.; Besancenot, J.-P.; Petit, J.-E.; Thibaudon, M.; Oliver, G.; et al. Variability and Geographical Origin of Five Years Airborne Fungal Spore Concentrations Measured at Saclay, France from 2014 to 2018. *Remote Sens.* **2019**, *11*, 1671. [CrossRef]

74. Millington, W.M.; Corden, M. Long term trends in indoor *Aspergillus/Penicillum* spore in Derby, UK form 1970 to 2003 and comparative study in 1994 and 1996 with indoor air of two local houses. *Aerobiologia* **2005**, *21*, 105–113. [CrossRef]

75. Burrows, S.M.; Elbert, W.; Lawrence, M.G.; Pöschl, U. Bacteria in the global atmosphere – Part 1: Review and synthesis of literature data for different ecosystems. *Atmos. Chem. Phys.* **2009**, *9*, 9263–9280. [CrossRef]

76. Tong, Y.; Lighthart, B. The annual bacterial particle concentration and size distribution in the ambient atmosphere in a rural area of the Willamette Valley, Oregon. *Aerosol Sci. Technol.* **2000**, *32*, 393–403. [CrossRef]

77. Negrin, M.M.; Del Panno, M.T.; Ronco, A.E. Study of bioaerosols and site influence in the La Plata area (Argentina) using conventional and DNA (fingerprint) based methods. *Aerobiologia* **2007**, *23*, 249–258. [CrossRef]

78. Bovallius, Å.; Bucht, B.; Roffey, R.; Ånäs, P. Three-year investigation of the natural airborne bacterial flora at four localities in Sweden. *Appl. Environ. Microbiol.* **1978**, *35*, 847–852. [CrossRef]

79. Di Giorgio, C.; Krempff, A.; Guiraud, H.; Binder, P.; Tiret, C.; Dumenil, G. Atmospheric pollution by airborne microorganisms in the city of Marseilles. *Atmos. Environ.* **1996**, *30*, 155–160. [CrossRef]

80. Bowers, R.M.; Clements, N.; Emerson, J.B.; Wiedinmyer, C.; Hannigan, M.P.; Fierer, N. Seasonal Variability in Bacterial and Fungal Diversity of the Near-Surface Atmosphere. *Environ. Sci. Technol.* **2013**, *47*, 12097–12106. [CrossRef]

81. Li, M.; Qi, J.; Zhang, H.; Huang, S.; Li, L.; Gao, D. Concentration and size distribution of bioaerosols in an outdoor environment in the Qingdao coastal region. *Sci. Total Environ.* **2011**, *409*, 3812–3819. [CrossRef]

82. Bertolini, V.; Gandolfi, I.; Ambrosini, R.; Bestetti, G.; Innocente, E.; Rampazzo, G.; Franzetti, A. Temporal variability and effect of environmental variables on airborne bacterial communities in an urban area of Northern Italy. *Appl. Microbiol. Biotechnol.* **2013**, *97*, 6561–6570. [CrossRef]

83. Brągoszewska, E.; Mainka, A.; Pastuszka, J.S. Concentration and Size Distribution of Culturable Bacteria in Ambient Air during Spring and Winter in Gliwice: A Typical Urban Area. *Atmos.* **2017**, *8*, 239. [CrossRef]

84. Striluk, M.L.; Aho, K.; Weber, C.F. The effect of season and terrestrial biome on the abundance of bacteria with plant growth-promoting traits in the lower atmosphere. *Aerobiologia* **2017**, *33*, 137–149. [CrossRef]

85. Tanaka, D.; Terada, Y.; Nakashima, T.; Sakatoku, A.; Nakamura, S. Seasonal variations in airborne bacterial community structures at a suburban site of central Japan over a 1-year time period using PCR-DGGE method. *Aerobiologia* **2015**, *31*, 143–157. [CrossRef]

86. Ravva, S.V.; Hernlem, B.J.; Sarreal, C.Z.; Mandrell, R.E. Bacterial communities in urban aerosols collected with wetted-wall cyclonic samplers and seasonal fluctuations of live and culturable airborne bacteria. *J. Environ. Monit.* **2012**, *14*, 473–481. [CrossRef] [PubMed]

87. Agarwal, S.; Mandal, P.; Majumdar, D.; Aggarwal, S.G.; Srivastava, A. Characterization of Bioaerosols and their Relation with OC, EC and Carbonyl VOCs at a Busy Roadside Restaurants-Cluster in New Delhi. *Aerosol Air Qual. Res.* **2016**, *16*, 3198–3211. [CrossRef]

88. Rajput, P.; Anjum, M.H.; Gupta, T. One year record of bioaerosols and particles concentration in Indo-Gangetic Plain: Implications of biomass burning emissions to high-level of endotoxin exposure. *Environ. Pollut.* **2017**, *224*, 98–106. [CrossRef]

89. Agarwal, S. Seasonal variability in size-segregated airborne bacterial particles and their characterization at different source-sites. *Environ. Sci. Pollut. Res.* **2017**, *24*, 13519–13527. [CrossRef]

90. Park, J.; Ichijo, T.; Nasu, M.; Yamaguchi, N. Investigation of bacterial effects of Asian dust events through comparison with seasonal variability in outdoor airborne bacterial community. *Sci. Rep* **2016a**, *6*, 35706:1–35706:8. [CrossRef]

91. Kallawicha, K.; Lung, S.-C.C.; Chuang, Y.-C.; Wu, C.-D.; Chen, T.-H.; Tsai, Y.-J.; Chao, H.J. Spatiotemporal Distributions and Land-Use Regression Models of Ambient Bacteria and Endotoxins in the Greater Taipei Area. *Aerosol Air Qual. Res.* **2015**, *15*, 1448–1459. [CrossRef]

92. Barrett, T.C.; Mok, W.W.K.; Murawski, A.M.; Brynildsen, M.P. Enhanced antibiotic resistance development from fluoroquinolone persisters after a single exposure to antibiotic. *Nat. Commun.* **2019**, *10*, 117:1–117:11. [CrossRef]

93. Knapp, C.W.; Dolfing, J.; Ehlert, P.A.; Graham, D.W. Evidence of increasing antibiotic resistance gene abundances in archived soils since 1940. *Environ. Sci. Technol.* **2009**, *44*, 580–587. [CrossRef]

94. Tiedje, J.M.; Wang, F.; Manaia, C.M.; Virta, M.; Sheng, H.J.; Ma, L.P.; Zhang, T.; Topp, E. Antibiotic resistance genes in the human-impacted environment: A One Health perspective. *Pedosphere* **2019**, *29*, 273–282. [CrossRef]

95. Xu, R.; Yang, Z.-H.; Zheng, Y.; Wang, Q.-P.; Bai, Y.; Liu, J.-B.; Zhang, Y.-R.; Xiong, W.-P.; Lu, Y.; Fan, C.-Z. Metagenomic analysis reveals the effects of long-term antibiotic pressure on sludge anaerobic digestion and antimicrobial resistance risk. *Bioresource Technol.* **2019**, *282*, 179–188. [CrossRef] [PubMed]

96. Ullah, R.; Yasir, M.; Bibi, F.; Abujamel, T.S.; Hashem, A.M.; Sartaj Sohrab, S.; Al-Ansari, A.; Al-Sofyani, A.A.; Al-Ghamdi, A.K.; Al-sieni, A.; et al. Taxonomic diversity of antimicrobial-resistant bacteria and genes in the Red Sea coast. *Sci. Total Environ.* **2019**, *677*, 474–483. [CrossRef] [PubMed]

97. Houndt, T.; Ochman, H. Long-term shifts in patterns of antibiotic resistance in enteric bacteria. *Appl. Environ. Microbiol.* **2000**, *66*, 5406–5409. [CrossRef] [PubMed]

98. Moellering, R.C., Jr.; Graybill, J.R.; McGowan, J.E.; Corey, L. Antimicrobial resistance prevention initiative—An update: Proceedings of an Expert Panel on Resistance. *Am. J. Med.* **2007**, *120*, S4–S25. [CrossRef] [PubMed]

99. Zhang, M.; Zuo, J.; Yu, X.; Shi, X.; Chen, L.; Li, Z. Quantification of multi-antibiotic resistant opportunistic pathogenic bacteria in bioaerosols in and around a pharmaceutical wastewater treatment plant. *J. Environ. Sci.* **2018**, *72*, 53–63. [CrossRef]

100. Cole, M.L.; Singh, O.V. Microbial occurrence and antibiotic resistance in ready-to-go food items. *J. Food Sci. Technol.* **2018**, *55*, 2600–2609. [CrossRef]

101. Mishra, M.; Arukha, A.P.; Patel, A.K.; Behera, N.; Mohanta, T.K.; Yadav, D. Multi-Drug Resistant Coliform: Water Sanitary Standards and Health Hazards. *Frontiers Pharmacol.* **2018**, *9*, 311:1–311:8. [CrossRef]

102. Singh, S.K.; Ekka, R.; Mishra, M.; Mohapatra, H. Association study of multiple antibiotic resistance and virulence: A strategy to assess the extent of risk posed by bacterial population in aquatic environment. *Environ. Monit. Assess.* **2017**, *189*, 320:1–320:12. [CrossRef]

103. Safatov, A.S.; Andreeva, I.S.; Buryak, G.A.; Vechkanov, V.A.; Vorobyeva, I.G.; Ol'kin, S.E.; Reznikova, I.K.; Solovyanova, N.A.; Teplyakova, T.V.; Arshinov, M.Y.; et al. The results of long-term monitoring of atmospheric aerosols biogenic components at altitudes 500–7000 m in the South of Western Siberia. In Proceedings of the Abstract European Aerosol Conference 2016, Tours, France, 4–9 September 2016; p. O2-AAS-AAP-08.

104. Blais Lecours, P.; Duchaine, C.; Thibaudon, M.; Marsolais, D. Health Impacts of Bioaerosol Exposure. In *Microbiology of Aerosols*; Delort, A.-M., Amato, P., Eds.; Wiley-Blackwell Published: Hoboken, NJ, USA, 2018; Chapter 4.1; pp. 251–268.

International Journal of
***Environmental Research
and Public Health***

Article

How Do Economic Growth, Urbanization, and Industrialization Affect Fine Particulate Matter Concentrations? An Assessment in Liaoning Province, China

Tuo Shi [1,2]**, Yuanman Hu** [1]**, Miao Liu** [1,*]**, Chunlin Li** [1,*]**, Chuyi Zhang** [1,2] **and Chong Liu** [1,2]

[1]   CAS Key Laboratory of Forest Ecology and Management, Institute of Applied Ecology,
     Chinese Academy of Sciences, No. 72, Wenhua Road, Shenyang 110016, China; tuoshi0411@163.com (T.S.);
     huym@iae.ac.cn (Y.H.); trullyzhang@163.com (C.Z.); liuchong@iae.ac.cn (C.L.)
[2]   College of Resources and Environment, University of Chinese Academy of Sciences, No. 19, Yuquan Road,
     Beijing 100049, China
*   Correspondence: lium@iae.ac.cn (M.L.); lichunlin@iae.ac.cn (C.L.)

Received: 2 July 2020; Accepted: 24 July 2020; Published: 28 July 2020

Abstract: With China's rapid development, urban air pollution problems occur frequently. As one of the principal components of haze, fine particulate matter ($PM_{2.5}$) has potential negative health effects, causing widespread concern. However, the causal interactions and dynamic relationships between socioeconomic factors and ambient air pollution are still unclear, especially in specific regions. As an important industrial base in Northeast China, Liaoning Province is a representative mode of social and economic development. Panel data including $PM_{2.5}$ concentration and three socio-economic indicators of Liaoning Province from 2000 to 2015 were built. The data were first-difference stationary and the variables were cointegrated. The Granger causality test was used as the main method to test the causality. In the results, in terms of the causal interactions, economic activities, industrialization and urbanization processes all showed positive long-term impacts on changes of $PM_{2.5}$ concentration. Economic growth and industrialization also significantly affected the variations in $PM_{2.5}$ concentration in the short term. In terms of the contributions, industrialization contributed the most to the variations of $PM_{2.5}$ concentration in the sixteen years, followed by economic growth. Though Liaoning Province, an industry-oriented region, has shown characteristics of economic and industrial transformation, policy makers still need to explore more targeted policies to address the regional air pollution issue.

Keywords: air pollution; fine particulate matter; economic growth; urbanization; industrialization; Granger causality test

---

1. Introduction

Since implementing the reform and opening-up policy in 1978, China has been experiencing a rapid process of social and economic development, attracting worldwide attention [1,2]. With the development of China's urbanization process, the population influx into cities, the consumption of resources and the transformation of the economic structure have caused a variety of social and environmental impacts [3,4]. Among them, air pollution is particularly prominent because it is closely associated with negative health effects [5,6].

In recent years, one of the primary pollutants most affecting China has been fine particulate matter ($PM_{2.5}$). $PM_{2.5}$ refers to small particles or droplets in the air less than 2.5 microns in aerodynamic diameter [7,8]. $PM_{2.5}$ easily binds to toxic and harmful substances due to its small size, long atmospheric residence time and extensive atmospheric transportation and seriously affects human health [9,10]. $PM_{2.5}$ exposure in 2015 was estimated to result in 8.9 million deaths globally, among which 28%

occurred in China [11]. To cope with severe and persistent $PM_{2.5}$ pollution and to meet pollutant concentration targets [12,13], it is urgent and necessary to explore the influence of human factors on $PM_{2.5}$ [14–16]. Hao and Liu [17] used a spatial lag model and spatial error model to investigate the socioeconomic influencing factors of urban $PM_{2.5}$ concentration in China. The results showed that the number of vehicles and the secondary industry had significantly positive effects on $PM_{2.5}$ concentration in cities. Wang et al. [18] found a positive correlation between $PM_{2.5}$ concentrations and urban area, and population and proportion of secondary industry, and determined the existence of an inverted U-shaped relationship between economic growth and $PM_{2.5}$ concentration. Existing studies have confirmed the contributions of socioeconomic factors to $PM_{2.5}$ pollution in China [13,19,20]; however, the dynamic relationships and causal interactions between them are still not well understood, especially in specific regions. The Granger causality test determines the causal relationships between variables based on the chronological order in which the events occurred [21]. The method has been widely used in the empirical analysis of the relationships between energy, environment, economic and social development, etc. [22–24]. As an important administrative unit of a country, "province" usually provides unified and periodic suggestions to the cities under its jurisdiction, but relevant studies at this scale were few. Understanding the principal environmental issues in each stage of development holds great significance for the formulation and implementation of pollution policy, and also for the improvement of public health in China with $PM_{2.5}$ as the primary pollutant.

In this paper, the panel data from 2000 to 2015 in Liaoning Province that combine a satellite derived $PM_{2.5}$ concentration data set and socioeconomic data were established. The panel Granger causality test was used as the main method to quantitatively test the causality among economic growth, urbanization, industrialization and $PM_{2.5}$ concentration. This study provides an idea for the formulation of regional periodic pollution control objectives which is significant to regional pollution control.

2. Materials and Methods

2.1. Study Area

Liaoning Province is located in Northeast China, covering an area of 148,000 km^2, including 14 prefecture-level cities (Figure 1). The population is 43.82 million, including 29.52 million urban residents. Liaoning Province is a region in Northeast China where cities characterized by heavy industry are concentrated.

In Liaoning Province, the secondary industry accounted for 48.12% of the total GDP in 2015, with the province ranking 5th among the 31 provinces in China. In April 2015, TomTom, the Dutch traffic navigation service provider, released a global traffic congestion ranking, and Shenyang, the capital of Liaoning Province, ranked 29th [25]. According to data from the China National Environmental Monitoring Centre, 11 out of 14 cities in Liaoning Province experienced severe air pollution in November 2015. Therefore, there is an urgent need to study the relationships and interactions between socioeconomic factors and ambient pollution in Liaoning Province

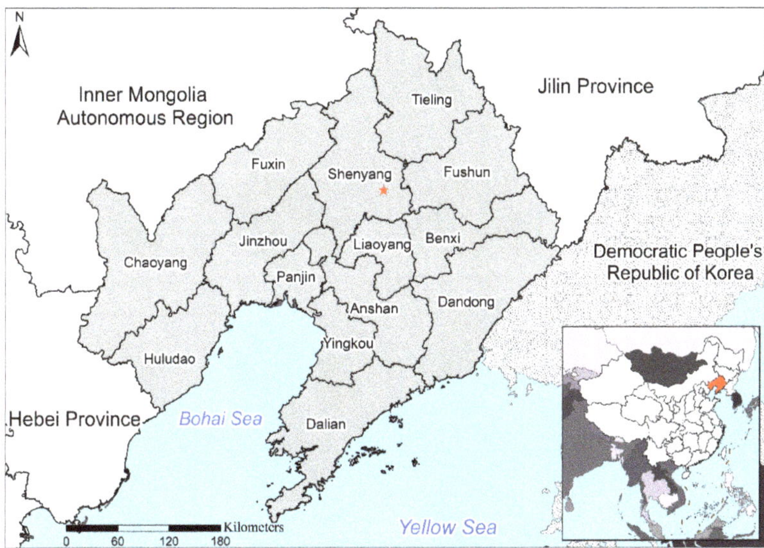

Figure 1. Location and cities in the study area.

2.2. Data

The annual PM$_{2.5}$ concentrations from 2000 to 2015 in the panel data were extracted from the global PM$_{2.5}$ concentration with a spatial resolution of 0.01° (http://fizz.phys.dal.ca/~{}atmos/martin/?page_id=140#V4.CH.02) [13,26–28]. The global PM$_{2.5}$ concentration data set was implemented by the atmospheric chemistry driven model GEOS-Chem. The algorithm in the model combines the aerosol optical depth obtained from multi-sensor products with the data from surface monitoring stations [13,29,30]. The correlation coefficient of the estimated and regulatory monitored PM$_{2.5}$ concentration was 0.81 [28]. To avoid uncertainty in the subsequent analysis caused by abnormal or missing values in the data, the three-year average was used as an annual average. The average PM$_{2.5}$ concentrations from 2000 to 2015 in 14 prefecture-level cities were extracted and calculated by city boundaries (Figure 2).

Referring to relevant studies, we selected GDP per capita (GDPPC), the proportion of urban impervious surface area (UIS) and the value added by industry as a percentage of GDP (IND) to represent the economic growth, urbanization and industrialization of each city, respectively [22]. The panel data on the economic growth and industrialization of the fourteen prefecture-level cities in Liaoning Province from 2000 to 2015 were collected from the China City Statistical Yearbook. Because China has cancelled the agricultural and non-agricultural household registration system since 2014, to avoid abnormal fluctuation of time series data, the proportion of urban artificial impervious surface area rather than the traditional proportion of urban population was used to express the urbanization level of each city [31]. The spatial resolutions of 30 m urban artificial impervious area data were obtained from Fine Resolution Observation and Monitoring of Global Land Cover (FROM-GLC, http://data.ess.tsinghua.edu.cn/urbanChina.html) [31,32]. The GDPPC data were converted to constant prices, and all data were logarithmically transformed to stabilize the time series data and reduce the heteroscedasticity when performing empirical tests (*ln*PM$_{2.5}$, *ln*GDPPC, *ln*UIS and *ln*IND).

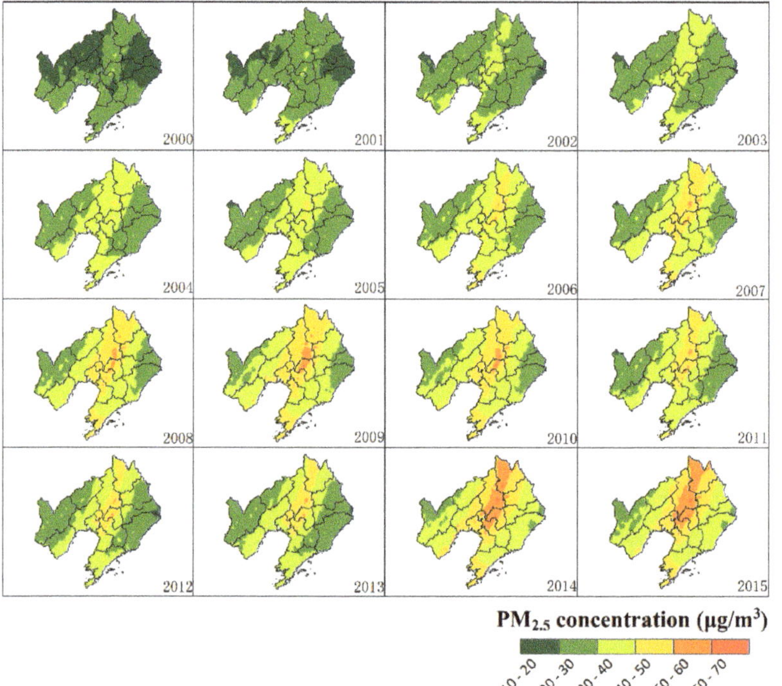

Figure 2. Spatial distribution of surface PM$_{2.5}$ concentrations in Liaoning province from 2000 to 2015.

2.3. Methodology

The procedure for estimating the causal relationships between PM$_{2.5}$ and the above socioeconomic factors using the panel data from 2000 to 2015 included five steps: the unit root test, panel cointegration test, panel fully modified least squares (FMOLS) regression, Granger causality test, variance decomposition and impulse response. The details are as follows:

A unit root test checks whether the unit root exists and if a time series variable is non-stationary [33]. If there is a unit root in the time series variable, it will lead to a pseudo-regression in subsequent regression analysis [34]. The null hypothesis is defined as the existence of a unit root, and the variables are non-stationary. In this study, the methods of Levin, Lin and Chu (LLC) and Im, Pesaran and Shin (IPS) were used for testing.

A panel cointegration test is used to test whether there is a long-term stable equilibrium relationship between variables. In this study, the Pedroni method was used to test the cointegration relationship between the socioeconomic variables and PM$_{2.5}$ concentrations [16].

The panel FMOLS regression designed by Phillips [35] is utilized to provide the optimal estimations of cointegrating regressions [36]. This method modifies least squares to account for the autocorrelation effects and the endogeneity in the regressors due to the existence of a cointegration relationship [35,37]. In this study, the panel FMOLS regression was used to explore the trends and directions of *ln*GDPPC, *ln*UIS and *ln*IND in *ln*PM$_{2.5}$ in the long term. The relationship between variables was expressed by the following equation, Equation (1):

$$ln\mathrm{PM}_{2.5it} = \alpha + \beta_1 ln\mathrm{GDPPC}_{it} + \beta_2 ln\mathrm{UIS}_{it} + \beta_3 ln\mathrm{IND}_{it} + \varepsilon_{it} \tag{1}$$

where i and t represent the city and the time indexes in the panel, as shown by subscripts i ($i = 1, \dots , 14$) and t ($t = 1, \dots , 16$), respectively. α is the intercept; βs are partial coefficients of lnGDPPC, lnUIS and lnIND; and εs refer to errors.

The panel vector error correction model (VECM) was used to investigate the direction and Granger causal relationships between the variables in the panel in the short or long term. In this study, short-term causality represented weak Granger causality because the dependent variable only responds to the short-term shocks of the stochastic environment (a stochastic environment refers to the agent's actions and does not uniquely determine the outcome), whereas long-term causality referred to the independent variable's response to the deviation from long-term equilibrium [22,38]. Generally, short-term causality affected 1–2 periods, while long-term causality represented the casual relationship of the whole period from 2000 to 2015 [22]. The short-term Granger causality depended on the χ^2-Wald statistics of the coefficient significances of the lagged terms of the explanatory variables [38]. The long-term Granger causality was determined by the error correction term (ECT) significance. If the variables are cointegrated, then the coefficients of the ECTs are expected to be at least one or all negative and significantly different from zero [22].

Variance decomposition explains the amount of information each endogenous variable contributes to the other variables in the autoregressions. The impulse response function indicates the effects of a shock to one innovation on current and future values of the endogenous variables [38,39]. The Cholesky decomposition technique was used in the VECM to determine the contribution of one variable on another and estimate how each variable responds to the changes in the other variables [22].

The above methods were realized in the software EViews 8.0 (IHS Global Inc., Englewood, CA, USA), and relevant statistical principles were followed according to the user guide [40,41].

3. Results

3.1. Data Description

The $PM_{2.5}$ concentrations data used in the study were extracted from the global data set provided by Van Donkelaar, Martin, Brauer and Boys [28]. In his study, sample points outside North America and Europe had precision with a correlation coefficient of 0.81 and a slope of 0.68. However, given the regional differences, the precision of the data involved in the study in Liaoning Province was yet to be verified.

Only in 2013 did the monitoring of particulate matter begin in various cities of China. Among them, cities in Liaoning Province started to have stable and continuous monitoring data from May 2014. Therefore, we selected the 76 regulatory stations that monitored $PM_{2.5}$ values in 2015 for verification, and the correlation coefficient was 0.7 (Figure 3). Additionally, Peng, Chen, Lü, Liu and Wu [29] compared 45 sample points values from published studies and the corresponding remote-sensing values in China, with 78.7% correlation. Therefore, it is reasonable to believe that the data can reflect the variation of $PM_{2.5}$ concentrations in the region and can be used for the following analysis.

The $PM_{2.5}$ concentration, GDPPC, UIS and IND of fourteen cities in Liaoning Province from 2000 to 2015 were selected; the descriptive statistics are summarized in Table 1.

Table 1. Description of the panel data from 2000 to 2015.

Variable	Obs.	Mean	Std. Dev	Min	Max
$PM_{2.5}$ (µg/m^3)	224	36.60	8.24	18.81	54.57
GDPPC (Yuan, RMB)	224	37,142.14	23,905.06	6184.72	121,457.46
UIS (%)	224	13.32	7.65	2.06	33.95
IND (%)	224	55.13	9.51	37.09	83.60

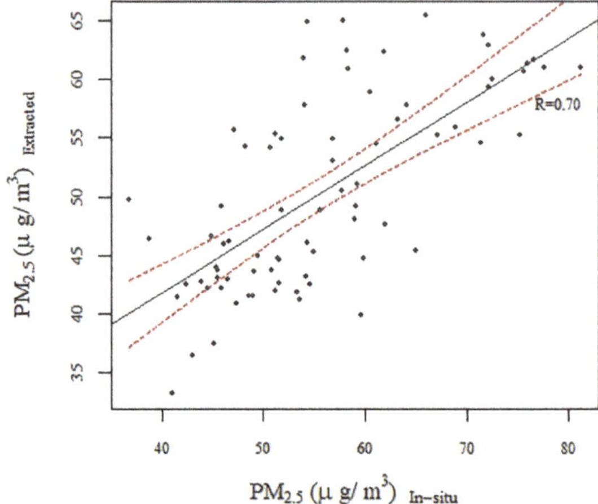

Figure 3. Scatter plot of regulatory stations that monitored PM$_{2.5}$ concentrations and remote-sensed PM$_{2.5}$ concentrations. Dashed red lines represent a 95% confidence interval of the fitting line.

Since 2000, PM$_{2.5}$ concentration has been on the rise in fourteen cities in Liaoning Province, except for a temporary decrease from 2009 to 2012, and after 2014, the concentration also weakened (Figure 4). Increasing trends also occurred in the GDPPC and UIS, but after 2013, the economic growth of most cities slowed down or even declined. The changes of UIS in fourteen cities were basically stable, and most cities showed faster increasing trends after 2009. Regarding IND, the proportions in all cities decreased after 2012, indicating a characteristic of industrial transformation, or that the contribution of industrialization to economic growth has declined.

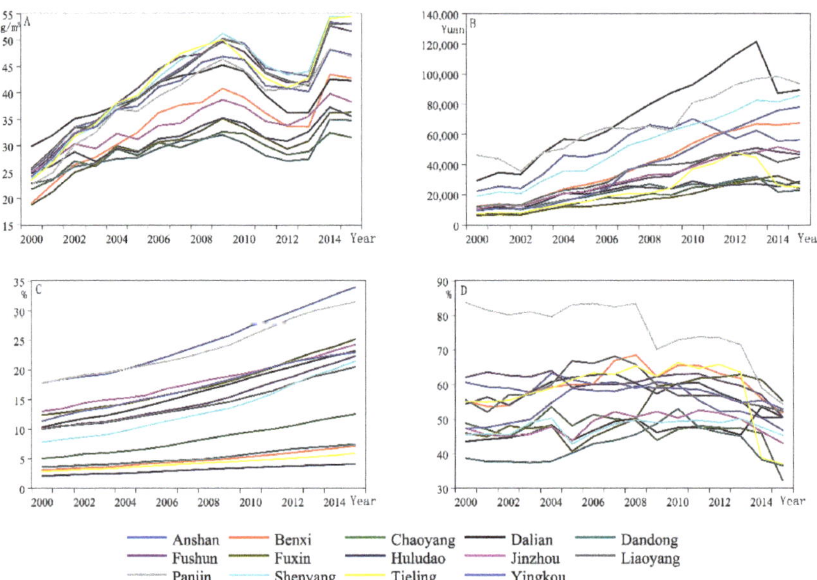

Figure 4. Data of PM$_{2.5}$ concentrations (**A**), GDP per capita (GDPPC) (**B**), proportion of urban impervious surface area (UIS) (**C**) and the value added by industry as a percentage of GDP (IND) (**D**) of fourteen cities in the panel that changed over the time series from 2000 to 2015.3.2. Panel Unit Root Test Results.

3.2. Panel Unit Root Test Results

The results (Table 2) showed that not all the variables in the panel were stationary at the levels; however, the four variables were basically stationary at the first difference. Therefore, we can reject the null hypothesis and assume the panel variables were stationary at the first difference.

Table 2. Panel unit root test results.

Variable	Level		1st Difference	
	Intercept	Intercept and Trend	Intercept	Intercept and Trend
	Levin, Lin and Chu (LLC)			
lnPM$_{2.5}$	−7.4320 ***	−2.6757 ***	−6.6609 ***	1.8893
lnUIS	−0.3350	−3.3226 ***	−6.8751 ***	−7.2374 ***
lnGDPPC	−13.618 ***	1.2149	−5.9671 ***	−17.066 ***
lnIND	1.4858	2.3637	−8.8109 ***	−6.2508 ***
	Im, Pesaran and Shin (IPS)			
lnPM$_{2.5}$	−3.9769 ***	−0.5582	−5.9219 ***	−5.0734 ***
lnUIS	5.0142	−1.0338	−5.2980 ***	−4.2049 ***
lnGDPPC	−5.7427 ***	4.40875	−4.7635 ***	−11.892 ***
lnIND	2.1611	4.9270	−6.2508 ***	−6.3221 ***

Note: Significance: * 0.1, ** 0.05, *** 0.01.

3.3. Panel Cointegration Test Results

The results (Table 3) showed that six statistics could significantly reject the null hypothesis that there was no cointegration relationship; that is, a long-term stable cointegration relationship between PM$_{2.5}$ concentration and explanatory variables existed in our panel data.

Table 3. Panel cointegration test results using the Pedroni methods.

		Alternative Hypothesis: Common AR Coefs. (Within-Dimension)			
		Statistic	Prob.	Weighted Statistic	Prob.
	Panel v-Statistic	1.2492	0.1058	1.0894	0.1380
	Panel rho-Statistic	0.0337	0.5134	−0.0132	0.4947
	Panel pp-Statistic	−1.9136 **	0.0278	−1.8940 **	0.0291
Pedroni	Panel ADF-Statistic	−2.1804 **	0.0146	−2.5478 ***	0.0054
		Alternative Hypothesis: Individual AR Coefs. (Between-Dimension)			
		Statistic		Prob.	
	Group rho-Statistic	1.6771		0.9532	
	Group pp-Statistic	−1.7092 **		0.0437	
	Group ADF-Statistic	−3.0995 ***		0.0010	

Note: Significance: * 0.1, ** 0.05, *** 0.01.

3.4. Panel Fully Modified Least Squares (FMOLS) Regression Results

The results are shown in Table 4, indicating that economic growth, urbanization and industrialization all had long-term positive effects on changes in PM$_{2.5}$ concentrations in the sixteen years.

Table 4. Panel fully modified least squares regression results.

Variable	Coefficient	Std. Error	t-Statistic
*ln*GDPPC	0.2620 ***	0.0025	104.2593
*ln*IND	0.2236 ***	0.0021	107.5758
*ln*UIS	0.0094 ***	0.0009	9.8713

$R^2 = 0.492128$, Adj. $R^2 = 0.487221$; Significance: * 0.1, ** 0.05, *** 0.01.

3.5. Panel Granger Causality Test Results

Table 5 showed that all the coefficients of ECT (-1) of variables were significant; that is, bidirectional and long-term causal relationships existed between both variables in the panel. According to the χ^2-Wald statistics, bidirectional short-term causal relationships between $PM_{2.5}$ concentrations and GDPPC were found in the structure. In addition, one-way short-term causalities were found from IND to $PM_{2.5}$ concentrations and UIS, from GDPPC to IND and UIS and from $PM_{2.5}$ concentrations to UIS. A more visual and clearer figure is shown (Figure 5) based on the above results.

Table 5. Panel Granger causality test results.

Dependent Variable	Independent Variables					
	Short-Run Causality (χ2-Wald Statistics)				Long-Run Causality	
	$\Delta ln PM_{2.5}$	ΔlnGDPPC	ΔlnUIS	ΔlnIND	ECT (-1)	*t-statistics*
$\Delta ln PM_{2.5}$		6.2655 **	1.7088	5.2909 *	−0.0665 ***	−5.6409
ΔlnGDPPC	12.0662 ***		0.9156	2.3951	−0.0704 ***	−3.2335
ΔlnUIS	6.1390 **	14.3349 ***		9.3067 ***	−0.0272 **	−2.3615
ΔlnIND	2.6420	14.4685 ***	1.0221		0.0072 ***	2.8231

Significance: * 0.1, ** 0.05, *** 0.01.

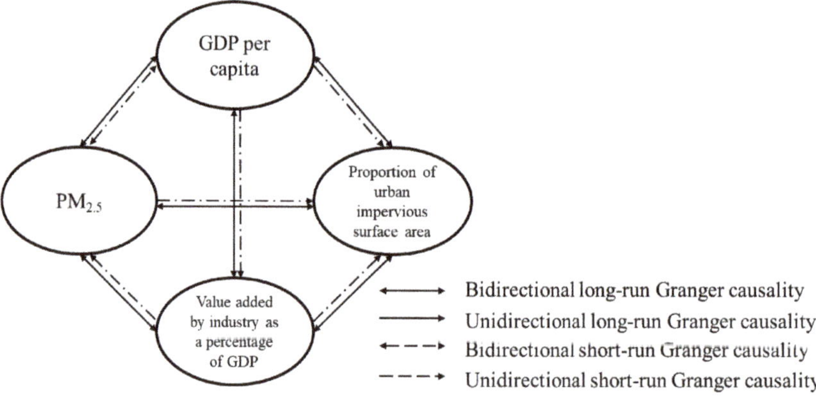

Figure 5. Diagram of the causal relationships between $PM_{2.5}$ concentrations, GDP per capita (GDPPC), the proportion of urban impervious surface area (UIS) and the value added by industry as a percentage of GDP (IND).

In the panel, all socioeconomic variables caused the variations of $PM_{2.5}$ concentrations in Liaoning Province, especially economic growth, which not only influenced changes in pollutant concentrations in the long and short term but also affected the changes in industrialization and urbanization in the long and short term. Additionally, industrialization directly caused changes in pollutant concentrations in the long and short run and caused variations in urbanization in the short and long term (Figure 5).

3.6. Variance Decomposition and Impulse Response Analysis Results

The results of the variance decomposition analysis in Table 6 compared the contribution of each variable to the changes in $PM_{2.5}$ concentration. In the panel, the variances of $PM_{2.5}$ concentration were mostly explained by its own standard shock (80.95%) in the 16-year period, while the contributions from the GDPPC, IND and UIS to the $PM_{2.5}$ concentration were 9.20%, 9.56% and 0.29%, respectively.

Table 6. Variance decomposition analysis results of pm$_{2.5}$ concentrations in the panel.

Period	S.E.	*ln*PM$_{2.5}$	*ln*GDPPC	*ln*IND	*ln*UIS
		Variance Decomposition of *ln*PM$_{2.5}$:			
1	0.068920	100.0000	0.000000	0.000000	0.000000
2	0.097998	98.17410	1.265334	0.488164	0.072398
3	0.114122	97.05163	2.419547	0.366743	0.162078
4	0.123131	95.67872	3.687611	0.410396	0.223269
5	0.128520	94.08125	4.921052	0.746677	0.251024
6	0.132068	92.35911	6.035212	1.349410	0.256273
7	0.134648	90.63484	6.970944	2.143034	0.251183
8	0.136672	89.00029	7.711078	3.044643	0.243994
9	0.138337	87.50652	8.266401	3.988203	0.238875
10	0.139744	86.17206	8.662361	4.928241	0.237340
11	0.140948	84.99492	8.928941	5.836583	0.239554
12	0.141987	83.96262	9.095029	6.697198	0.245149
13	0.142889	83.05865	9.185900	7.501821	0.253625
14	0.143674	82.26610	9.222504	8.246885	0.264513
15	0.144360	81.56939	9.221616	8.931577	0.277415
16	0.144962	80.95495	9.196351	9.556689	0.292006

The impulse responses result presented in Figure 6 showed that the responses of the $PM_{2.5}$ concentration to itself decreased because of shocks from decreasing UIS and IND in the first two years. Then, from the fifth year, the response of the $PM_{2.5}$ concentration continued to decrease because of shocks from decreasing GDPPC and decreasing IND in the latest seven years.

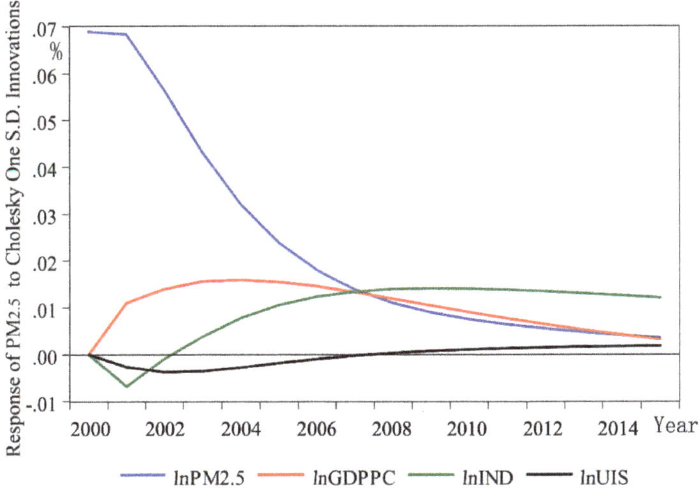

Figure 6. Results of the impulse response of *ln*PM$_{2.5}$ to Cholesky one S.D. innovations of the variables.

4. Discussion

4.1. The Analysis of Relationships between PM$_{2.5}$ and Socio-Economic Development in Liaoning Province

Studies have shown that changes in fine particulate pollution concentrations in China are influenced by natural factors and human activities [42,43]. Therefore, to explore the impacts of urban socioeconomic factors on PM$_{2.5}$ concentrations, we selected three indicators: urban GDP per capita, the proportion of urban impervious surface area and the value added by industry as a percentage of GDP, representing economic growth, urbanization and industrialization, respectively, which were assumed to be the most significant socioeconomic factors affecting PM$_{2.5}$ concentrations in China. In our results, all selected socioeconomic variables were long-term causalities of the changes of PM$_{2.5}$ concentrations, and economic growth and industrialization also significantly affected the variations in PM$_{2.5}$ concentrations in the short term. The variance decomposition results showed that industrialization was the determinate factor affecting PM$_{2.5}$ concentration variations in Liaoning Province, which was basically the same with the results found by Li, Fang, Wang and Sun [22], but only five cities in Liaoning Province were included in their industry-oriented panel, and the study period and indicators were different from ours. This further confirmed the attribute of Liaoning Province as a socio-economic mode of industry-oriented development.

Liaoning Province is an area in Northeast China where cities characterized by heavy industry are concentrated. Equipment manufacturing, the coal industry, the metallurgy industry and commodity production are the strengths of Liaoning Province [44]. For a long time period, heavy industry had been the main driving force of economic growth of most cities in Liaoning Province, promoting the rapid urbanization process. The concentrating population and developing economies would also motivate the urban industrial activities [45]. However, with the popularization and development of technology, the pressure of market competition increases. As a result, the supply of products in Liaoning Province far exceeds the market demand, and the problem of overcapacity is becoming increasingly serious [44]. Following the third scientific and technological revolution, the new science and technology industry, represented by electronics, computers, biological engineering, etc., seriously impacted traditional industries, resulting in a decline in the proportion of primary and secondary industries and leading to the rise of emerging industries such as the internet industry. However, in Liaoning Province, the tertiary industry only accounted for 38.7% of GDP in 2013, 5.8% less than the national average [46]. In 2015, the tertiary industry as a percentage of GDP rose to 46.06%, with major growth, basically equal to the national average. The slowdown in economic growth (Figure 4B) and the increase in the proportion of the tertiary industry indicated that adhering to the transformation of economic structure and industrial structure is a policy with both opportunities and difficulties. However, in recent years, the PM$_{2.5}$ concentration has declined (Figure 4A), further proving the validity of industrial structure transformation.

4.2. The Analysis of Environmental Kuznets Curve (EKC)

Although industrialization contributed the most to the PM$_{2.5}$ concentration changes in the sixteen years in Liaoning Province, the contribution of economic growth dominated a longer period (Table 6). Moreover, some relationships between the economic growth and PM$_{2.5}$ concentration changes were also noteworthy, such as the feedback effects in the Granger causality test (Figure 5) and fluctuations in the impulse response of shocks (Figure 6). Therefore, we constructed a regression model based on the Environmental Kuznets Curve (EKC) theory to study the relationship between economic growth and PM$_{2.5}$ pollution. Grossman and Krueger [47] found that an inverted U-shaped relationship existed between economic growth and environmental pollution [48]. With a low level of economic development in a country or region, the degree of environmental pollution is relatively low, and with an improved economic level, the degree of environmental pollution intensifies. However, when economic development reaches a certain level, that is to say, reaches an "inflection point", environmental quality gradually improves thenceforth with the increase in income.

Our result of the EKC regression between GDPPC and $PM_{2.5}$ is shown in Figure 7. According to the model equation, when the GDPPC was equal to CNY 74.8 thousand, the pollution reached the inflection point, and a decreasing trend appeared. Referring to the panel data, we found that the data of GDPPC higher than the turning point mainly appeared in the later periods of the time series, and the value added by industry as a percentage of GDP declined. The EKC result further proved that economic growth did not always increase $PM_{2.5}$ concentrations in Liaoning Province, suggesting that changing economic growth mode was a correct choice for pollution control.

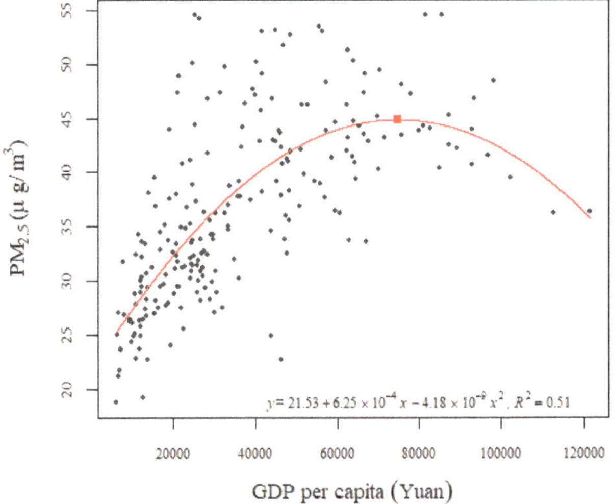

Figure 7. Scatter plot and Environmental Kuznets Curve (EKC) fitting line between $PM_{2.5}$ concentrations and GDP per capita.

4.3. Implications for Regional Air Pollution Management

Through the study on the relationships between socioeconomic factors and $PM_{2.5}$ concentration changes in Liaoning Province from 2000 to 2015, we found that the industrialization and economic growth were the main causes affecting the $PM_{2.5}$ concentration changes from the perspective of short-term impacts and long-term contributions. As the traditional pillar industry of economic growth in Liaoning Province, the contributions of the secondary industry to regional pollution is predictable. According to the above data and results, we also found that the dependence of economic growth on the secondary industry in Liaoning Province was weakened, and the EKC curve also showed that economic growth did not always lead to the increase in $PM_{2.5}$ concentrations. In 2014, the number of days of severe pollution (150–250 μg/m^3) in Shenyang reached 22 days; in 2018, the number of days of severe pollution was only 2 days. Although there is still a big gap between China's pollution level and the world standard, the improvement of atmospheric environment is obvious. This informs us that the transformation of economic structure is effective for the management of atmospheric pollution. However, improving energy efficiency and developing and utilizing clean energy is the key direction of taking into account both economic growth and environmental protection [49,50].

Among the socioeconomic variables, the urbanization process only showed the long-term impact on $PM_{2.5}$ concentration changes, and the contribution was weak. In other words, the urban expansion and population growth had little direct effects on the changes of $PM_{2.5}$ concentration, but indirectly affected the changes through economic growth and the industrialization process [19]. The causality diagram (Figure 5) showed that $PM_{2.5}$ changes, industrialization and economic growth also affected the urbanization process in both the short and long term. As the level of urbanization in each period is closely related to the pollution exposure [51], the goal of "new-type urbanization" is not only to emphasize the rapid urbanization, but also to meet the health needs of residents [52]. Therefore, the study on relationships between regional environment and socioeconomic factors is necessary for the phased management of regional pollution, and more variables may be added according to the data availability and research objectives.

4.4. Limitations

The study results have explained the impacts of socioeconomic development on $PM_{2.5}$ concentrations and the causal relationships among them to a large extent in Liaoning Province; however, there are still some limitations. For example, the surface $PM_{2.5}$ data used in this study are the longest time series pollutant data available at present, but there is also a possibility that the lower spatial resolution of the data has affected the accuracy of the assessment results. If better data could be obtained (i.e., higher spatial resolution and longer time series), it would be beneficial to further explore the causes of regional and internal pollution differences in the future. On the other hand, complex coupling relationships among economic growth, urbanization, industrialization and $PM_{2.5}$ concentrations were observed in this study. Determining how to decouple these relationships to further develop targeted solutions that tackle the pollution issue remains a challenging and urgent task. In the future, it is also necessary and meaningful to study and compare the relationships between policies and environment in other regions such as agriculture- or service-oriented areas and comprehensive areas.

5. Conclusions

In the panel data used in this study, the variables were all cointegrated. The Granger causality test results showed that economic growth, industrialization and urbanization were all long-term causalities of the changes of $PM_{2.5}$ concentrations, and economic growth and industrialization also significantly affected changes in $PM_{2.5}$ concentrations in the short term. The results of variance decomposition and the impulse response analysis showed that industrialization was the most important variable affecting $PM_{2.5}$ concentrations. However, controlling only one socioeconomic factor to slow pollution growth is not feasible because there are either long-term or short-term and either bidirectional or unidirectional relationships among them. Though Liaoning Province has shown characteristics of economic and industrial transformation, it is also necessary to formulate more targeted policies to solve the problem of regional air pollution.

Author Contributions: Conceptualization, Y.H. and M.L.; formal analysis, T.S.; funding acquisition, Y.H; methodology, T.S., C.Z. and C.L. (Chunlin Li); project administration, Y.H.; validation, M.L. and C.L. (Chong Liu); writing—original draft, T.S.; writing—review and editing, M.L. and C.L. (Chunlin Li). All authors have read and agreed to the published version of the manuscript.

Funding: This study was financially supported by the National Natural Science Foundation of China (No. 41730647, 41671184 and 41671185).

Acknowledgments: The authors would like to thank Aaron van Donkelaar and Randall Martin of the Atmospheric Composition Analysis Group of Dalhousie University in Canada who offered the data used in this research.

Conflicts of Interest: The authors declare no conflict of interest.

References

1. Zhang, K.H.; Song, S.F. Rural-urban migration and urbanization in China: Evidence from time-series and cross-section analyses. *China Econ. Rev.* **2003**, *14*, 386–400. [CrossRef]
2. Liu, M.; Hu, Y.M.; Li, C.L. Landscape metrics for three-dimensional urban building pattern recognition. *Appl. Geogr.* **2017**, *87*, 66–72. [CrossRef]
3. Jiang, Z.; Lin, B. China's energy demand and its characteristics in the industrialization and urbanization process. *Energy Policy* **2012**, *49*, 608–615. [CrossRef]

4. Cheng, Z.; Li, L.; Liu, J. Identifying the spatial effects and driving factors of urban $PM_{2.5}$ pollution in China. *Ecol. Indic.* **2017**, *82*, 61–75. [CrossRef]

5. Lave, L.B.; Seskin, E.P. *Air Pollution and Human Health*; RFF Press: New York, NY, USA; Milton Park, Abingdon: Oxon, UK, 2013. [CrossRef]

6. Kampa, M.; Castanas, E. Human health effects of air pollution. *Environ. Pollut.* **2008**, *151*, 362–367. [CrossRef]

7. Davidson, C.I.; Phalen, R.F.; Solomon, P.A. Airborne particulate matter and human health: A review. *Aerosol Sci. Technol.* **2005**, *39*, 737–749. [CrossRef]

8. Buser, M.; Parnell, C.; Shaw, B.; Lacey, R. Particulate matter sampler errors due to the interaction of particle size and sampler performance characteristics: Ambient $PM_{2.5}$ samplers. *Trans. ASABE* **2007**, *50*, 241–254. [CrossRef]

9. Kelly, F.J.; Fussell, J.C. Size, source and chemical composition as determinants of toxicity attributable to ambient particulate matter. *Atmos. Environ.* **2012**, *60*, 504–526. [CrossRef]

10. Lu, F.; Xu, D.; Cheng, Y.; Dong, S.; Guo, C.; Jiang, X.; Zheng, X. Systematic review and meta-analysis of the adverse health effects of ambient $PM_{2.5}$ and PM_{10} pollution in the Chinese population. *Environ. Res.* **2015**, *136*, 196–204. [CrossRef]

11. Burnett, R.; Chen, H.; Szyszkowicz, M.; Fann, N.; Hubbell, B.; Pope, C.A.; Apte, J.S.; Brauer, M.; Cohen, A.; Weichenthal, S.; et al. Global estimates of mortality associated with long-term exposure to outdoor fine particulate matter. *Proc. Natl. Acad. Sci. USA* **2018**, *115*, 9592–9597. [CrossRef]

12. Sheehan, P.; Cheng, E.; English, A.; Sun, F. China's response to the air pollution shock. *Nat. Clim. Chang.* **2014**, *4*, 306. [CrossRef]

13. Shi, T.; Liu, M.; Hu, Y.M.; Li, C.L.; Zhang, C.Y.; Ren, B.H. Spatiotemporal Pattern of Fine Particulate Matter and Impact of Urban Socioeconomic Factors in China. *Int. J. Environ. Res. Public Health* **2019**, *16*, 1099. [CrossRef] [PubMed]

14. Zhang, L.; LeGates, R.; Zhao, M. *Understanding China's Urbanization: The Great Demographic, Spatial, Economic, and Social Transformation*; Edward Elgar Publishing: Cheltenham, UK, 2016.

15. Han, L.J.; Zhou, W.Q.; Li, W.F.; Li, L. Impact of urbanization level on urban air quality: A case of fine particles ($PM_{2.5}$) in Chinese cities. *Environ. Pollut.* **2014**, *194*, 163–170. [CrossRef] [PubMed]

16. Zhao, X.; Zhou, W.; Han, L.; Locke, D. Spatiotemporal variation in $PM_{2.5}$ concentrations and their relationship with socioeconomic factors in China's major cities. *Environ. Int.* **2019**, *133*, 105145. [CrossRef] [PubMed]

17. Hao, Y.; Liu, Y.-M. The influential factors of urban $PM_{2.5}$ concentrations in China: A spatial econometric analysis. *J. Clean. Prod.* **2016**, *112*, 1443–1453. [CrossRef]

18. Wang, S.; Zhou, C.; Wang, Z.; Feng, K.; Hubacek, K. The characteristics and drivers of fine particulate matter ($PM_{2.5}$) distribution in China. *J. Clean. Prod.* **2017**, *142*, 1800–1809. [CrossRef]

19. Jiang, P.; Yang, J.; Huang, C.H.; Liu, H.K. The contribution of socioeconomic factors to $PM_{2.5}$ pollution in urban China. *Environ. Pollut.* **2018**, *233*, 977–985. [CrossRef]

20. Fang, C.L.; Liu, H.M.; Li, G.D.; Sun, D.Q.; Miao, Z. Estimating the Impact of Urbanization on Air Quality in China Using Spatial Regression Models. *Sustainability* **2015**, *7*, 15570–15592. [CrossRef]

21. Granger, C.W.J. Testing for Causality—A Personal Viewpoint. *J. Econ. Dyn. Control* **1980**, *2*, 329–352. [CrossRef]

22. Li, G.; Fang, C.; Wang, S.; Sun, S. The Effect of Economic Growth, Urbanization, and Industrialization on Fine Particulate Matter ($PM_{2.5}$) Concentrations in China. *Environ. Sci. Technol.* **2016**, *50*, 11452–11459. [CrossRef]

23. Adedoyin, F.F.; Alola, A.A.; Bekun, F.V. An assessment of environmental sustainability corridor: The role of economic expansion and research and development in EU countries. *Sci. Total Environ.* **2020**, *713*. [CrossRef] [PubMed]

24. Li, G.; Zakari, A.; Tawiah, V. Does environmental diplomacy reduce CO_2 emissions? A panel group means analysis. *Sci. Total Environ.* **2020**, *722*, 137790. [CrossRef] [PubMed]

25. Mi, X. Top 10 most congested cities in China. China: China Internet Information Center. 2015. Available online: http://www.china.org.cn/top10/2015-04/14/content_35314976.htm (accessed on 14 April 2015).

26. Boys, B.; Martin, R.; Van Donkelaar, A.; MacDonell, R.; Hsu, N.; Cooper, M.; Yantosca, R.; Lu, Z.; Streets, D.; Zhang, Q. Fifteen-year global time series of satellite-derived fine particulate matter. *Environ. Sci. Technol.* **2014**, *48*, 11109–11118. [CrossRef]

27. Van Donkelaar, A.; Martin, R.V.; Li, C.; Burnett, R.T. Regional Estimates of Chemical Composition of Fine Particulate Matter Using a Combined Geoscience-Statistical Method with Information from Satellites, Models, and Monitors. *Environ. Sci. Technol.* **2019**, *53*, 2595–2611. [CrossRef] [PubMed]

28. Van Donkelaar, A.; Martin, R.V.; Brauer, M.; Boys, B.L. Use of Satellite Observations for Long-Term Exposure Assessment of Global Concentrations of Fine Particulate Matter. *Environ. Health Perspect.* **2015**, *123*, 135–143. [CrossRef]

29. Peng, J.; Chen, S.; Lü, H.; Liu, Y.; Wu, J. Spatiotemporal patterns of remotely sensed PM$_{2.5}$ concentration in China from 1999 to 2011. *Remote Sens. Environ.* **2016**, *174*, 109–121. [CrossRef]

30. Han, L.J.; Zhou, W.Q.; Li, W.F. City as a major source area of fine particulate (PM$_{2.5}$) in China. *Environ. Pollut.* **2015**, *206*, 183–187. [CrossRef]

31. Gong, P.; Li, X.; Zhang, W. 40-Year (1978–2017) human settlement changes in China reflected by impervious surfaces from satellite remote sensing. *Sci. Bull.* **2019**, *64*, 756–763. [CrossRef]

32. Gong, P.; Wang, J.; Yu, L.; Zhao, Y.; Zhao, Y.; Liang, L.; Niu, Z.; Huang, X.; Fu, H.; Liu, S. Finer resolution observation and monitoring of global land cover: First mapping results with Landsat TM and ETM+ data. *Int. J. Remote Sens.* **2013**, *34*, 2607–2654. [CrossRef]

33. Du, G.; Liu, S.; Lei, N.; Huang, Y. A test of environmental Kuznets curve for haze pollution in China: Evidence from the penal data of 27 capital cities. *J. Clean. Prod.* **2018**, *205*, 821–827. [CrossRef]

34. Guangyue, X.; Deyong, S. An empirical study on the environmental Kuznets curve for China's carbon emissions: Based on provincial panel data. *Chin. J. Popul. Resour. Environ.* **2011**, *9*, 66–76. [CrossRef]

35. Phillips, P.C. Fully modified least squares and vector autoregression. *Econometrica* **1995**, *63*, 1023–1078. [CrossRef]

36. Borrero, H.; Garza, N. Growth and distribution endogenously determined: A theoretical model and empirical evidence. *Braz. J. Political Econ.* **2019**, *39*, 344–361. [CrossRef]

37. Shahbaz, M. A Reassessment of Finance-Growth Nexus for Pakistan: Under the Investigation of FMOLS and DOLS Techniques. *IUP J. Appl. Econ.* **2009**, *8*, 65–80.

38. Chen, J.; Zhou, C.; Wang, S.; Li, S. Impacts of energy consumption structure, energy intensity, economic growth, urbanization on PM$_{2.5}$ concentrations in countries globally. *Appl. Energy* **2018**, *230*, 94–105. [CrossRef]

39. Xu, J.J.; Yip, T.L.; Liu, L. A directional relationship between freight and newbuilding markets: A panel analysis. *Marit. Econ. Logist.* **2011**, *13*, 44–60. [CrossRef]

40. EViews. *EViews 8 User's Guide I*; IHS Global Inc.: Irvine, CA, USA, 2013.

41. EViews. *EViews 8 User's Guide II*; IHS Global Inc.: Irvine, CA, USA, 2013.

42. Lin, G.; Fu, J.; Jiang, D.; Hu, W.; Dong, D.; Huang, Y.; Zhao, M. Spatio-temporal variation of PM$_{2.5}$ concentrations and their relationship with geographic and socioeconomic factors in China. *Int. J. Environ. Res. Public Health* **2014**, *11*, 173–186. [CrossRef]

43. Guan, D.; Su, X.; Zhang, Q.; Peters, G.P.; Liu, Z.; Lei, Y.; He, K. The socioeconomic drivers of China's primary PM$_{2.5}$ emissions. *Environ. Res. Lett.* **2014**, *9*, 024010. [CrossRef]

44. Wang, R. Reasons for Negative Growth of Economic Growth in Liaoning Province and Countermeasure Analysis. *Economy* **2016**, *9*, 25.

45. Sadorsky, P. Do urbanization and industrialization affect energy intensity in developing countries? *Energy Econ.* **2013**, *37*, 52–59. [CrossRef]

46. Li, W.; Luo, J. Problems and Countermeasures in Liaoning's Economic Development. *Chin. Bus. Trade* **2014**, 177–178. [CrossRef]

47. Grossman, G.M.; Krueger, A.B. Environmental Impacts of a North American Free Trade Agreement. *Soc. Sci. Electron. Publ.* **1991**, *8*, 223–250.

48. Lin, B.Q.; Zhu, J.P. Changes in urban air quality during urbanization in China. *J. Clean. Prod.* **2018**, *188*, 312–321. [CrossRef]

49. Byrne, J.; Shen, B.; Li, X. The challenge of sustainability: Balancing China's energy, economic and environmental goals. *Energy Policy* **1996**, *24*, 455–462. [CrossRef]
50. Zhang, X.; Wu, L.; Zhang, R.; Deng, S.; Zhang, Y.; Wu, J.; Li, Y.; Lin, L.; Li, L.; Wang, Y.; et al. Evaluating the relationships among economic growth, energy consumption, air emissions and air environmental protection investment in China. *Renew. Sustain. Energy Rev.* **2013**, *18*, 259–270. [CrossRef]
51. Aunan, K.; Wang, S. Internal migration and urbanization in China: Impacts on population exposure to household air pollution (2000–2010). *Sci. Total Environ.* **2014**, *481*, 186–195. [CrossRef]
52. Chen, T.; Hui, E.C.-M.; Lang, W.; Tao, L. People, recreational facility and physical activity: New-type urbanization planning for the healthy communities in China. *Habitat Int.* **2016**, *58*, 12–22. [CrossRef]

International Journal of
Environmental Research and Public Health

Article

Passive Exposure to Pollutants from a New Generation of Cigarettes in Real Life Scenarios

Joseph Savdie [1], Nuno Canha [1,2], Nicole Buitrago [1] and Susana Marta Almeida [1,*]

[1]   Center for Nuclear Sciences and Technologies (C2TN), Instituto Superior Técnico, University of Lisbon, Estrada Nacional 10, 2695-066 Bobadela-LRS, Portugal; jsavdie@gmail.com (J.S.); nunocanha@ctn.tecnico.ulisboa.pt (N.C.); nbuitrago21@gmail.com (N.B.)

[2]   Centre for Environmental and Marine Studies (CESAM), University of Aveiro, Campus Universitário de Santiago, 3810-193 Aveiro, Portugal

*   Correspondence: smarta@ctn.tecnico.ulisboa.pt

Received: 18 April 2020; Accepted: 13 May 2020; Published: 15 May 2020

Abstract: The use of electronic cigarettes (e-cigarettes) and heat-not-burn tobacco (HNBT), as popular nicotine delivery systems (NDS), has increased among adult demographics. This study aims to assess the effects on indoor air quality of traditional tobacco cigarettes (TCs) and new smoking alternatives, to determine the differences between their potential impacts on human health. Measurements of particulate matter (PM_1, $PM_{2.5}$ and PM_{10}), black carbon, carbon monoxide (CO) and carbon dioxide (CO_2) were performed in two real life scenarios, in the home and in the car. The results indicated that the particle emissions from the different NDS devices were significantly different. In the home and car, the use of TCs resulted in higher PM_{10} and ultrafine particle concentrations than when e-cigarettes were smoked, while the lowest concentrations were associated with HNBT. As black carbon and CO are released by combustion processes, the concentrations of these two pollutants were significantly lower for e-cigarettes and HNBT because no combustion occurs when they are smoked. CO_2 showed no increase directly associated with the NDS but a trend linked to a higher respiration rate connected with smoking. The results showed that although the levels of pollutants emitted by e-cigarettes and HNBT are substantially lower compared to those from TCs, the new smoking devices are still a source of indoor air pollutants.

Keywords: indoor air quality; e-cigarettes; heat-not-burn tobacco; traditional smoking products; tobacco smoke; passenger cars

---

1. Introduction

There is a scientific and medical consensus that cigarette smoking is causally related to lung cancer, heart disease, emphysema and other serious diseases in smokers [1]. Every year, about 8 million people worldwide die from tobacco use [2], and its consumption has been consistently declared as the leading cause of morbidity and mortality in the world [3]. Tobacco smoke is a complex mixture of numerous toxic and carcinogenic substances, containing more than 8000 chemicals produced by distillation, pyrolysis and combustion reactions when tobacco is burnt during both the smoldering and puffing of a cigarette [4].

Convincing scientific evidence has been available for a long time from experimental and epidemiological studies demonstrating that exposure to environmental tobacco smoke (ETS), called secondhand smoke (SHS) or passive smoke, also causes respiratory and heart diseases including lung cancer in adult nonsmokers [5]. In 2017, 1.22 million deaths were caused by SHS [2] (approximately 15% of the deaths linked to tobacco). In children, SHS interferes with lung development, promotes allergic sensitization and asthma, and increases the risk of sudden infant death syndrome [6,7]. The International Agency for Research on Cancer has classified ETS as carcinogenic [5].

Following smoking bans introduced in many countries prohibiting tobacco smoking in public spaces to minimize exposure, the tobacco industry initiated major investments in promoting new (sometimes unregulated) products for consumers. These products were advertised as more appealing than traditional cigarettes (TCs) in terms of social tolerance and health risks. Beliefs that these new products are useful as cessation tools are associated with elevated odds of use in locations where TCs are prohibited [8].

Examples of new smoking products are electronic cigarettes (e-cigarettes), which are battery-powered devices that produce an aerosol from a water-based solution, and heat not-burn tobacco (HNBT), which has been described as a hybrid between TCs and e-cigarettes.

Investigations (some of them developed by the tobacco industry) concluded that although these products are still not entirely safe, they can be considered harmless compared to TCs and, if regulated and controlled, a method to quit addiction to TCs [9–11].

Despite these claims, some research results suggest that inhaling propylene glycol-containing e-cigarette aerosols may have adverse health effects, especially in the respiratory and cardiovascular systems [12,13]. Vaping indoors can also release vegetable glycerin, nicotine, aldehydes and heavy metals at levels that may pose a health risk to others [14,15]. In the United States, during 2019, more than 2000 people developed serious lung damage in a poisoning outbreak associated with the use of vaping devices, and 39 people have died from the condition. The United States Centers for Disease Control and Prevention has recently identified vitamin E acetate, an ingredient added to illicit vaping liquids, as the main cause. Recent research has also shown that HNBT produces toxic compounds (e.g., formaldehyde), which are inhaled together with the aerosol [16]. It is also unclear if these new products reduce or increase nicotine addiction [17]. It has been suggested that they can change the epidemiological perception of smoking and likely attract adolescents into smoking dependence [18–20].

Due to the increasing popularity of e-cigarettes and HNBT as alternatives to TCs, the World Health Organization (WHO) recognized the importance of monitoring and closely following the evolution of new tobacco products, including products with potentially "modified risks". There is a need for further documentation and research about the emissions, impacts on indoor air quality, potential health risks for passive smokers and benefits of the new devices [21]. This study evaluated the levels of particles, black carbon, carbon monoxide and carbon dioxide during the smoking of e-cigarettes, HNBT and TCs in homes and cars to assess the potential exposure of smokers and non-smokers.

2. Materials and Methods

2.1. Sampling Sites Description

Home measurements were performed in the sitting room of an occupied flat located in Lisbon, Portugal (Figure 1). The sitting room had a volume of 73 m^3 and was decorated with typical home furniture. During the experiments, the room was occupied by two people. The air quality monitoring equipment was placed 1.5 m away from the smoker with probes and absorption tubes pointed upwards, at a height of approximately 1 m from the floor. Subjects were told to smoke as usual and not to blow directly onto the equipment.

Car measurements were performed inside a medium volume car (Diesel Opel Corsa, from 2007) traveling on a low traffic intensity route of 4.95 km at a mean speed of 34 km/h. The route was located in the municipality of Loures, Portugal, between the neighborhoods of Bobadela and São João de Talha (Figure 1). The real time monitors were placed in the back seat of the car, in open boxes that were fastened with seatbelts to prevent their slipping. The probes or absorption tubes of the various devices were positioned in the area corresponding to the breathing zone of a child. The study was carried out with two occupants in the car: a driver (the smoker) and a non-smoking passenger seated in the front passenger seat.

Figure 1. Top: measurement locations within the Lisbon metropolitan area (Portugal)—car route in Loures municipality (**right**, **top**) and location of the studied flat in Lisbon municipality (**right**, **middle**). Bottom: arrangement of the measuring instruments in the home (**left** and **center**) and car (**right**).

2.2. Smoking Devices

Three different types of NDS were used in this work, all used by volunteer smokers:

Traditional cigarettes (TC) are comprised of a blend of dried and cured tobacco leaves which are rolled into a thin rolling paper for smoking. TCs burn at temperatures of around 800 °C, generating smoke that contains harmful chemicals. This work used two types of cigarette of a commonly smoked brand in Portugal, Chesterfield blue (TC1) and Chesterfield menthol (blue caps) (TC2).

E-cigarettes are battery-powered devices that produce an aerosol, from a water-based solution, containing a mixture of nicotine, glycerin, propylene glycol and flavoring chemicals, differing depending on the commercial brand. This work used two different types of e-cigarette: the one most common in the USA (JUUL: Slate JUUL, 4.5V, 8W, 5% nicotine pods) and that in Europe (Vape: IStick TC40W, nicotine free liquid).

Heat-not-burn tobacco (HNBT) is comprised of a small cigarette made of elements that include a tobacco plug, hollow acetate tube, polymer-film filter, cellulose-acetate mouthpiece filter, and outer and mouth-end papers. It is equipped with electronics that heat specially prepared and blended tobacco, just enough to release a flavorful nicotine-containing vapor but without burning the tobacco. HNBT is heated up to temperatures below 350 °C in an effort to produce lower amounts of air toxicants [22]. This work used the iQOS from Philip Morris International, which is the most popular brand in Europe and America.

2.3. Measurement Equipment and Protocol

Continuous measuring portable monitors were used to carry out measurements of indoor concentrations of smoking related pollutants:

The **DustTrack DRX monitor** (8533 model, TSI, Dallas, TX, USA) was used to measure the concentration of particles in a size range between 0.1 to 15 μm. It is a multi-channel, battery-operated, data-logging device, which uses a light-scattering laser photometer that allows the simultaneous measurement of size-segregated mass fraction concentrations corresponding to PM_1, $PM_{2.5}$, respirable, PM_{10}, and total PM size fractions. The resolution of the equipment is $\pm 0.1\%$ of the reading or 0.001 mg/m^3.

The **CPC TSI 3007** was used to measure the number concentration of ultrafine particles (UFP) with a size range between 0.01 and 1.0 μm ($PM_{0.01-1}$). It operates by drawing an aerosol sample continuously through a heated saturator, in which alcohol is vaporized and diffused into the sample stream. Together, the aerosol sample and alcohol vapor pass into a cooled condenser where the alcohol vapor becomes supersaturated. Here, particles grow quickly into larger alcohol droplets and pass through an optical detector where they are counted. The accuracy of the equipment is $\pm 20\%$, and the resolution is 0.001 $\mu g/m^3$.

The **MicroAethalometer AE51** (AethLabs, San Francisco, CA, USA) was used to measure black carbon. In the AE51, the air sample is collected by a T60 filter medium (Teflon coated glass fiber). During operation, the microprocessor makes optical measurements, measures and stabilizes the airflow, and calculates the mass concentration of black carbon. The measurement is performed at 880 nm, and the concentration is obtained by the rate of change in the absorption of the transmitted light due to the continuous deposition of black carbon in the filter and the determination of the attenuation of the source light. The measurement precision is ± 0.1 $\mu g/m^3$, at a 150 ml/min flow rate, and the resolution is 0.001 μg.

The **TSI 7545** (7545 model, TSI, Dallas, TX, USA) was used to simultaneously measure and log CO, using an electro-chemical sensor, and CO_2, with a non-dispersive infrared sensor. The accuracy of the CO and CO_2 concentrations is $\pm 3\%$ of the reading, and the resolution is 0.1 ppm for CO and 1 ppm for CO_2.

In homes, an initial non-smoking scenario was recorded for 2 hours and used as a control. Afterwards, each NDS was continuously measured for 2 hours divided into eight 15-minute intervals. Each interval consisted of NDS being smoked with 10 "puffs" for 5 minutes leaving a 10-minute decay period between smokes.

In cars, the measurement for each NDS was made by completing three repetitions composed of three different individual laps (Figure 2). Lap A consisted of a "cleaning lap" where all windows were open and there was no smoking; Lap B was a "blank/control lap" where all windows were closed except for the driver's, which was opened halfway, with no smoking; and Lap C consisted of a "smoking lap", which replicated the conditions of the blank/control lap (all windows closed except for the driver's) with smoking. During Lap C, measurements were registered separately for the complete lap (measurements C1), which included the pollutants' decay, and only during the smoking period within the lap, beginning when the cigarette was lit until it was turned off (measurements C2). Each lap lasted between 8 and 10 minutes in which 10 "puffs" were taken per NDS, for an average smoke time of 3 minutes and with a 7-minute decay period. To maintain the external conditions, the study test drives took place outside of the traffic peak period.

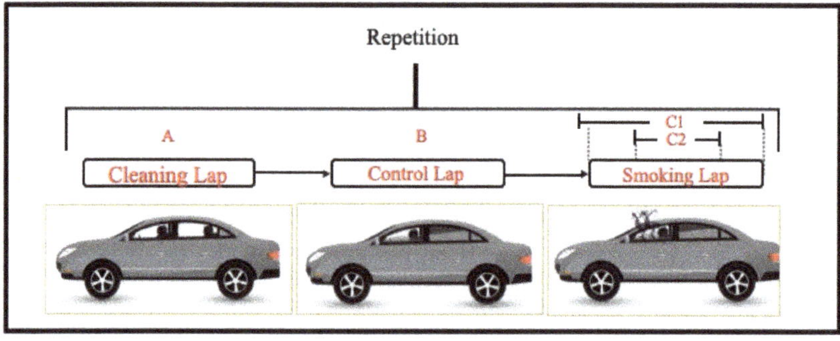

Figure 2. Car measurement methodology.

2.4. Emission Factors

Emission factors for the air pollutants emitted in homes were calculated using Equation (1) [23]:

$$EF = (C_{ave} * ACH * V)/(n_{ave}), \tag{1}$$

where EF is the emission factor of TCs, e-cigarettes or HNBT in $\mu g/h$; C_{ave} is the timed-average pollutant indoor concentrations during the smoking session ($\mu g/m^3$); n_{ave} is the number of TCs, e-cigarettes or HNBT being smoked during the average unit smoking time; ACH is the air change per hour (h^{-1}); and V is the room volume (m^3).

Black carbon concentrations were used to calculate the ACH as it is a conservative and stable pollutant, according to Equation (2) [23]:

$$ACH = (lnC_{ini} - lnC_{end})/t, \tag{2}$$

where C_{ini} is the initial concentration of black carbon (ng/m^3), C_{end} is the final concentration of black carbon (ng/m^3), t is the total time (h) and ACH is the air change per hour (h^{-1}).

2.5. Statistical Analysis

The analysis of the variance of the results was performed by non-parametric statistics for a significance level of 0.05. The Mann–Whitney U test was used to test whether two independent groups are likely to derive from the same population, considering the null hypothesis that the two samples have the same median. Therefore, this test assessed whether observations in one sample tend to be larger than observations in the other, such as in the case of air pollutant concentrations associated with the different types of smoking product, the air pollutant levels for the background and smoking periods, and the contribution of the particles' sizes to the PM_{10} for the different NDS. The statistical calculations were performed using the Statistica software.

3. Results and Discussion

3.1. Home Scenario

A comprehensive evaluation of the levels of smoking related pollutants in a home while TCs and new smoking products (e-cigarettes and HNBT) were being smoked was performed. The concentrations of the measured indoor air pollutants are summarized in Table 1, and the basic statistics are summarized in Table S1 of the Supplementary Materials.

Table 1. Air pollutant average concentrations and emission factors for traditional cigarettes (TC), e-cigarettes and heat-not-burn tobacco (HNBT) in the home. NDS, nicotine delivery systems; UFP, ultrafine particles; BC, black carbon.

	NDS	PM_1 (μg·m^{-3})	$PM_{2.5}$ (μg·m^{-3})	PM_{10} (μg·m^{-3})	UFP (particles·cm^{-3})	BC (μg·m^{-3})	CO (mg·m^{-3})	CO_2 (mg·m^{-3})
	Control	21.0	22.6	25.4	4690	0.21	1.66	1810
Concentrations	TC	3470	3480	3480	110,000	13.2	4.16	2220
	e-cigarette	1350	1370	1380	37,800	4.30	1.00	2890
	HNBT	80.6	81.6	87.8	35,700	1.18	1.29	2640

	NDS	PM_1 (μg·min^{-1})	$PM_{2.5}$ (μg·min^{-1})	PM_{10} (μg·min^{-1})	UFP (particles·min^{-1})	BC (μg·min^{-1})	CO (mg·min^{-1})	CO_2 (mg·min^{-1})
	TC	844	845	846	2.46×10^9	3.37	0.92	604
Emission Factors	e-cigarette	419	424	427	9.89×10^8	1.10	0.26	836
	HNBT	21.9	22.2	23.7	1.20×10^9	0.36	0.33	720

3.1.1. Particulate Matter

Figure 3 depicts the contribution of each particle size fraction (PM_1, $PM_{1-2.5}$, $PM_{2.5-10}$) to the PM_{10} for the studied NDS and control. The Mann–Whitney test showed that there was a significant difference between the contributions of the three particle size ranges to the PM_{10} in the non-smoking and NDS trials. PM_1 was the dominant size fraction for TCs (98.6%), e-cigarettes (91.1%) and HNBT (92.1%) followed by $PM_{2.5-10}$ (TCs: 1.2%, e-cigarettes: 6.5% and HNBT: 6.8%), whereas in the control, the contribution of the coarsest particles to the PM_{10} mass increased to 43.9%.

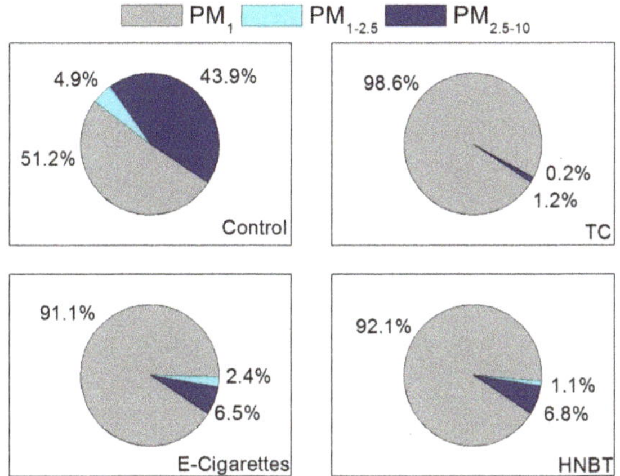

Figure 3. Contribution of each particle size fraction (PM_1, $PM_{1-2.5}$, $PM_{2.5-10}$) to the PM_{10} in the home discriminated by NDS.

The use of TCs led to the highest increase in PM_1 (3470 ± 1570 µg·m^{-3}), $PM_{2.5}$ (3480 ± 1570 µg·m^{-3}) and PM_{10} (3480 ± 1570 µg·m^{-3}) concentrations, followed by the e-cigarettes (PM_1: 1350 ± 1510 µg·m^{-3}; $PM_{2.5}$: 1370 ± 1520 µg·m^{-3}; PM_{10}: 1380 ± 1520 µg·m^{-3}) and HNBT (PM_1: 80.6 ± 51.3 µg·m^{-3}; $PM_{2.5}$: 81.6 ± 51.3 µg·m^{-3}; PM_{10}: 87.8 ± 51.7 µg·m^{-3}). The Mann–Whitney test showed that the concentrations were significantly different between all types of cigarettes and that PM_{10} concentrations measured during the smoking of TCs, e-cigarettes and HNBT were significantly higher than the levels measured in the non-smoking period (165, 64 and 4 times higher, respectively).

Another study on smoke exposure [23] also described higher PM concentrations for TCs than for e-cigarettes and HNBT. However, during the smoking of TCs, Ruprecht et al. [23] obtained PM_1, $PM_{2.5}$ and PM_{10} concentrations 10, 23 and 2 times lower than those measured in the present study, respectively. Schober et al. [24] also measured lower $PM_{2.5}$ levels associated with the smoking of e-cigarettes (197 µg·m^{-3}) than those in the present study.

The differences between the NDS are likely caused by the fact that in TCs, there is a combustion at a temperature <800 °C, which is lower than the temperature needed for complete combustion (around 1300 °C), while e-cigarettes and HNBT are only heated. According to Jiang et al. [25], heating tobacco or e-liquids result in 95% less substances emitted than those produced by the combustion that occurs in TCs. Schober et al. [26] showed that the vaping of the e-cigarettes releases more particles than the use of HNBT. E-cigarette aerosols contain fine and ultrafine liquid particles that are formed from supersaturated propylene glycol vapor, which can penetrate into the respiratory system and cause oxidative stress and inflammatory reactions [27]. Pisinger and Dossing [28] mentioned the irritation of the respiratory tract, evidence of an inflammatory process, a dry cough and an impairment of lung function as short term effects of vaping.

The guidelines defined by the World Health Organization and the limit values according to the Portuguese legislation for indoor air quality ($PM_{2.5}$: 25 µg·m^{-3}; PM_{10}: 50 µg·m^{-3}) were exceeded for

TCs (139 and 70 times higher for $PM_{2.5}$ and PM_{10}, respectively), e-cigarettes (54 and 27 times higher) and HNBT (3.2 and 1.7 times higher).

Figure 4 shows the temporal trends of PM_{10} levels measured during TC, e-cigarette, and HNBT consumption. The PM_{10} concentrations associated with the TC and e-cigarette trials presented a rapid increase above the background, while for HNBT, the increment was less pronounced but still visible. PM_{10} peaks of more than 8000 $\mu g \cdot m^{-3}$ were reached for e-cigarettes and TCs. For TCs, PM_{10} levels showed a long decay period, causing an accumulation for each additional cigarette smoked, whereas for both e-cigarettes and HNBT, PM_{10} showed a faster decay and no sign of accumulation. Protano et al. [29] described a similar behavior, since a 1 hour time interval after each smoking each TC was not enough to allow the PM concentration to decrease to the background levels. According to Martuzevicius et al. [30], e-cigarette aerosols have been shown to have a half-life 100 times shorter than TC emissions. The rapid evaporation of liquid droplets from e-liquids is the main reason for the quick decay and the lack of atmospheric accumulation of PM during the use of e-cigarettes.

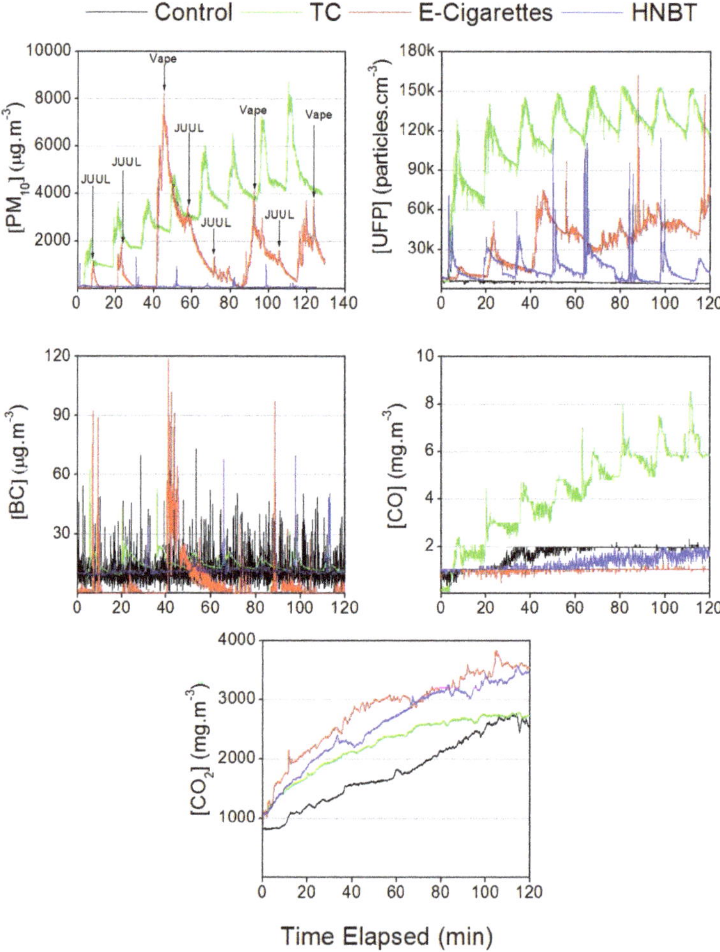

Figure 4. Temporal trends for the PM_{10}, UFP, BC, CO and CO_2 measured during Traditional Cigarettes (TC), e-cigarette and Heat-not-burn tobacco (HNBT) consumption in the home.

The PM_1, $PM_{2.5}$ and PM_{10} emission factors were the highest for TCs, followed by e-cigarettes and HNBT. The emission factors calculated by Ruprecht et al. [23] for TCs were lower for PM_1 (320 ± 132 µg·min^{-1}) than those calculated in this study (844 µg·min^{-1}), but higher for $PM_{2.5}$ and PM_{10} (1480 ± 570 and 1540 ± 570 µg·min^{-1}, respectively) than the ones calculated here (845 and 846 µg·min^{-1} for $PM_{2.5}$ and PM_{10}, respectively). The same study found that both the e-cigarette and HNBT emission factors were non-detectable, significantly differentiating themselves from the elevated values obtained for the present work.

3.1.2. Ultrafine Particles

The Mann–Whitney test showed that the UFP number concentrations were significantly higher during all the smoking sessions than during the background, but the levels for TCs ($110,000 \pm 36,000$ particles·cm^{-3}) stood out compared with those for e-cigarettes ($37,800 \pm 19,000$ particles·cm^{-3}) and HNBT ($35,700 \pm 11,500$ particles·cm^{-3}). The levels for TCs, e-cigarettes and HNBT were 23.4, 8.1 and 7.6 times higher than background, respectively. The UFP concentrations were higher when combustion was occurring i.e., during TC use [25]. This fact explains why both e-cigarettes and HNBT showed lower UFP concentrations compared to TCs.

Atmospheric UFP are mainly composed of organic compounds, trace metal oxides and elemental carbon [31]. Ruprecht et al. [23] found that for selected metals, trace elements and organic compound emission factors varied between TCs, e-cigarettes and HNBT. This means that the type of NDS used highly influences the UFP number concentration. Avino [32] also showed that during a TC test, the increase in the particle number concentration is due to the emission during the smoking activity of particles with a mode of roughly 100 nm, while the e-cigarettes emit particles sized with a mode of about 30 nm.

The UFP number concentrations for TCs and HNBT were similar to those measured by Ruprecht et al. [23] ($123,000 \pm 37,000$ and $27,700 \pm 10,300$ particles·m^{-3}, respectively). For e-cigarettes, Ruprecht et al. [23] measured concentrations 4.4 times lower (8660 ± 560 particles·m^{-3}) and Schober et al. [24] obtained concentrations 1.6 times higher ($61,700 \pm 16,000$ particles·m^{-3}) than in the present study. The discrepancies found are likely due to high variability in emissions due to the types of equipment and e-liquid being used. Schober et al. [24] used a Red Kiwi (second generation e-cigarette), which is larger and has more wattage than the Elips Series C (second generation e-cigarette) used by Ruprecht et al. [23] and smaller than the third generation e-cigarettes used in this study. Moreover, Zhao et al. [33] also showed that the heating coil temperature, puff duration and puff flow rate in e-cigarettes influence the number concentration of the particles.

The real-time UFP number concentration plot presented in Figure 4 shows an initial cumulative behavior in TCs that reaches a plateau at around 150,000 particles per cm^3. The UFP temporal pattern for e-cigarettes and HNBT shows a behavior similar to the one obtained by Protano et al. [29], which is characterized by non-accumulation and rapid decay.

The UFP emission factors were the highest for TCs, followed by e-cigarettes and HNBT. The emission factors obtained for TCs, e-cigarettes and HNBT by Ruprecht et al. [23] (130×10^{10}, 1.1×10^{10} and 5.3×10^{10} particles per min) were much higher than those obtained in the present study.

3.1.3. Black Carbon

The highest black carbon concentrations were measured while TCs were being smoked (13.2 ± 5.2 µg·m^{-3}), followed by e-cigarettes (4.3 ± 10.4 µg·m^{-3}) and HNBT (1.2 ± 0.7 µg·m^{-3}), which are values approximately 63, 20 and 5.6 times higher than those in the non-smoking trials, considering the values presented in Table 1.

Black carbon particles are produced due to the incomplete combustion of carbon-containing materials [34]. As the tobacco or tobacco-derived products within TCs are burned at temperatures below the 1300 °C threshold needed for complete combustion to occur [25], these NDS have been directly identified as black carbon emission sources [35,36].

On the other hand, probably due to the fact that they evaporate a liquid charge rather than combusting it, studies conducted by van Drooge et al. [37] and Ruprecht et al. [23] have shown no connection between the use of e-cigarettes and black carbon emissions. According to van Drooge et al. [37], the difference between the black carbon concentrations recorded during the non-smoking and e-cigarette smoking scenarios are directly linked to outdoor black carbon concentrations, thus indicating that black carbon is not an emission of the e-cigarette vapor. This is the reason why in the study by Ruprecht et al. [23], the temporal patterns show lower black carbon concentrations during e-cigarette smoking than in the control test, similarly to in the present work.

Figure 4 shows that the black carbon measured during the TC smoking trials presented an initial cumulative behavior, reaching a plateau around 20.0 $\mu g \cdot m^{-3}$. Both e-cigarettes and HNBT had non-cumulative effects and rapid decays, besides the high spikes observed.

3.1.4. Carbon Monoxide

The Mann–Whitney test shows that the use of TCs led to a significant increase in CO levels in homes to 4.2 ± 1.8 mg·m^{-3}, a concentration 2.5 above background levels without smoking. The smoking of HNBT and e-cigarettes had no effect on the CO concentration, as already demonstrated by previous studies [9,24,37], because CO is a byproduct of the incomplete combustion of carbonaceous matter that occurs in TCs [38]. The real time CO concentration plotted in Figure 4 shows that both e-cigarettes and HNBT had a steady, non-cumulative behavior, unlike the TCs, which had a cumulative and incremental behavior without reaching a plateau.

None of the NDS surpassed the guidelines defined by the World Health Organization nor the limit values according to Portuguese legislation (10 mg·m^{-3} for 8 h; 30 mg·m^{-3} for 1 h).

3.1.5. Carbon Dioxide

The CO_2 concentrations were 2890 ± 660 mg·m^{-3} for e-cigarettes, 2640 ± 680 mg·m^{-3} for HNBT and 2220 ± 520 mg·m^{-3} for TCs; approximately 1.6, 1.5 and 1.2 times higher than control levels, respectively. All the NDS as well as the control scenario (also with two occupants) exceeded the recommended World Health Organization CO_2 maximum concentration (1800 mg·m^{-3}).

The real time CO_2 measurements (Figure 4) show similarities in the incremental behavior of all the NDS and the control. The concentrations steadily increased, reaching almost double their initial values after one hour and roughly thrice after two hours, indicating that exhalations during NDS use did not increase CO_2 concentrations in peak increments as with other pollutants. A study conducted by Sadjadi and Minai [39] states that this increase in CO_2 concentrations is related to an increase in the respiration rate of smokers as a response to inflammation in order to compensate for the decrease in oxygen inhalation during smoking rather than to the emissions originating from NDS use.

3.2. Car Scenario

Smoking in the interior of cars is of particular concern for the smoker and other non-smoking passengers, principally for the most susceptible such as children and pregnant woman, because the concentrations of potentially harmful substances are expected to be high due to the reduced volume of the cabin. The mean concentrations measured during the test drives are summarized in Table 2, and for the basic statistics, Table S2 from the Supplementary Materials can be consulted.

Table 2. Air pollutant average concentrations measured in the car for traditional cigarettes (TC1 and TC2), e-cigarettes (JUUL and Vape) and heat-not-burn tobacco (HNBT).

NDS	Lap	PM$_1$ (μg.m^{-3})	PM$_{2.5}$ (μg.m^{-3})	PM$_{10}$ (μg.m^{-3})	UFP (particles.cm^{-3})	BC (μg.m^{-3})	CO (mg.m^{-3})	CO$_2$ (mg.m^{-3})
TC1	Control	46.2	49.5	57.2	31,733	0.83	0.81	1059
	Smoking	963	967	973	141,000	2.11	3.02	1130
TC2	Control	43.4	45.3	49.7	42,700	1.46	1.10	1090
	Smoking	905	907	912	142,000	6.11	4.12	11,900
JUUL	Control	19.2	21.1	24.5	28,500	0.57	0.43	883
	Smoking	129	131	134	47,800	1.15	0.82	982
Vape	Control	21.0	21.8	23.3	17,600	0.59	0.43	956
	Smoking	1150	1170	1170	56,300	0.70	1.09	1090
HNBT	Control	14.5	15.9	18.3	7940	0.61	0.45	925
	Smoking	23.3	24.7	26.7	22,100	0.46	0.74	1020

3.2.1. Particulate Matter

Figure 5 depicts the contribution of each particle size fraction (PM$_1$, PM$_{1-2.5}$, PM$_{2.5-10}$) to the PM$_{10}$ for the studied NDS during the different laps. Although no difference was observed between the cleaning and control laps, the Mann–Whitney test indicated a significant difference between the non-smoking and the NDS trials. For the NDS, PM$_1$ was the dominant size fraction for TC1 (98.3%), TC2 (99.2%), JUUL (95.3%), vape (97.9%) and HNBT (87.9%), with negligible contributions from the other two fractions. In the control, the two coarser fractions (PM$_{1-10}$) have a significantly higher contributions during the smoking periods, representing between 9.8% and 21.5% of the PM$_{10}$ mass.

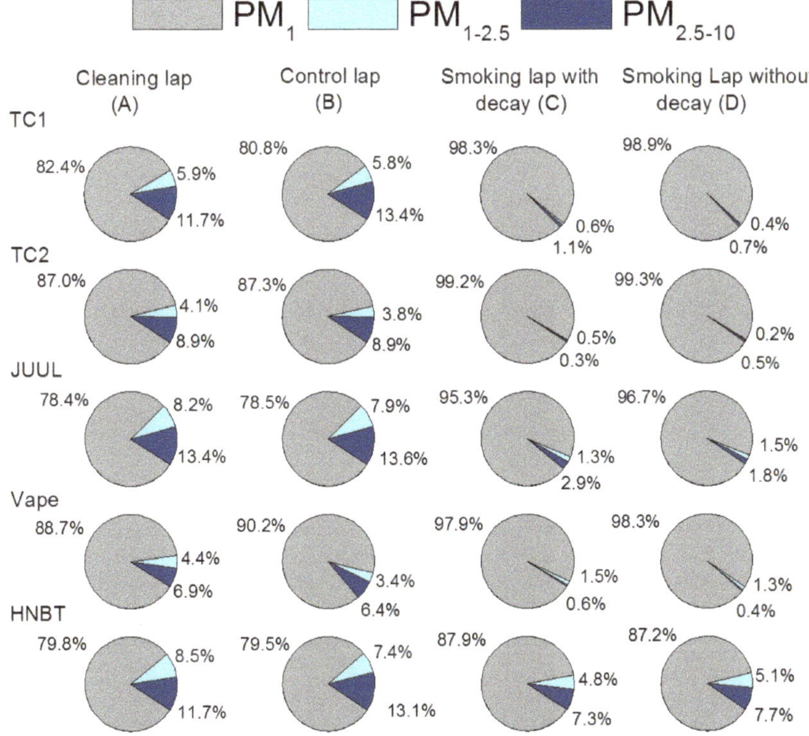

Figure 5. Contribution of each particle size fraction (PM$_1$, PM$_{1-2.5}$, PM$_{2.5-10}$) to the PM$_{10}$ in the car discriminated by NDS.

The highest PM_{10} concentrations were measured while the vape was smoked (1170 ± 1160), followed by TC1 (973 ± 597 $\mu g \cdot m^{-3}$), TC2 (912 ± 881 $\mu g \cdot m^{-3}$), JUUL (134 ± 190 $\mu g \cdot m^{-3}$) and HNBT (26.7 ± 22.7 $\mu g \cdot m^{-3}$). The Mann–Whitney test showed that the PM_{10} concentrations were significantly different for all the types of cigarette except for TC1 and TC2, between which significant differences were not observable.

Figure 6 shows the temporal evolution of the PM_{10} concentrations. There is an incremental and cumulative behavior for TC1 and TC2, reaching a plateau at around 1000 $\mu g \cdot m^{-3}$ before the concentrations start to slowly decrease back to control levels. The JUUL, vape and HNBT time patterns show significant concentration spikes during use but then rapid decreases in concentration.

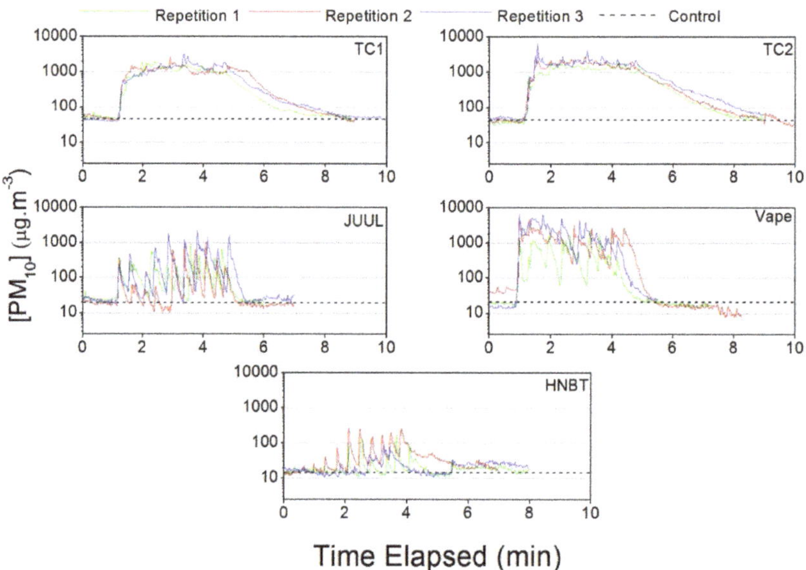

Figure 6. PM_{10} concentrations measured in the car during traditional cigarette (TC1 and TC2), e-cigarette (JUUL and Vape) and heat-not-burn tobacco (HNBT) consumption.

Geiss et al. [40] measured PM in the vehicle cabins and obtained an average $PM_{2.5}$ concentration in the cars of 26.9 $\mu g \cdot m^{-3}$, similar to those found in the control level measurements in the present study. Schober et al. [26] studied NDS emissions in seven different vehicles and observed higher mean $PM_{2.5}$ concentrations for TCs (64–1990 $\mu g \cdot m^{-3}$) when compared to vape (8–490 $\mu g \cdot m^{-3}$), HNBT (6–34 $\mu g \cdot m^{-3}$) and control (4–11 $\mu g \cdot m^{-3}$). In the present study, the e-cigarette vape showed the highest mean levels of $PM_{2.5}$ and PM_{10}, even when comparing with TCs.

3.2.2. Ultrafine Particles

The highest UFP concentrations were measured while TC2 were being smoked (142,000 ± 42,000 particles·cm⁻³), followed by TC1 (141,000 ± 56,000 particles·cm⁻³), vape (56,300 ± 39,700 particles·cm⁻³), JUUL (47,800 ± 12,700 particles·cm⁻³) and HNBT (22,100 ± 16,800 particles·cm⁻³). These values are 3.3, 4.4, 3.2, 1.7 and 2.8 times higher than those in the control scenario, respectively.

TC1 and TC2 showed a longer decay period than the other NDS. Clear spikes were observed for JUUL, HNBT and vapes when "puffs" were taken, but the patterns did not show accumulation and had rapid decays (Figure 7).

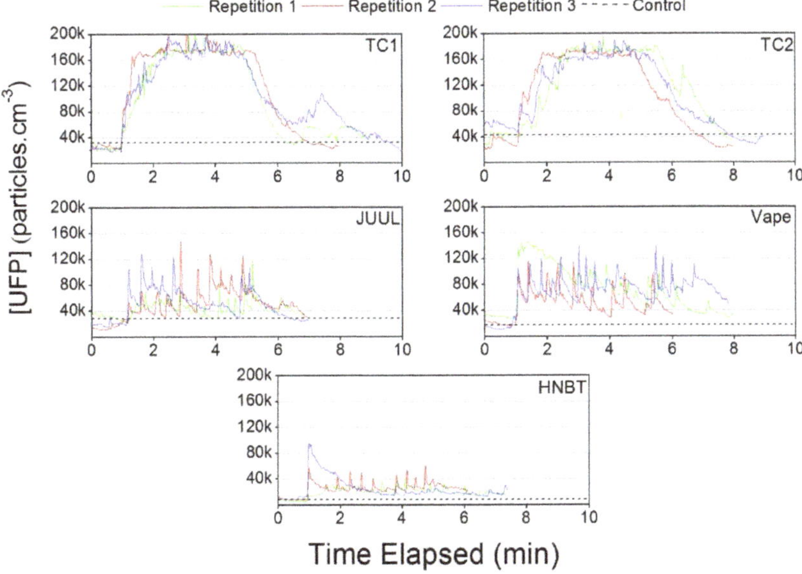

Figure 7. Ultrafine particle concentrations measured in the car during traditional cigarette (TC1 and TC2), e-cigarette (JUUL and Vape) and heat-not-burn tobacco (HNBT) consumption.

The UFP concentrations measured during TC1 and TC2 consumption were significantly higher than for the other NDS, likely due to the combustion that occurred. As previously stated, TCs burn at temperatures of 800 °C, which leads to incomplete combustion, while vape, JUUL and HNBT are only heated. TC1 and TC2 also contain heavy metals and hydrocarbons [41], both of which can be found in the chemical composition of atmospheric UFP [31].

The study developed by Schober et al. [26] showed that TCs also presented the highest UFP levels (ranging from 24,300 to 236,000 particles·cm⁻³), but with HNBT (mean value of 37,900 ± 38,100 particles·cm⁻³, ranging from 16,700 to 124,000 particles·cm⁻³) having higher UFP levels than e-cigarettes (mean value of 31,000 ± 24,100 particles·cm⁻³, ranging from 10,200 to 74,000 particles·cm⁻³) in 71% of the cases.

3.2.3. Black Carbon

The black carbon concentrations were the highest for TC2 (6.1 ± 4.0 µg·m⁻³), followed by TC1 (2.1 ± 0.9 µg·m⁻³), JUUL (1.2 ± 0.6 µg·m⁻³), vape (0.7 ± 1.0 µg·m⁻³) and HNBT (0.5 ± 0.3 µg·m⁻³), representing levels 4.2, 2.5, 2.0, 0.4 and 0.7 times higher than those in the control scenario, respectively. The incomplete combustion that occurs in TCs explains the comparably higher concentrations obtained for this type of NDS.

The real time black carbon concentrations presented in Figure 8 show an incremental behavior during the use of TC1 and TC2 and a steady decrease after smoking. The JUUL presented a non-cumulative effect, a rapid decay and spikes in concentrations during its use. Both the vape and HNBT patterns showed a non-cumulative effect and rapid decay like the pattern for JUUL, but no spikes in concentrations were observed.

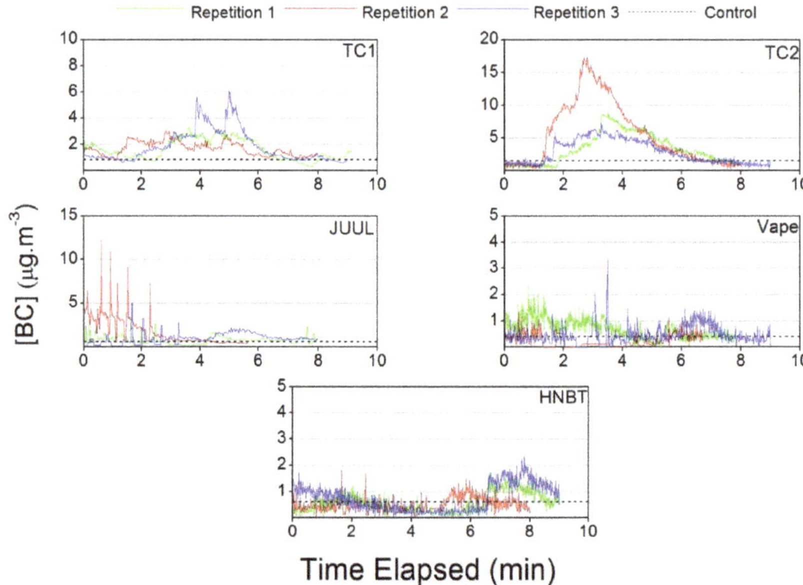

Figure 8. Black carbon concentrations measured in the car during traditional cigarette (TC1 and TC2), e-cigarette (JUUL and Vape) and heat-not-burn tobacco (HNBT) consumption.

The concentrations in the present study were lower than the black carbon concentrations measured, in vehicles from non-smokers, by Lee et al. [42] (1.9 $\mu g \cdot m^{-3}$), Cunha-Lopes et al. [43] (5.1 ± 7.3 $\mu g \cdot m^{-3}$) and Correia et al. [44] (5.5 ± 5.9 $\mu g \cdot m^{-3}$), except for TC1 and TC2. Onat et al. [45] measured a set of indoor pollutants in different commuting vehicles in Istanbul and registered, for cars, an average black carbon concentration of 2.3 ± 1.3 $\mu g \cdot m^{-3}$ with closed windows, similar to the results obtained in this study for TC1. Fruin et al. [46] showed that driving behind vehicles in traffic with open windows has a significant effect on the black carbon exposure. This work measured very high levels of black carbon in cars driving behind transit buses reaching up to 92 $\mu g \cdot m^{-3}$. This would mean that black carbon concentrations in vehicles can be much more related to the outdoor environment rather than to indoor sources, even with a significant emitting source such as an NDS.

3.2.4. Carbon Monoxide

Statistical tests showed that the CO concentrations for TC1 (3.0 ± 1.5 $mg \cdot m^{-3}$) and TC2 (4.1 ± 1.6 $mg \cdot m^{-3}$) were significantly higher than for vape (1.1 ± 0.3 $mg \cdot m^{-3}$), JUUL (0.8 ± 0.1 $mg \cdot m^{-3}$) and HNBT (0.7 ± 0.3 $mg \cdot m^{-3}$). Figure 9 shows an incremental and cumulative behavior for TC1 and TC2. E-Cigarettes, JUUL and HNBT show a steady behavior regarding concentrations, with no increases or accumulation occurring during their use. The observed differences are likely linked to the incomplete combustion processes in TC1 and TC2.

Northcross et al. [47] measured CO concentrations in cars during the smoking of TCs and obtained an average concentration of 2.8 ± 1.0 $mg \cdot m^{-3}$ when all windows were half open, while a study conducted by Dirks et al. [48] measured CO concentrations in vehicles ranging from 0.7 to 3.2 $mg \cdot m^{-3}$, depending on the window conditions and the ventilation setting in the car.

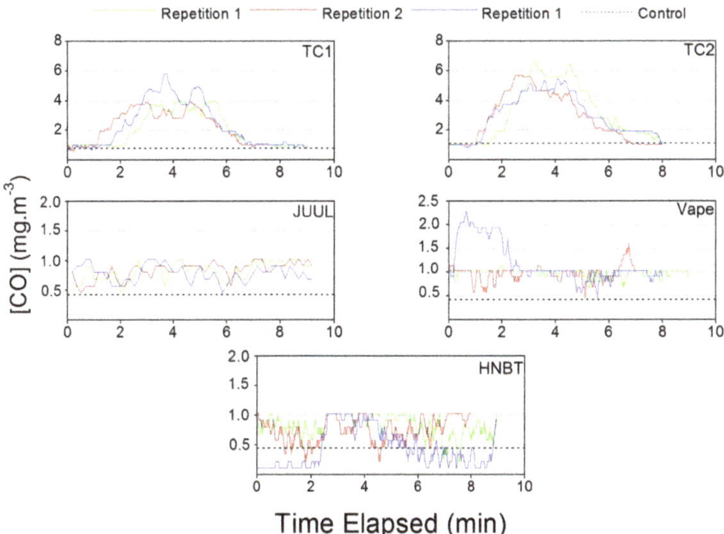

Figure 9. Carbon monoxide concentrations measured in the car during traditional cigarette (TC1 and TC2), e-cigarette (JUUL and Vape) and heat-not-burn tobacco (HNBT) consumption.

3.2.5. Carbon Dioxide

The CO_2 concentrations were the highest during TC2 consumption (1190 ± 50 mg·m^{-3}), followed by TC1 (1130 ± 90 mg·m^{-3}), vape (1090 ± 60 mg·m^{-3}), HNBT (1020 ± 60 mg·m^{-3}) and JUUL (982 ± 43 mg·m^{-3}).

Smoking is linked with an increase in respiration rate, which increases CO_2 concentrations in indoor environments (Figure 10).

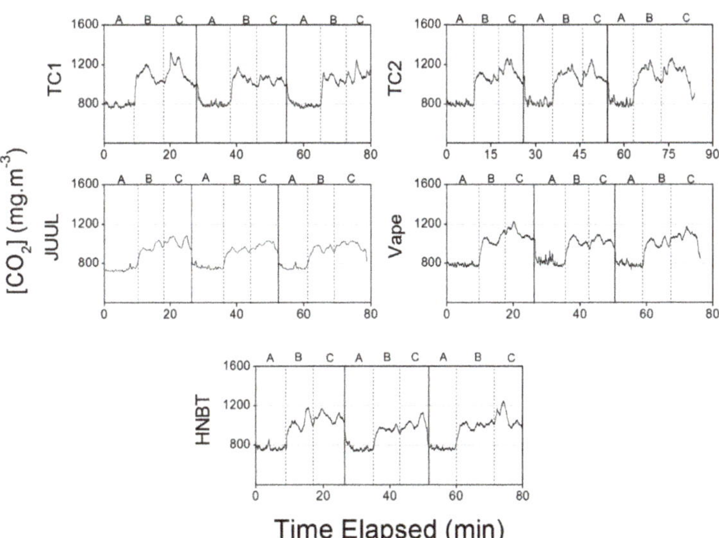

Figure 10. Carbon dioxide concentrations measured in the car during traditional cigarette (TC1 and TC2), e-cigarette (JUUL and Vape) and heat-not-burn tobacco (HNBT) consumption.

Goh et al. [49] measured CO_2 concentrations ranging between 810 and 1080 mg·m^{-3} in cars, similar to the results for the cleaning laps in the present study. The same study obtained CO_2 concentrations for two occupants (with all the windows closed) of 2160 mg·m^{-3} nine minutes after the beginning of the experiment. Even without smoking, these values are almost twice the levels measured in the present study for TC2 (1190 mg·m^{-3}).

4. Conclusions

Although traditional tobacco smoking has been in decline since the 1980s, newer generations of NDS have been steadily increasing in popularity ever since they were introduced into the market in 2013. This accelerated growth, together with their recent appearance, has led to an impendent need for studies to be developed measuring the effects of such.

The present study allowed the evaluation of the concentrations of smoke pollutants, more specifically, the particulate matter and gases originating from different types of NDS in real life scenarios where smoking is still common among electronic nicotine delivery systems users, which consider these a safer option than TCs.

The results showed that although the levels of pollutants emitted by e-cigarettes and HNBT are substantially lower compared to those from TCs, the new smoking devices are still a source of indoor air pollutants. All smoking options are avoidable sources of indoor pollutants, and to protect the health of smokers and non-smokers, they should not be used in homes and cars.

The presented results pertain to a single brand of HNBT and specific brands of e-cigarettes and may not represent the possible variability among different brands or manufacturers. Additionally, the configurations of the equipment as well as the e-liquid charges used for each e-cigarette may not represent other brands or configurations of these devices.

Supplementary Materials: The following are available online at http://www.mdpi.com/1660-4601/17/10/3455/s1, Table S1: Concentrations of air pollutants measured in the home for traditional cigarettes (TC1 and TC2), e-cigarettes (JUUL and Vape), and HNBT, Table S2: Concentrations of air pollutants measured in the car for traditional cigarettes (TC1 and TC2), e-cigarettes (JUUL and Vape), and HNBT.

Author Contributions: Conceptualization, J.S., N.C. and S.M.A.; methodology, J.S., N.C. and S.M.A.; validation, N.C. and S.M.A; investigation, J.S., N.B., N.C. and S.M.A.; resources, S.M.A.; data curation, J.S., N.C. and S.M.A.; writing—original draft preparation, J.S.; writing—review and editing, N.C. and S.M.A.; supervision, N.C. and S.M.A.; project administration, S.M.A.; funding acquisition, S.M.A. All authors have read and agreed to the published version of the manuscript.

Acknowledgments: This work was supported by LIFE Index-Air project (LIFE15 ENV/PT/000674). This work reflects only the authors' view and EASME is not responsible for any use that may be made of the information it contains. Authors also acknowledge the support of the Portuguese Foundation for Science and Technology (FCT, Portugal) through the Postdoc Grant SFRH/BPD/102944/2014, the contract IST-ID/098/2018, and the strategic projects UIDB/04349/2020+UIDP/04349/2020 and UIDB/50017/2020+UIDP/50017/2020.

Conflicts of Interest: The authors declare no conflict of interest.

References

1. U.S. Department of Health and Human Services. *The Health Consequences of Smoking: 50 Years of Progress;* A Report of the Surgeon General. U.S. Department of Health and Human Services; Centers for Disease Control and Prevention, National for Chronic Disease Prevention and Health Center Promotion, Office on Smoking and Health: Atlanta, Greece, 2014.

2. Ng, M.; Freeman, M.K.; Fleming, T.D.; Robinson, M.; Dwyer-Lindgren, L.; Thomson, B.; Wollum, A.; Sanman, E.; Wulf, S.; Lopez, A.D.; et al. Smoking prevalence and cigarette consumption in 187 countries, 1980–2012. *JAMA J. Am. Med. Assoc.* **2014**, *311*, 183–192. [CrossRef] [PubMed]

3. Feigin, V.L.; Roth, G.A.; Naghavi, M.; Parmar, P.; Krishnamurthi, R.; Chugh, S.; Mensah, G.A.; Norrving, B.; Shiue, I.; Ng, M.; et al. Global burden of Stroke and Risk Factors in 188 Countries, During 1990–2013: A Systematic Analysis for the Global Burden of Disease Study 2013. *Lancet Neurol.* **2016**, *15*, 913–924. [CrossRef]

4. Food and Drug Administration. Harmful and potentially harmful constituents in tobacco products and tobacco smoke; established list. *Fed. Regist.* **2012**, *77*, e20034–e20037.

5. International Agency for Research on Cancer (IARC). *Monographs on the Evaluation of the Carcinogenic Risk of Chemicals to Humans*; Tobacco Smoke and Involuntary Smoking; IARC: Lyon, France, 2004; Volume 83.

6. McCarville, M.; Sohn, M.-W.; Oh, E.; Weiss, K.; Gupta, R. Environmental tobacco smoke and asthma exacerbations and severity: The difference between measured and reported exposure. *Arch. Dis. Child.* **2013**, *98*, 510–514. [CrossRef]

7. Feleszko, W.; Ruszczyński, M.; Jaworska, J.; Strzelak, A.; Zalewski, B.M.; Kulus, M. Environmental tobacco smoke exposure and risk of allergic sensitisation in children: A systematic review and meta-analysis. *Arch. Dis. Child.* **2014**, *99*, 985–992. [CrossRef]

8. Dunbar, Z.R.; Giovino, G.; Wei, B.; O'connor, R.J.; Goniewicz, M.L.; Travers, M.J. Use of electronic cigarettes in smoke-free spaces by smokers: Results from the 2014–2015 population assessment on tobacco and health study. *Int. J. Environ. Res. Public Health* **2020**, *17*, 978. [CrossRef]

9. Mitova, M.I.; Campelos, P.B.; Goujon-Ginglinger, C.G.; Maeder, S.; Mottier, N.; Rouget, E.G.R.; Tharin, M.; Tricker, A.R. Comparison of the impact of the Tobacco Heating System 2.2 and a cigarette on indoor air quality. *Regul. Toxicol. Pharmacol.* **2016**, *80*, 91–101. [CrossRef]

10. Shahab, L.; Goniewicz, M.L.; Blount, B.C.; Brown, J.; McNeill, A.; Alwis, K.U.; Feng, J.; Wang, L.; West, R. Nicotine, Carcinogen, and Toxin Exposure in Long-Term E-Cigarette and Nicotine Replacement Therapy Users: A Cross-sectional Study. *Ann. Intern. Med.* **2017**, *166*, 390–400. [CrossRef]

11. Polosa, R.; Farsalinos, K.; Prisco, D. Health impact of electronic cigarettes and heated tobacco systems. *Intern. Emerg. Med.* **2019**, *14*, 817–820. [CrossRef]

12. Carnevale, R.; Sciarretta, S.; Violi, F.; Nocella, C.; Loffredo, L.; Perri, L.; Peruzzi, M.; Marullo, A.G.; De Falco, E.; Chimenti, I.; et al. Acute impact of tobacco vs electronic cigarette smoking on oxidative stress and vascular function. *Chest* **2016**, *150*, 606–612. [CrossRef]

13. Vlachopoulos, C.; Ioakeimidis, N.; Abdelrasoul, M.; Terentes-Printzios, D.; Georgakopoulos, C.; Pietri, P.; Stefanadis, C.; Tousoulis, D. Electronic cigarette smoking increases aortic stiffness and blood pressure in young smokers. *J. Am. Coll. Cardiol.* **2016**, *67*, 2802–2803. [CrossRef] [PubMed]

14. Logue, J.M.; Sleiman, M.; Montesinos, V.N.; Russell, M.L.; Litter, M.I.; Benowitz, N.L.; Gundel, L.A.; Destaillats, H. Emissions from Electronic Cigarettes: Assessing Vapers' Intake of Toxic Compounds, Secondhand Exposures, and the Associated Health Impacts. *Environ. Sci. Technol.* **2017**, *51*, 9271–9279. [CrossRef] [PubMed]

15. Li, L.; Lin, Y.; Xia, T.; Zhu, Y. Effects of Electronic Cigarettes on Indoor Air Quality and Health. *Annu. Rev. Public Health* **2020**, *41*, 363–380. [CrossRef] [PubMed]

16. Davis, B.; Williams, M.; Talbot, P. iQOS: Evidence of pyrolysis and release of a toxicant from plastic. *Tobac. Contr.* **2019**, *28*, 34–41.

17. Palazzolo, D.L. Electronic cigarettes and vaping: A new challenge in clinical medicine and public health. A literature review. *Front. Public Heal.* **2013**, *1*, 56. [CrossRef]

18. Barrington-Trimis, J.L.; Urman, R.; Berhane, K.; Unger, J.B.; Cruz, T.B.; Pentz, M.A.; Samet, J.M.; Leventhal, A.M.; McConnell, R. E-Cigarettes and Future Cigarette Use. *Pediatrics* **2016**, *138*, e20160379. [CrossRef]

19. Jenssen, B.P.; Boykan, R. Electronic Cigarettes and Youth in the United States: A Call to Action (at the Local, National and Global Levels). *Children* **2019**, *6*, 30. [CrossRef]

20. Zainol Abidin, N.; Zainal Abidin, E.; Zulkifli, A.; Karuppiah, K.; Syed Ismail, S.N.; Amer Nordin, A.S. Electronic cigarettes and indoor air quality: A review of studies using human volunteers. *Rev. Environ. Health* **2017**, *32*, 235–244. [CrossRef]

21. Callahan-Lyon, P. Electronic cigarettes: Human health effects. *Tob. Control* **2014**, *23*. [CrossRef]

22. Auer, R.; Concha-Lozano, N.; Jacot-Sadowski, I. Heat-Not-Burn Tobacco Cigarettes. *JAMA Intern. Med.* **2017**, *177*, 1050–1052. [CrossRef]

23. Ruprecht, A.A.; De Marco, C.; Saffari, A.; Pozzi, P.; Mazza, R.; Veronese, C.; Angellotti, G.; Munarini, E.; Ogliari, A.C.; Westerdahl, D.; et al. Environmental pollution and emission factors of electronic cigarettes, heat-not-burn tobacco products, and conventional cigarettes. *Aerosol Sci. Technol.* **2017**, *51*, 674–684. [CrossRef]

24. Schober, W.; Szendrei, K.; Matzen, W.; Osiander-Fuchs, H.; Heitmann, D.; Schettgen, T.; Jörres, R.A.; Fromme, H. Use of electronic cigarettes (e-cigarettes) impairs indoor air quality and increases FeNO levels of e-cigarette consumers. *Int. J. Hyg. Environ. Health* **2014**, *217*, 628–637. [CrossRef] [PubMed]

25. Jiang, Z.; Ding, X.; Fang, T.; Huang, H.; Zhou, W.; Sun, Q. Study on heat transfer process of a heat not burn tobacco product flow field. *J. Phys. Conf. Ser.* **2018**, *1064*, 012011. [CrossRef]

26. Schober, W.; Fembacher, L.; Frenzen, A.; Fromme, H. Passive exposure to pollutants from conventional cigarettes and new electronic smoking devices (IQOS, e-cigarette) in passenger cars. *Int. J. Hyg. Environ. Health* **2019**, *222*, 486–493. [CrossRef] [PubMed]

27. Cervellati, F.; Muresan, X.M.; Sticozzi, C.; Gambari, R.; Montagner, G.; Forman, H.J.; Torricelli, C.; Maioli, E.; Valacchi, G. Comparative effects between electronic and cigarette smoke in human keratinocytes and epithelial lung cells. *Toxicol. Vitr.* **2014**, *28*, 999–1005. [CrossRef]

28. Pisinger, C.; Døssing, M. A systematic review of health effects of electronic cigarettes. *Prev. Med. (Baltim).* **2014**, *69*, 248–260. [CrossRef]

29. Protano, C.; Manigrasso, M.; Avino, P.; Vitali, M. Second-hand smoke generated by combustion and electronic smoking devices used in real scenarios: Ultrafine particle pollution and age-related dose assessment. *Environ. Int.* **2017**, *107*, 190–195. [CrossRef]

30. Martuzevicius, D.; Prasauskas, T.; Setyan, A.; O'Connell, G.; Cahours, X.; Julien, R.; Colard, S. Characterization of the Spatial and Temporal Dispersion Differences Between Exhaled E-Cigarette Mist and Cigarette Smoke. *Nicotine Tob. Res.* **2019**, *21*, 1371–1377. [CrossRef]

31. Cass, G.R.; Hughes, L.A.; Bhave, P.; Kleeman, M.J.; Allen, J.O.; Salmon, L.G. The chemical composition of atmospheric ultrafine particles. *Philos. Trans. R. Soc. A Math. Phys. Eng. Sci.* **2000**, *358*, 2581–2592. [CrossRef]

32. Avino, P.; Scungio, M.; Stabile, L.; Cortellessa, G.; Buonanno, G.; Manigrasso, M. Second-hand aerosol from tobacco and electronic cigarettes: Evaluation of the smoker emission rates and doses and lung cancer risk of passive smokers and vapers. *Sci. Environ.* **2018**, *642*, 137–147. [CrossRef]

33. Zhao, T.; Shu, S.; Guo, Q.; Zhu, Y. Effects of design parameters and puff topography on heating coil temperature and mainstream aerosols in electronic cigarettes. *Atmos. Environ.* **2016**, *134*, 61–69. [CrossRef]

34. Niranjan, R.; Thakur, A.K. The toxicological mechanisms of environmental soot (black carbon) and carbon black: Focus on Oxidative stress and inflammatory pathways. *Front. Immunol.* **2017**, *8*, 1–20. [CrossRef] [PubMed]

35. Dautzenberg, B. Heated tobacco: Technology and nature of emissions. In Proceedings of the 3rd ENSP-Cnpt International Conference on Tobacco Control 2018, Madrid, Spain, 14–16 June 2018.

36. You, R.; Lu, W.; Shan, M.; Berlin, J.M.; Samuel, E.L.G.; Marcano, D.C.; Sun, Z.; Sikkema, W.K.A.; Yuan, X.; Song, L.; et al. Nanoparticulate carbon black in cigarette smoke induces DNA cleavage and Th17-mediated emphysema. *Elife* **2015**, *4*, 1–20. [CrossRef] [PubMed]

37. van Drooge, B.L.; Marco, E.; Perez, N.; Grimalt, J.O. Influence of electronic cigarette vaping on the composition of indoor organic pollutants, particles, and exhaled breath of bystanders. *Environ. Sci. Pollut. Res.* **2019**, *26*, 4654–4666. [CrossRef]

38. Malmgren, A.; Riley, G. Biomass Power Generation. *Compr. Renew. Energy* **2012**, *5*, 27–53.

39. Sadjadi, K.; Minai, C. Comparison of Vital Lung Capacity between Smokers and Non-Smokers. *Saddleback J. Biol.* **2010**, *8*, 51–52.

40. Geiss, O.; Barrero-Moreno, J.; Tirendi, S.; Kotzias, D. Exposure to particulate matter in vehicle cabins of private cars. *Aerosol Air Qual. Res.* **2010**, *10*, 581–588. [CrossRef]

41. Rodgman, A.; Perfetti, T.A. *The Chemical Components of Tobacco and Tobacco Smoke*; CRC Press: Boca Raton, FL, USA, 2013; ISBN 9781420078831.

42. Lee, K.; Sohn, H.; Putti, K. In-vehicle exposures to particulate matter and black carbon. *J. Air Waste Manag. Assoc.* **2010**, *60*, 130–136. [CrossRef]

43. Cunha-Lopes, I.; Martins, V.; Faria, T.; Correia, C.; Almeida, S.M. Children's exposure to sized-fractioned particulate matter and black carbon in an urban environment. *Build. Environ.* **2019**, *155*, 187–194. [CrossRef]

44. Correia, C.; Martins, V.; Cunha-Lopes, I.; Faria, T.; Diapouli, E.; Eleftheriadis, K.; Almeida, S.M. Particle exposure and inhaled dose while commuting in Lisbon. *Environ. Pollut.* **2020**, *257*, 113547. [CrossRef]

45. Onat, B.; Şahin, Ü.A.; Uzun, B.; Akın, Ö.; Özkaya, F.; Ayvaz, C. Determinants of exposure to ultra fi ne particulate matter, black carbon, and $PM_{2.5}$ in common travel modes in Istanbul. *Atmos. Environ.* **2019**, *206*, 258–270. [CrossRef]

46. Fruin, S.A.; Winer, A.M.; Rodes, C.E. Black carbon concentrations in California vehicles and estimation of in-vehicle diesel exhaust particulate matter exposures. *Atmos. Environ.* **2004**, *38*, 4123–4133. [CrossRef]

47. Northcross, A.L.; Trinh, M.; Kim, J.; Jones, I.A.; Meyers, M.J.; Dempsey, D.D.; Benowitz, N.L.; Katharine Hammond, S. Particulate mass and polycyclic aromatic hydrocarbons exposure from secondhand smoke in the back seat of a vehicle. *Tob. Control* **2014**, *23*, 14–20. [CrossRef] [PubMed]
48. Dirks, K.N.; Talbot, N.; Salmond, J.A.; Costello, S.B. In-cabin vehicle carbon monoxide concentrations under different ventilation settings. *Atmosphere* **2018**, *9*, 338. [CrossRef]
49. Goh, C.C.; Kamarudin, L.M.; Shukri, S.; Abdullah, N.S.; Zakaria, A. Monitoring of carbon dioxide (CO2) accumulation in vehicle cabin. In Proceedings of the 2016 3rd International Conference on Electronic Design (ICED), Phuket, Thailand, 11–12 August 2016; pp. 427–432. [CrossRef]

International Journal of
*Environmental Research
and Public Health*

Article

Evaluation on Air Purifier's Performance in Reducing the Concentration of Fine Particulate Matter for Occupants according to its Operation Methods

Hyungyu Park [1], Seonghyun Park [2] and Janghoo Seo [3,*]

[1] Department of Architecture, Graduated School, Kookmin University, Seoul 02707, Korea; sang911008@kookmin.ac.kr

[2] Department of Industry-Academic Cooperation Foundation, Kookmin University, Seoul 02707, Korea; marine86@kookmin.ac.kr

[3] School of Architecture, Kookmin University, Seoul 02707, Korea

* Correspondence: seojh@kookmin.ac.kr; Tel.: +82-02-910-4593

Received: 29 April 2020; Accepted: 29 July 2020; Published: 1 August 2020

Abstract: Fine particulate matter entering the body through breathing cause serious damage to humans. In South Korea, filter-type air purifiers are used to eliminate indoor fine particulate matter, and there has been a broad range of studies on the spread of fine particulate matter and air purifiers. However, earlier studies have not evaluated an operating method of air purifiers considering the inflow of fine particulate matter into the body or reduction performance of the concentration of fine particulate matter. There is a limit to controlling the concentration of fine particulate matter of the overall space where an air purifier is fixed in one spot as the source of indoor fine particulate matter is varied. Accordingly, this study analyzed changes in the concentration of indoor fine particulate matter through an experiment according to the discharging method and location of a fixed air purifier considering the inflow route of fine particulate matter into the body and their harmfulness. The study evaluated the purifiers' performance in reducing the concentration of fine particulate matter in the occupants' breathing zone according to the operation method in which a movable air purifier responds to the movement of occupants. The results showed the concentration of fine particulate matter around the breathing zone of the occupants had decreased by about 51 $\mu g/m^3$ compared to the surrounding concentration in terms of the operating method in which an air purifier tracks occupants in real-time, and a decrease of about 68 $\mu g/m^3$ in terms of the operating method in which an air purifier controls the zone. On the other hand, a real-time occupant tracking method may face a threshold due to the moving path of an air purifier and changes in the number of occupants. A zone controlling method is deemed suitable as an operating method of a movable air purifier to reduce the concentration of fine particulate matter in the breathing zone of occupants.

Keywords: particulate matters (PM); air purifier; experiment; real-time monitoring unit; transfer unit; occupant; breathing zone

---

1. Introduction

Recently, high levels of fine particulate matter are observed in the atmosphere of the East Asian regions, including South Korea, regardless of the season due to rapid economic growth and industrialization, causing adverse effects on the body [1,2]. The inflow of outdoor particulate matter into the indoor space is causing deterioration of the indoor air quality. According to the World Health Organization, 3.8 million people a year die prematurely from illness attributable to household air pollution [3]. Particulate matter is classified into PM_{10}, which has particulate matter less than 10 μm and $PM_{2.5}$, which has particulate matter less than 2.5 μm [4]. Particles, in general, are filtered using the cilia or respiratory tract and cannot enter the lung when they are under the body, but fine particles of small size can penetrate the lung and accumulate in the alveoli. Particles accumulated in the alveoli cause an inflammatory response and generate active oxygen, causing necrotized tissues. When particles accumulate in the bronchial tubes, they cause phlegm and cough as well as drying of the bronchial mucosa, which can easily allow penetration of germs and leave people with chronic lung conditions vulnerable to infectious diseases like pneumonia [5,6]. In particular, fine particulate matter less than 2.5 μm have been reported to cause diseases such as angina, stroke, and heart attack [7–10]. This indicates that fine and ultrafine particulate matter penetrating the body through breathing can cause serious damage to the body.

Filter-type air purifiers are used to control indoor fine particulate matter in South Korea, and there has been a broad range of studies on air purifiers and fine particulate matter. However, previous studies focused on the purification effects caused by the airflow rate and filter grade of air purifiers and have not evaluated the operating method of air purifiers that considers the inflow path of fine particulate matter into the body and the breathing zone of occupants or the performance in reducing the concentration of particulate matter [11–19]. As various sources cause indoor fine particulate matter [20], operating a fixed air purifier can reduce the average concentration of indoor fine particulate matter, but the reduction effect against fine particulate matter in local areas such as the breathing zone of occupants is inadequate. Contrarily, we thought that delivering purified air discharged from the air purifier to the breathing zone of occupants by changing the discharge angle of the air purifier and attaching a transfer unit can reduce the concentration of fine particulate matter in the air that occupants breathe regardless of the average concentration of particulate matter in indoor spaces.

Accordingly, this study considered the inflow of fine particulate matter into the body and their harmfulness and analyzed changes in the concentration of indoor fine particulate matter according to the discharge method of a fixed air purifier and changes in the concentration of indoor fine particulate matter through the experiment, and the study evaluated the reduction performance of fine particulate matter in the breathing zone of occupants according to the operation method in which a movable air purifier responds to the movement of occupants.

2. Materials and Methods

2.1. Experiment Equipment and Method

The experiment was conducted indoor after sunset to minimize changes in indoor temperature caused by solar radiation. Figure 1 shows the experimental space and installation locations of the experiment's equipment. The size of the experiment space is 7.0 m (L) × 10 m (W) × 2.8 m (H), and the volume is 196 m^3. The experiment space was divided into four artificial zones, and the particulate matter (PM) concentration measuring devices were installed in the center of each zone. Pollutants were set to be generated from a height of 1.2 m from the center of the hallway wall. Table 1 shows the experiment devices used in this study. The target pollutant was a Polystyrene Latex (PSL) 2 μm standard solution, which was mixed with distilled water and sprayed in the space with the aerosol generator. The scanning electron microscope [21] was used to observe and analyze the particle shape of PSL, and the result showed the particle had an even spherical shape of 2.0 μm (SD ± 3%), as shown in Figure 2. The TES-5322 model was used for concentrating the PM concentration, and the measurement accuracy of the $PM_{2.5}$ concentration is below ± 5μg/m^3 when less than 50 μg/m^3, and below ± 10% when greater than 50 μg/m^3. The H13 grade filter that can filter out particles greater than 0.3 μm by 99.95% was applied.

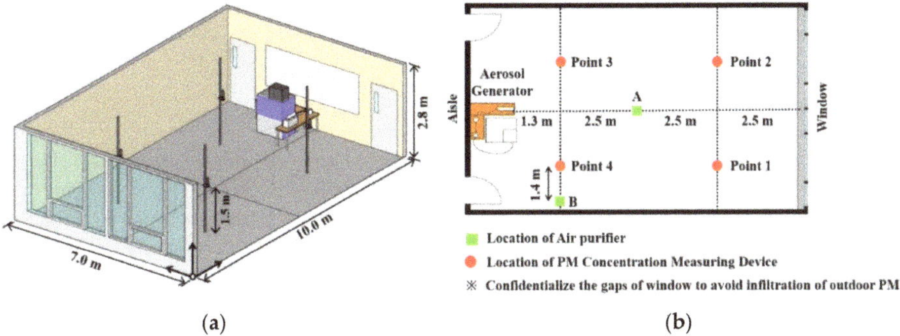

(a)   (b)

Figure 1. Experimental space and location of experiment equipment. (**a**) Experimental space; (**b**) Location of experiment equipment.

Table 1. Equipment and materials for experiment.

Purpose	Model	Quantity
Air purifier	Mi Air 2	1
PM measuring device	TES-5322	5
Thermo-hygrometer	TR-72WF	4
Anemometer	TSI 9565	1
$PM_{2.5}$ generating device	Aerosol generator	1
Standard particle solution	Polystyrene Latex (Standard particle 15 mL)	3
Air supply and flow control of aerosol generator	Compressor (AM 400D)	1
Water removal of aerosol air supply	Dryer (TX15K)	1

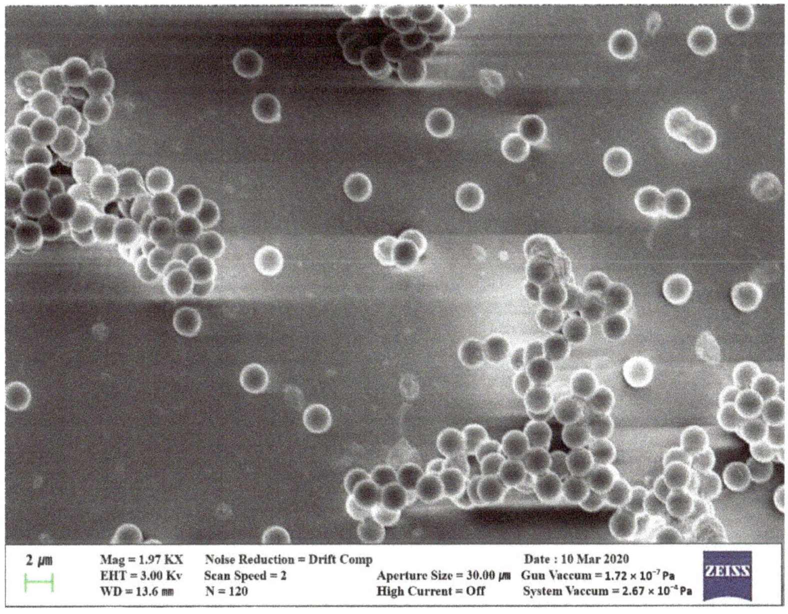

Figure 2. Expanded picture of standard solution through a scanning electron microscope (SEM).

The height of the PM concentration measuring device for measuring changes in the concentration of fine particulate matter in the breathing zone of occupants was set to 1.5 m by considering the breathing area of occupants while standing.

2.2. Arrangement of the Air Purifier Considering Draft

There are methods to control the concentration of various indoor pollutants such as fine particulate matter, including generation control, elimination control, and dilution control. If fine particulate matter is the target pollutant, the concentration can be controlled by eliminating the source or collecting fine particulate matter dispersed in the space. Therefore, the most effective method for reducing the concentration of indoor fine particulate matter is to locate and operate an air purifier close to the source of pollutants. However, outdoor fine particulate matter can flow in through openings or cracks in the building, and fine particulate matter are generated from a variety of activities by occupants, including cooking, exercising, and ventilation. Thus, the source of pollutants cannot be easily characterized. Furthermore, many measurement sensors are required to detect the source of the indoor fine particulate matter early [22–25]. This study selected a method of delivering air discharged from the air purifier toward the breathing zone of occupants to improve the reduction performance of the concentration fine particulate matter by operating the air purifier. In this case, occupants could inhale purified air regardless of the source. Contrarily, occupants could feel discomfort from drafts if the velocity of air currents is extremely high, and therefore, this needs to be considered. Gong et al. performed the study on the allowable wind velocity range needed for finding human recognition for the local airflow, under isothermal and non-isothermal conditions, and designed individual ventilation of the tropical regions through the experiment, and suggested that a wind velocity of minimum 0.3 m/s and maximum 0.9 m/s is acceptable based on the comfort of occupants [26].

This study set the installation height of the $PM_{2.5}$ measuring device to 1.5 m by considering the breathing area of occupants while standing. In the case where purified air is discharged toward the breathing zone of occupants, the air purifier was arranged to maintain the velocity of the discharged air that reaches the 1.5 m-high measuring device at 0.8 m/s, as shown in Figure 3. The speed set here takes the precedence over the performance of reducing fine dust concentrations in the breathing zone and not over the human thermal comfort. The air purifier used in this study inhales air from the bottom and discharges from the top, and if the discharge angle is changed toward the breathing zone of the occupants, the velocity may change due to the pressure loss. Accordingly, a separate discharge outlet was made with a 3D printer to form the same face velocity regardless of the discharge angle, and it is shown in Figure 4.

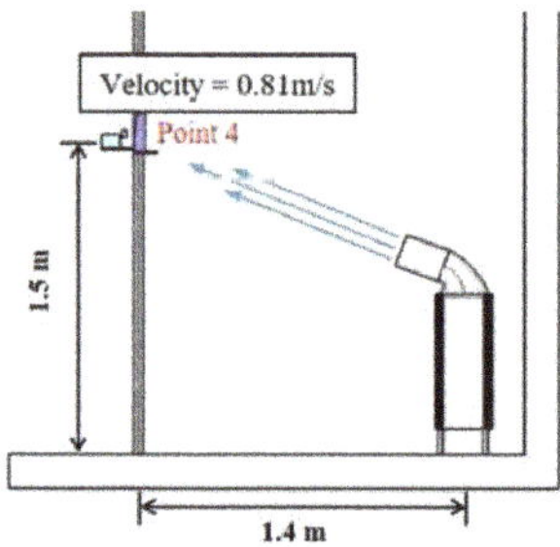

Figure 3. Air purifier of discharge direction toward breathing.

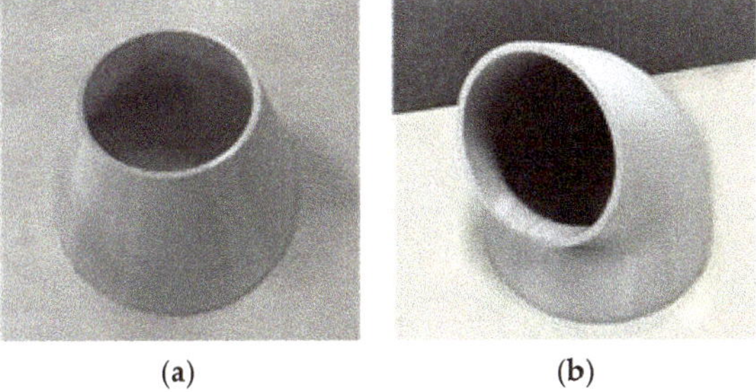

(a)   (b)

Figure 4. Discharge outlet made with a 3D print.: (**a**) Discharged upward (**b**) Discharged toward occupants.

2.3. Real-time Monitoring and Transfer Unit Production Using an Arduino Board

To reduce the concentration of fine particulate matter in the breathing zone of occupants through the operation of an air purifier, it is necessary to respond to the movement of occupants or monitor real-time changes of the concentration of fine particulate matter, as well as changes in the discharge angle of the air purifier. A data communication module for real-time monitoring of the measurement sensor for indoor fine particulate matter and a remote transfer unit for the air purifier were produced. Figure 5 shows the overview of the data communication module system, and Figure 6 shows the diagram of the remote transfer unit.

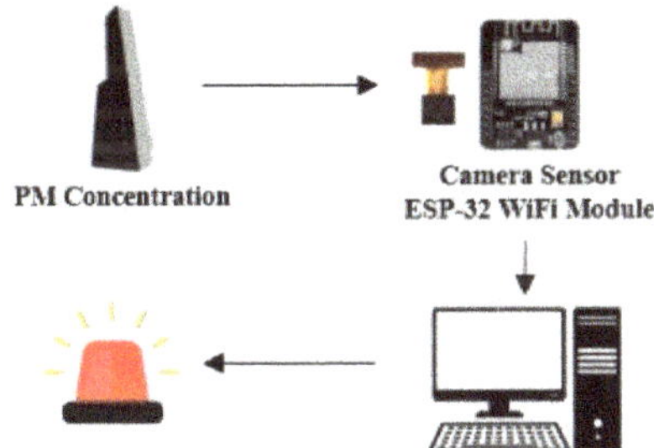

Figure 5. Overview of the data communication module system.

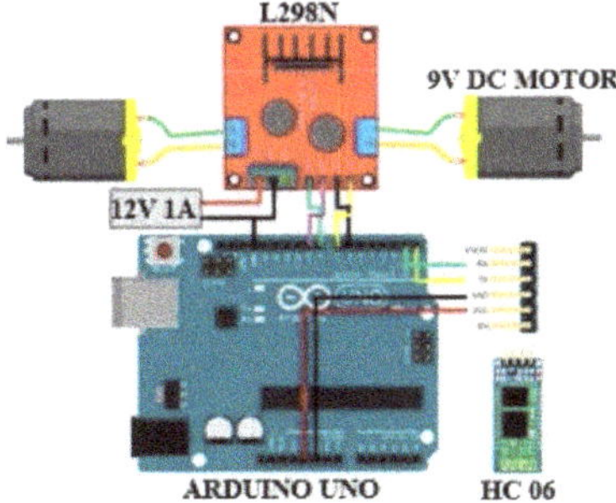

Figure 6. Diagram of the Arduino-based remote transfer unit.

The $PM_{2.5}$ concentration value monitored through the camera sensor is transferred to the server via ESP 32 WiFi module. Whether the air purifier needs to be moved is determined based on the $PM_{2.5}$ concentration data received, and if its movement is necessary, the remote transfer unit sends a signal to the Arduino board to move it.

Figure 7 shows the data communication module and remote transfer unit for real-time monitoring of the movable air purifier made for the experiment of this study.

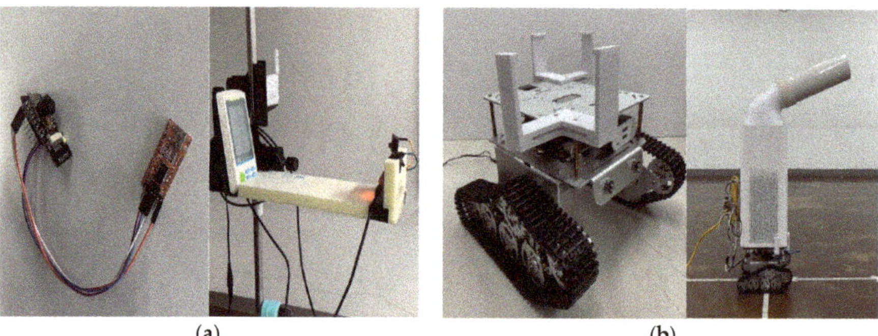

(a)   (b)

Figure 7. Data communication module and remote transfer unit used in the experiment: (**a**) Data communication module, (**b**) Remote transfer unit.

3. Measurement Experiment with Fixed Air Purifier

3.1. Case Setting

This experiment was conducted to evaluate the air purifiers' performance in reducing the concentration of fine particulate matter in the breathing zone of occupants according to the location of the fixed air purifier and the direction of the discharged air. The air purifier was operated to make the background PM$_{2.5}$ concentration of the experimental space to be less than 30 µg/m^3. As shown in Figure 8, the experiment generated pollutants through the aerosol generator after 10 min from the start of measurement, and the air purifier was operated for 150 min through a remote control after 30 min from the operation of the aerosol generator. The generation of pollutants was discontinued after 90 min.

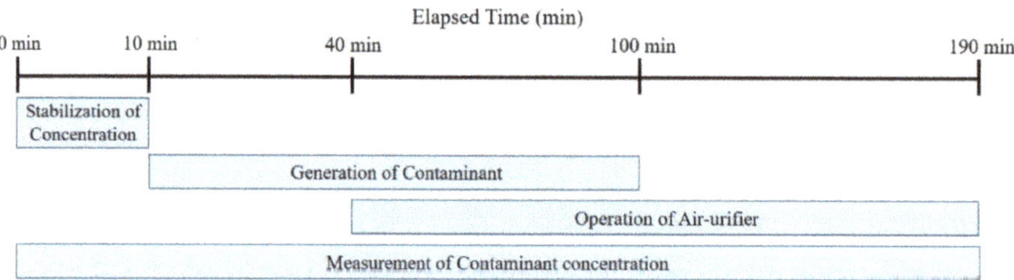

Figure 8. Experiment progress.

The case conditions according to the location of the air purifier and changes in the direction of discharge are shown in Table 2. In Case 1, the air purifier of upward discharge is located in the center, and it is the control group to compare with other cases in which the location and discharge direction of the air purifier were changed. In Case 2, the air purifier of the upward discharge was located on the nearby wall that was 1.4 m away from the P 4 (Point 4) measuring device and was set to analyze the reduction of the concentration of fine particulate matter in the breathing zone of occupants by installing the purifier near occupants regardless of the location of the pollutant source.

Table 2. Experiment case conditions.

Cases Variables	1	2	3	4
Location of air purifier	A	B	A	B
Direction of discharge	Upward		Toward point 4	
Initial PM concentration	27 μg/m^3	26 μg/m^3	29 μg/m^3	30 μg/m^3
Temperature/Relative Humidity	$25 \pm 2.2\,°C/28 \pm 1.8\%$			

Case 3 and 4 were set to analyze the reduction of the fine particulate matter in local areas according to changes in the discharge direction of the air purifier. Case 3 set the direction of the discharge from the central location toward the P 4 measuring device as in Case 1, and the wind velocity of the air current reaching the P 4 measurement point was 0.23 m/s. Case 4 installed the air purifier near the occupants while changing the direction of the discharge, and the wind velocity of the air current reaching the P 4 measurement point was 0.81 m/s. The wind velocity value was the average value measured every 10 s for a total of 60 times with TSI 9565 (TSI Inc., Shoreview, MN, USA; Velocity).

3.2. Experiment Results

Figure 9 shows changes in the PM$_{2.5}$ concentration according to the time by measurement point of each case. The changes in the concentration of fine particulate matter in the local areas according to the location of the air purifier were compared and analyzed with the results of Case 1 and Case 2. In all the measurement points of Case 1 and 2, the PM$_{2.5}$ concentration has increased according to the pollutants, and the rising curve of the PM$_{2.5}$ concentration was maintained consistently. When the generation of pollutants was discontinued, the PM$_{2.5}$ concentration was decreased by the air purifier. In Case 1, the average PM$_{2.5}$ concentration difference in P 1 (Point 1), P 2 (Point 2), P 3 (Point 3), and P 4 (Point 4) was about 25 μg/m^3 from 40 min to 100 min. This is due to the distance from the source of pollutants to the measurement points. In Case 2, the PM$_{2.5}$ concentration in P 4 that is close to the air purifier of the upward discharge was higher than the PM$_{2.5}$ of the other measurement points. This was due to the movement of pollutants to the discharge outlet of the air purifier. In Case 3, P 4 also had the highest PM$_{2.5}$ concentration and this meant that the P 4 point assumed as the breathing zone of the occupants did not fall within the influence of purified air discharged from the air purifier. The concentration of PM at points other than point P 4 was lower than that of P 4, but compared to Case 1, the PM concentration in P 3 decreased by 21 μg/m^3, whereas points P 1 and P 2 maintained similar levels.

Meanwhile, the average PM$_{2.5}$ concentration in P 4 from 40 min to 100 min was about 77 μg/m^3 lower than other measurement points. In particular, the PM$_{2.5}$ concentration in P 4 was maintained at less than 50 μg/m^3 since the operation of the air purifier, and it reached 30 μg/m^3 within 20 min after discontinuing the operation of the aerosol generator. It could be believed that the concentration of PM at point P 3 increased due to the diffusion of pollutants from the pollution source, and points P 1 and P 2 were similar to that of Case 1. Thus, the air purifier must be located near occupants, and the breathing zone of occupants must be located within the influence of purified air discharged from the air purifier to reduce the concentration of fine particulate matter in the breathing zone of occupants.

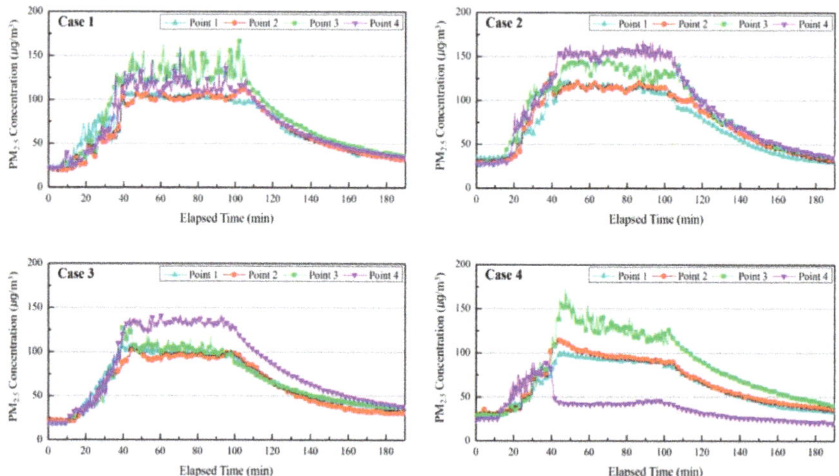

Figure 9. Comparing PM$_{2.5}$ concentrations by measurement points (**Case 1–4**).

4. Measurement Experiment with Movable Air Purifier

4.1. Operating Method of Movable Air Purifier

In this section, the real-time monitoring device and the remote transfer unit were applied to allow the air purifier, which delivers purified air to the breathing zone of occupants, to respond to the movement of occupants and the reduction performance of fine particulate matter in the breathing zone of occupants was evaluated. In addition, the experiment was conducted through the real-time occupant tracking method and the zone controlling method by considering the threshold according to the moving path of the air purifier.

4.1.1. Real-Time Occupant Tracking Method

Firstly, a controlling method of tracking occupants to reduce the concentration of fine particulate matter in the breathing zone of occupants, regardless of the surrounding concentration was set. The performance evaluation of this method was conducted through the measurement experiment.

The target experimental space and location for measuring the PM$_{2.5}$ concentrations were the same as those in the experiment conditions of Section 2, but an additional device for measuring the concentration of fine particulate matter was installed at the P 5 (Point 5), which was an assumed breathing zone of the occupant. The measuring device for the concentration of fine particulate matter and the movable air purifier were fixed in the location, which was assumed as the breathing zone of the occupants. Figure 10 displays an implementation of a controlling method of tracking occupants in real-time. For the experiment, the aerosol generator and the air purifier were operated at the same time after 10 min from starting the recording of each measurement device, and the operation of the aerosol generator was discontinued after 190 min. The movement of the measuring device in the breathing zone of the occupants was set to move to random points every 15, 20, and 30 min to analyze the reduction effect of the concentration of fine particulate matter in the breathing zone of the occupants in terms of controlling the real-time tracking of occupants.

4.1.2. Zone Controlling Method

Meanwhile, a method of controlling occupant tracking of a movable air purifier has limitations in its actual use. such as the moving path of the air purifier or changes in the number of occupants. Accordingly, the study conducted a measurement experiment for performance verification by setting the moving locations of the air purifier by zone, which allows the air purifier to move to the set location with the high concentration of fine particulate matter or the area where occupants are located and reduce the concentration of fine particulate matter in the breathing zone of occupants within that area. The experimental space and the locations of the device for measuring the concentration of fine particulate matter are the same as those in the measurement experiment for the occupant tracking method.

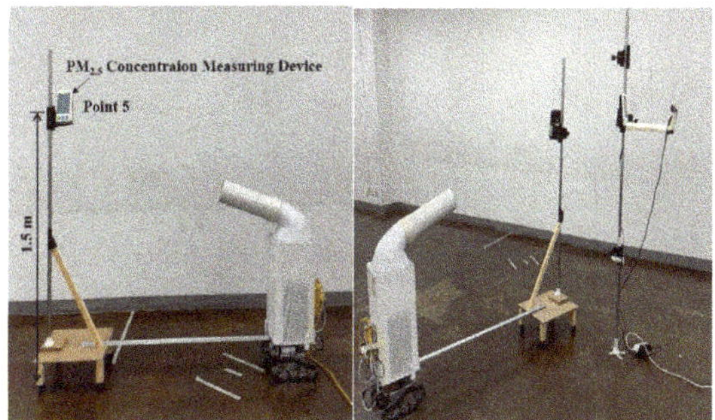

Figure 10. A method of implementing the air purifier that tracks occupants in real-time.

As shown in Figure 11, the experiment generated pollutants after 10 min from the start of measurement, and the air purifier was operated at the same time. The generation of pollutants was discontinued after 190 min. Figure 12 shows the moving path and location of the air purifier, and the air purifier was set to be located near the walls of each zone using the moving path of the wall.

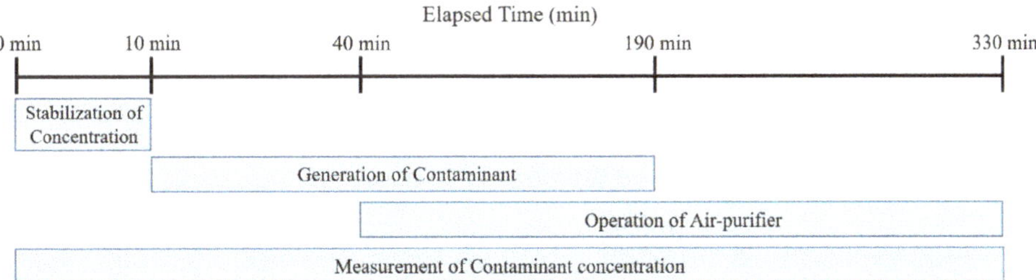

Figure 11. Experiment Progress (Real-Time Occupant Tracking Method and Zone Controlling Method).

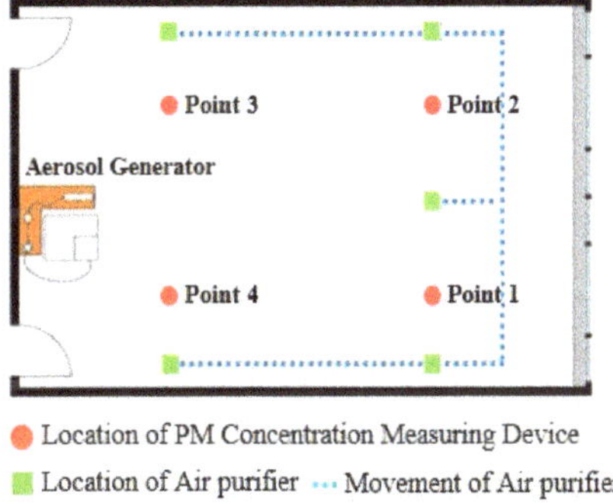

Figure 12. Moving path of the air purifier that controls zones.

The air purifier was relocated every 30 min to analyze changes in the concentration of indoor fine particulate matter, according to the moving path and location (Figure 13).

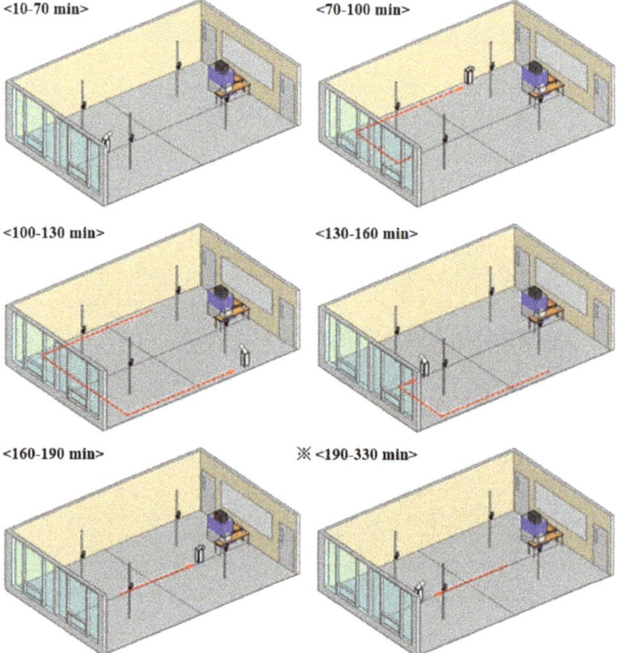

Figure 13. Location of the air purifier at each time and moving path from the previous location (Zone Controlling Method).

4.2. Experiment Results

Figure 14 shows changes in the $PM_{2.5}$ concentration by measurement point according to the operation of the air purifier that tracks occupants in real-time.

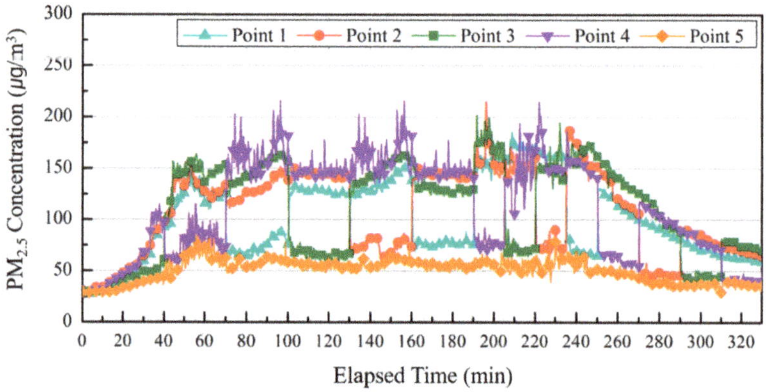

Figure 14. Comparing $PM_{2.5}$ concentrations by measurement points according to the operation of the air purifier that tracks occupants.

The PM2.5 concentrations were compared and analyzed at P 5 (Point 5), which is considered as the breathing zone of occupants and continuously received purified air currents discharged from the air purifier and other measurement points. The $PM_{2.5}$ concentrations at all measurement points had increased after 10 min from the start of the measurement and reached about 100 $\mu g/m^3$ at P 1, P 2, and P 4 at 40 min. The upward slopes of P 5 and P 3, which were closest to the discharge outlet of the air purifier, were relatively low compared to those of other points and showed concentrations of 40 $\mu g/m^3$ and 55 $\mu g/m^3$, respectively. During the overall experiment time, the average concentration at P 5 was reduced by 51 $\mu g/m^3$ compared to the average concentration at other measurement points, which demonstrated that the operation of an air purifier that tracks occupants in real-time could reduce the concentration of fine particulate matter in the breathing zone of occupants even if the concentration of fine particulate matter in the surrounding area is high. In addition, the concentration of fine particulate matter in measurement points near the discharge outlet was lower than other measurement points. This means the operation of an air purifier that tracks occupants in real-time could reduce the concentration of fine particulate matter in the breathing zone of occupants regardless of the surrounding concentration of fine particulate matter. The result of measuring the $PM_{2.5}$ concentration at measurement points near the discharge outlet proves that the concentration of fine particulate matter could be reduced if the location is close to the air purifier and is within the influence of the purified air discharged from the air purifier.

Figure 15 shows changes in the $PM_{2.5}$ concentration according to the time of each measurement point with the operation of the air purifier that controls zones through the set moving path.

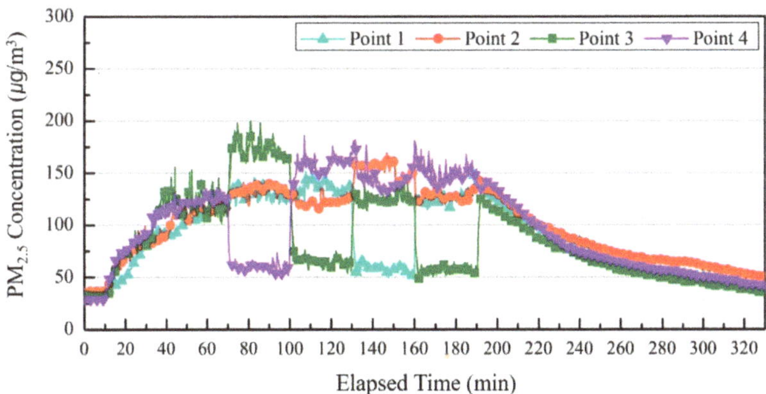

Figure 15. Comparing $PM_{2.5}$ concentrations by measurement points according to the operation of the air purifier that controls zones.

The $PM_{2.5}$ concentration in all the measurement points between 10–70 min located in the back of the room without moving an air purifier had increased to a similar level, reaching an average of 128 $\mu g/m^3$ concentration at 70 min.

At 70–100 min, when the moved air purifier was located at the P 4 wall point, the average $PM_{2.5}$ concentration at P 4 was about 56 $\mu g/m^3$ for 30 min, and the $PM_{2.5}$ concentration at P 1, P 2, and P 3 were 129, 133, and 172 $\mu g/m^3$, respectively. The reason for the high $PM_{2.5}$ concentration in P 3 was because pollutants generated from the aerosol generator were spread to the P 3 point due to the direction of the air current discharged from the air purifier. At 100–130 min, when the air purifier was located by the nearby wall of P 3 using the same moving path, a similar result for 70–100 min was observed. The average $PM_{2.5}$ concentration in the P 3 point was about 60 $\mu g/m^3$ for 30 min due to controlling of purified air discharged from the air purifier. P 4 showed the highest level of 161 $\mu g/m^3$. At 130–160 min when the air purifier was located at the nearby wall of P 1, the $PM_{2.5}$ concentration at P 1, which was affected by the purified air discharged from the air purifier, was about 56 $\mu g/m^3$ for

30 min, and the $PM_{2.5}$ concentrations at the P 2, P 3, and P 4 points were 153, 124, and 145 $\mu g/m^3$ for 30 min, respectively. At 160–190 min, the $PM_{2.5}$ concentration in P 3, which was controlled by purified air discharged from the air purifier, was 55 $\mu g/m^3$ for 30 min, and the $PM_{2.5}$ concentrations at P 1, P 2, and P 4 were 125, 129, and 151 $\mu g/m^3$ for 30 min, respectively. The average concentration of the fine particulate matter from 70 min to 190 min in measurement points within the influence of purified air discharged from the air purifier was reduced by about 68 $\mu g/m^3$ compared to other measurement points. This is due to the decreased removal rate of fine particulate matter in the breathing zone of the occupants for the corresponding time because the air purifier for controlling zones moves around.

5. Conclusions

This study performed measurement experiments to evaluate the reduction performance of fine particulate matter concentration in the breathing zone for a method of delivering purified air discharged from an air purifier to the human breathing zone. The results are as follows:

The method of installing a fixed air purifier at a location adjacent to occupants without changing the discharged direction cannot improve the performance in reducing the concentration of fine particulate matter within the breathing zone of occupants; the method of delivering purified air discharged from an air purifier can better reduce the concertation of fine particulate matter in the breathing zone of occupants compared to the concentration of fine particulate matter in the surrounding area, but if the distance from the air purifier to the controlling point is distant and so the velocity of airflow is not sufficient, there is no effect in reducing the fine particulate matter concentration.

In the case of a mobile air purifier, the real-time occupant tracking method was effective in terms of reducing the concentration of fine particulate matter in the breathing zone of occupants by 51 $\mu g/m^3$ compared to the surrounding PM concentration, but there are limits in actual use regarding the moving path of the air purifier or the change in the number of occupants. On the contrary, the operation of the movable air purifier showed that the fine PM concentration of the occupant's respiratory zone to be 68 $\mu g/m^3$ lower than other measurement points. Thus, it is more effective to divide the target space by zone and move an air purifier around considering the number of occupants and the mobility of an air purifier.

This study compared and evaluated an air purifier's performance in reducing the concentration of fine particulate matter in the breathing zone of occupants against measurement points of other zones by setting measurement points of each zone in one place when delivering purified air discharged from an air purifier. However, the level of reduction performance of fine particulate matter could differ according to the range of influencing purified air currents within the same zone. In addition, in the case of the zone controlling method, there are also limitations regarding the moving path and the distance between the occupants. Accordingly, a follow-up study on the range of purified air currents discharged from an air purifier, method of expanding the range, method of installing air purifier on the ceiling, and changing control mode according to the number of occupants will be conducted in future.

Author Contributions: Conceptualization, H.P.; methodology, H.P. and S.P.; writing-original draft preparation, H.P.; writing—review and editing, H.P. and S.P.; Supervision, J.S.; All authors have read and agreed to the published version of the manuscript.

Funding: This work was supported by the National Research Foundation of Korea (NRF) grant funded by the Korea government (MSIT) (No. 2018R1A2B2007165).

Conflicts of Interest: The authors declare no conflict of interest.

References

1. Dockery, D.W. Health effects of particulate air pollution. *Ann. Epidemiol.* **2009**, *19*, 257–263. [CrossRef] [PubMed]

2. Yamaguchi, N.; Ichijo, T.; Sakotani, A.; Baba, T.; Nasu, M. Global dispersion of bacterial cells on Asian dust. *Sci. Rep.* **2012**, *2*, 525–529. [CrossRef] [PubMed]

3. World Health Organization. Household Air Pollution and Health. Available online: http://www.who.int/news-room/fact-sheets/detail/household-air-pollution-and-health (accessed on 24 January 2020).
4. United States Environmental Protection Agency. Particulate Matter (PM) pollution. Available online: https://www.epa.gov/pm-pollution (accessed on 24 January 2020).
5. Kim, K.; Kabir, E.; Kabir, S. A review on the human health impact of airborne particulate matter. *Environ. Int.* **2015**, *74*, 136–143. [CrossRef] [PubMed]
6. Franck, U.; Odeh, S.; Wiedensohler, A.; Wehner, B.; Herbarth, O. The effect of particle size on cardiovascular disorders—The smaller the worse. *Sci. Total Environ.* **2011**, *409*, 4217–4221. [CrossRef]
7. Ristovski, Z.D.; Miljevic, B.; Surawski, N.C.; Morawska, L.; Fong, K.M.; Goh, F.; Yang, I.A. Respiratory health effects of diesel particulate matter. *Respirology* **2012**, *17*, 201–212. [CrossRef]
8. Brook, R.D.; Rajagopalan, S.; Pope, C.A.; Brook, J.R.; Bhatnagar, A.; Diez-Roux, A.V.; Holguin, F.; Hong, Y.; Luepker, R.V.; Mittleman, M.A.; et al. Particulate matter air pollution and cardiovascular disease: An update to the scientific statement from the American heart association. *Circulation* **2010**, *121*, 2331–2378. [CrossRef]
9. Hoek, G.; Krishnan, R.M.; Beelen, R.; Peters, A.; Ostro, B.; Brunekreef, B. Long-term air pollution exposure and cardio-respiratory mortality: A review. *Environ. Health* **2013**, *12*, 43. [CrossRef]
10. Shah, A.S.; Lee, K.K.; Mcallister, D.A.; Hunter, A.; Nair, H.; Whiteley, W.; Langlish, J.P.; Newby, D.E.; Mills, N.L. Short term exposure to air pollution and stroke: Systematic review and meta-analysis. *BMJ* **2015**, *350*, h1295. [CrossRef]
11. Fisk, W.J.; Faulkner, D.; Palonen, J.; Sepanen, O. Performance and costs of particle air filtration technologies. *Indoor Air* **2002**, *12*, 223–234. [CrossRef]
12. Cha, D.W.; Kim, S.H.; Cho, S.Y. Radon reduction efficiency of the air cleaner equipped with a Korea carbon filter. *J. Odor Indoor Environ.* **2017**, *16*, 364–368. [CrossRef]
13. Urso, P.; Cattaneo, A.; Garramone, G.; Peruzzo, C.; Cavallo, D.M.; Carrer, P. Identification of particulate matter determinants in residential homes. *Build. Environ.* **2015**, *86*, 61–69. [CrossRef]
14. Liuliu, D.; Batterman, S.; Parker, E.; Godwin, C.; Chin, J.; O'Toole, A.; Robins, T.; Brakefield-Caldwel, W.; Lewis, T. Particle concentrations and effectiveness of free-standing air filters in bedrooms of children with asthma in Detroit, Michigan. *Build. Environ.* **2011**, *46*, 2303–2313. [CrossRef]
15. Ma, H.; Shen, H.; Shui, T.; Li, Q.; Zhou, L. Experimental study on ultrafine particle removal performance of portable air cleaners with different filters in an office room. *Int. J. Environ. Res. Public Health* **2015**, *13*, 102. [CrossRef] [PubMed]
16. Fermo, P.; Comite, V.; Falciola, L.; Guglielmi, V.; Miani, A. Efficiency of an air cleaner device in reducing aerosol particulate matter (PM) in indoor environments. *Int. J. Environ. Res. Public Health* **2020**, *17*, 18. [CrossRef] [PubMed]
17. Park, J.H.; Lee, T.J.; Park, M.J.; Oh, H.; Jo, Y.M. Effect of air cleaners and school characteristic on classroom concentration of particulate matter in 34 elementary schools in Korea. *Build. Environ.* **2019**, *167*, 106437. [CrossRef] [PubMed]
18. Oh, H.J.; Nam, I.; Yun, H.; Kim, J.; Yang, J. Characterization of indoor air quality and efficiency of air purifier in childcare centers, Korea. *Build. Environ.* **2014**, *82*, 203–214. [CrossRef]
19. Ciuzas, D.; Prasauskas, T.; Krugly, E.; Jurelionis, A.; Seduikyte, L.; Martuzevicius, D. Indoor air quality management by combined ventilation and air cleaning: An experimental study. *Aerosol Air Qual. Res.* **2016**, *16*, 2550–2559. [CrossRef]
20. Li, Z.; Wen, Q.; Zhang, R. Sources, health effects and control strategies of indoor fine particulate matter (PM2.5): A review. *Sci. Total Environ.* **2017**, *586*, 610–622. [CrossRef]
21. Goldstein, J.I.; Newbury, D.E.; Michael, J.R.; Ritchie, N.W.M.; Scott, J.H.J.; Joy, D.C. *Scanning Electron Microscopy and X-ray Microanalysis*; Springer: Berlin, Germany, 2017.
22. Nazaroff, W.W. Indoor particle dynamics. *Indoor Air* **2004**, 175–183. [CrossRef]
23. Liu, D.; Nazaroff, W.W. Modeling pollutant penetration across building envelopes. *Atmos. Environ.* **2001**, *35*, 4451–4462. [CrossRef]
24. Wallace, L.A.; Emmerich, S.J.; Howard-reed, C. Continuous measurements of air change rates in an occupied house for 1 year: The effect of temperature, wind, fans, and windows. *J. Expo. Anal. Environ. Epidemiol.* **2002**, *12*, 296–306. [CrossRef] [PubMed]

25. Long, C.M.; Suh, H.H.; Catalano, P.J.; Koutrakis, P. Using time-and size-resolved particulate data to quantify indoor penetration and deposition behavior. *Environ. Sci. Technol.* **2001**, *35*, 2089–2099. [CrossRef] [PubMed]
26. Gong, N.; Tham, K.W.; Melikov, A.K.; Wyon, D.P.; Sekhar, S.C.; Cheong, K.W. The acceptable air velocity range for local air movement in the tropics. *HVAC&R Res.* **2006**, *12*, 1065–1076. [CrossRef]

International Journal of
Environmental Research and Public Health

Article

Quantifying the Health Burden Misclassification from the Use of Different PM$_{2.5}$ Exposure Tier Models: A Case Study of London

Vasilis Kazakos, Zhiwen Luo * and Ian Ewart

School of Built Environment, University of Reading, Reading RG6 6DF, UK; v.kazakos@pgr.reading.ac.uk (V.K.); i.j.ewart@reading.ac.uk (I.E.)
* Correspondence: z.luo@reading.ac.uk

Received: 27 December 2019; Accepted: 4 February 2020; Published: 9 February 2020

Abstract: Exposure to PM$_{2.5}$ has been associated with increased mortality in urban areas. Hence, reducing the uncertainty in human exposure assessments is essential for more accurate health burden estimates. Here, we quantified the misclassification that occurred when using different exposure approaches to predict the mortality burden of a population using London as a case study. We developed a framework for quantifying the misclassification of the total mortality burden attributable to exposure to fine particulate matter (PM$_{2.5}$) in four major microenvironments (MEs) (dwellings, aboveground transportation, London Underground (LU) and outdoors) in the Greater London Area (GLA), in 2017. We demonstrated that differences exist between five different exposure Tier-models with incrementally increasing complexity, moving from static to more dynamic approaches. BenMap-CE, the open source software developed by the U.S. Environmental Protection Agency, was used as a tool to achieve spatial distribution of the ambient concentration by interpolating the monitoring data to the unmonitored areas and ultimately estimating the change in mortality on a fine resolution. Indoor exposure to PM$_{2.5}$ is the largest contributor to total population exposure concentration, accounting for 83% of total predicted population exposure, followed by the London Underground, which contributes approximately 15%, despite the average time spent there by Londoners being only 0.4%. After incorporating housing stock and time-activity data, moving from static to most dynamic metric, Inner London showed the highest reduction in exposure concentration (i.e., approximately 37%) and as a result the largest change in mortality (i.e., health burden/mortality misclassification) was observed in central GLA. Overall, our findings showed that using outdoor concentration as a surrogate for total population exposure but ignoring different exposure concentration that occur indoors and time spent in transit, led to a misclassification of 1174–1541 mean predicted mortalities in GLA. We generally confirm that increasing the complexity and incorporating important microenvironments, such as the highly polluted LU, could significantly reduce the misclassification of health burden assessments.

Keywords: PM$_{2.5}$; population exposure; tier-models; health burden misclassification; BenMap-CE

---

1. Introduction

There is growing evidence that air pollution and specifically fine particulate matter (PM$_{2.5}$) contribute significantly to health burden and further, there is a close relationship between long-term air pollution exposure and adverse health effects in urban populations [1,2]. The assessment of Global Burden of Disease (GDB) indicated that PM$_{2.5}$ contributed 4.24 million deaths globally in 2015 [3]. Assessments of human health effects attributed to an air pollutant are dependent on the magnitude of human exposure to that pollutant. Thus, the accuracy of a health burden assessment is determined by the uncertainty of predicted population exposure. Quantifying the population exposure to air pollution is subject to several challenges:

The spatiotemporal variability of ambient concentration is strongly influenced by emissions dynamics, predominantly from road transport, (such as peaks in traffic-related pollution during rush hours), meteorological conditions, which determine the transport and dilution of air pollutants and local conditions such as the urban form (e.g., the presence of high buildings can reduce the dispersion of the pollutants), which are the most important factors leading to significant variation of air pollutants in urban areas.

The proportion of outdoor air infiltrated to indoor microenvironments (MEs) is influenced by different housing designs and patterns of behaviour inside the building.

The spatiotemporal variability of people's activity (population time–activity patterns) in various MEs [4].

Around 75% of European populations live in cities, with a highly variable range of activities carried out at different times and in different places [5]. The quality of data, or the absence of key components within an epidemiological exposure assessment, is likely to affect the magnitude and significance of the prediction misclassification in a health burden assessment (Figure 1).

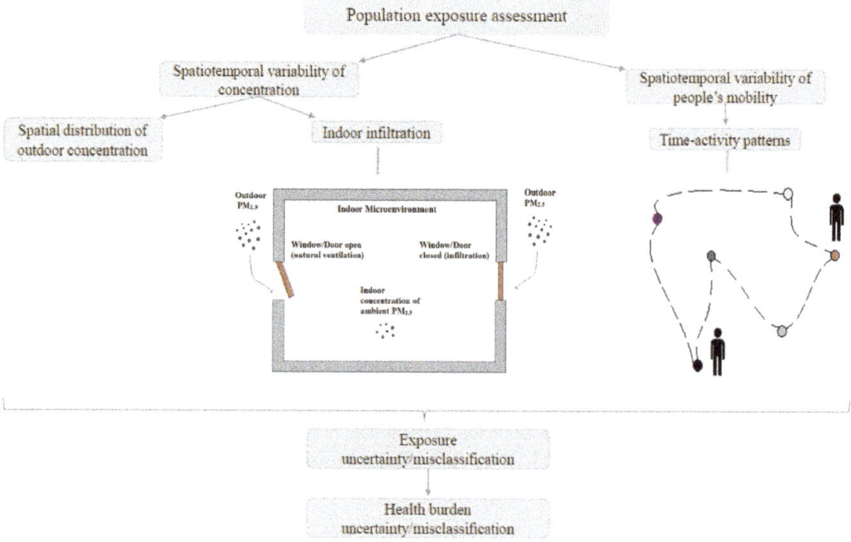

Figure 1. Schematic diagram of an exposure assessment structure for health burden misclassification.

Traditionally, epidemiological studies relied on centralized ambient concentration measurements of limited monitoring sites [6–10]. This is likely to lead to an exposure error, since several monitoring studies have suggested that air pollution data from a single site can represent only a small surrounding area especially in urban environments, due to pollutants' spatial heterogeneity [11,12]. Ambient air pollutant concentration can be estimated in several ways such as through field observations, statistical modelling such as land-use regression (LUR) and air quality dispersion models (AQM) that can use various spatial resolutions [13]. Willers et al. [14] indicated that using air quality data measured at a single site and assuming that exposure across cities was the same, could cause considerable misclassification of exposure. In their study, they examined the difference in mortality risk between neighborhoods in the city of Rotterdam and found that the mortality risks between neighborhoods had a difference of up to 7%. By utilizing land use regression techniques and air quality models, several studies have managed to demonstrate that an increased spatial resolution of the exposure concentration could lead to significantly different exposure or health burden estimates [15–18]. Similarly, Punger and West [19] assessed the effect of spatial resolution to population-weighted $PM_{2.5}$ concentrations in the

U.S. by utilizing the Community Multiscale Air Quality (CMAQ) model. They found that population exposures, maximum concentrations and standard deviations all reduced at coarser resolutions. At 408 km resolution, exposure and maximum concentration were 27% and 71% lower, respectively than those at 12km resolution. Attributable mortality also reduced as the resolution became coarser. Several studies have shown that coarse resolutions might result in lower mortality attributed to $PM_{2.5}$ [20]. Fenech et al. [21] concluded that total mortality estimates were sensitive to model resolution up to ±5% across Europe, whereas Korhonen et al. [22] found that, considering only local sources of primary $PM_{2.5}$, the mortality reduced by 70% in the whole country (Finland) and 74% in urban areas when the resolution changed from 250 m to 50 km.

Apart from the exposure misclassification due to the different levels of spatiotemporal resolution of outdoor concentration, there are other significant contributors, in particular the infiltration of outdoor pollutants to indoor MEs and different time-activity patterns in MEs. As particles infiltrate and persist indoors, where people living in urban areas spent over 80% of their daily time [23], most of the exposure to $PM_{2.5}$ actually occurred in the indoor microenvironments [23–25]. The fraction of ambient $PM_{2.5}$ that infiltrates indoor microenvironments can vary due to particle size, building characteristics, meteorological conditions and human activities [26]. Consequently, relying on outdoor measurements alone can therefore lead to exposure misclassification. Moreover, variations in the time spent in various MEs (e.g., outdoors, indoors, vehicles, subway) also influence population exposure to outdoor-generated $PM_{2.5}$ due to the spatial variability of both outdoor concentrations and the indoor transport of ambient $PM_{2.5}$. Baxter et al. [27] compared four different approaches to $PM_{2.5}$ exposure prediction, where each model was of a different complexity. In their study they focused on the heterogeneity in exposures but did not investigate the influence on health effect predictions. They suggested that geographic heterogeneity in both housing stock (and thus a relatively consistent Air Change Rate) and human activity patterns contribute to significant heterogeneity in ambient $PM_{2.5}$ exposure both within and between cities that is not demonstrated by stationary monitors. Ma et al. [28] compared three different types of $PM_{2.5}$ exposure estimates to illustrate the differences in exposure levels between estimates obtained from different approaches. They found that the daily average $PM_{2.5}$ exposures for residents with different activity patterns may vary significantly even when they were living in the same neighborhood. Several studies have also investigated the correlation between outdoor $PM_{2.5}$ and mortality, although their results are skewed by the fact that people spend the majority of the time indoors. Ji and Zhao [29] used existing epidemiological data on ambient $PM_{2.5}$-related mortality to estimate mortality associated with indoor exposure to outdoor-generated PM. This was the first attempt to quantify that relationship and their results indicated that outdoor PM had substantial effects on health caused by exposure within indoor MEs. Recently, Fenech and Aquilina [30] used the annual mean $PM_{2.5}$ concentrations derived from local fixed monitoring stations to estimate the $PM_{2.5}$-related mortality in the Maltese Islands. They found that the attributable fraction of all-cause mortality associated with long-term $PM_{2.5}$ exposure ranged from 5.9% to 11.8%, indicating that $PM_{2.5}$ concentration is a major component of attributable deaths. Azimi and Stephens [31] used a modified version of the common exposure-response function and developed a framework for estimating the total U.S. mortality burden attributed to exposure to $PM_{2.5}$ of both indoor and outdoor origins. They found that residential exposure to outdoor-generated $PM_{2.5}$ accounted for 36% to 48% of total exposure, indicating that efforts to mitigate mortality associated with exposure to $PM_{2.5}$ should consider indoor pollution control as well.

That of particular importance is how different exposure approaches impact long-term health burden/mortality predictions and the magnitude of the resultant impact. We made multiple comparisons between refined ambient $PM_{2.5}$ exposure surrogates (that account for important factors such as the infiltration and time-activity) and the fixed-site monitor $PM_{2.5}$ concentrations to indicate the importance of including more dynamic data to epidemiological studies and to demonstrate how more complex modelling approaches modify mortality predictions. By using BenMap-CE we were able to provide the spatial distribution of health outcomes influenced by the exposure misclassification. While a number of studies have already investigated exposure misclassification when using different approaches and others have estimated health effects based on specific exposure metrics, the aim of this work is to move one step further and answer the question: how much is the misclassification that occurs when using different exposure approaches to predict health burden?

2. Materials and Methods

This work aims to quantify the long-term health burden misclassification that occurs when different $PM_{2.5}$ exposure metrics are utilized. An ecologic design was used to generate associations between air pollution exposure and health outcomes. We investigated the Greater London Area (GLA), building on recent exposure studies that have explicitly estimated London population exposure using hybrid dynamic models [32]. Here, we have described five different exposure Tier-models of incrementally increased complexity are considered by gradually including data of important MEs, such as infiltration rates of the different dwelling types and the London Underground, where London's population spend most or part of their daily time. The London Travel Demand Survey (LTDS) space-activity data were categorized into three major ME groups. The analysis estimated the magnitude of the change (i.e., avoided or incurred) in mortalities when moving from the central-site monitored concentrations as a surrogate for population exposure (Tier-model 1) to more refined exposure Tier-models. The original ambient $PM_{2.5}$ concentrations were based on average hourly data measured by 23 monitoring stations located in the GLA [33] and the examined MEs were: i) indoors (i.e., home-indoor), ii) aboveground transportation iii) the London Underground and iv) outdoors. The following sections describe the structure of the methodology and the development of each component.

2.1. Developing Tier Models to Estimate Human Exposure

To capture different exposure assessment methods that have been used in epidemiology, we developed five different Tier models of increased complexity, moving from static to more dynamic approaches (Table 1). This method was separated into two parts: i) The microenvironments and time-activity patterns were classified and calculated based on the derived information; ii) the time-activity information was matched with corresponding microenvironmental concentrations to estimate the dynamic time-weighted exposure. The exposure time was considered costly and the metrics estimated the annual hourly-average $PM_{2.5}$ exposures, which were then used as an input for BenMap-CE [34].

Table 1. Tier models for assessing the time-weighted exposure.

Tier Models	Exposure Equation	Approach
Tier model 1	$E = C_{out}$	Outdoors only
Tier model 2	$E = C_{ind}$	Indoor only
Tier model 3	$E = \sum C_{out}*F_i *x_i,$	Indoor only (dwellings)
Tier model 4	$E = (C_{out}*t_{out}) + (\sum C_{out}*F_i *x_i)*t_{ind} + (\sum C_{out}* F_j)* t_{abg} + (C_{undg}*t_{undg})$	Outdoor + Indoor + Transportation (abg. and undg.)
Tier model 5	$E = (C_{out}* t_{out}) + [(\sum C_{out}*F_i *x_i)*t_{ind}] + (\sum C_{out}* F_j)* t_{abg} + (C_{undg\text{-}hvac}*t_{undg\text{-}hvac}) + (C_{deep\text{-}undg} *t_{deep\text{-}undg})$	Outdoor + Indoor + Transportation (abg., deep-line + subsurface undg)

The Tier-model stages and the respective approaches are briefly described below.
Tier model 1: Outdoor

$$E = C_{out},\tag{1}$$

where E is mean exposure and C_{out} is mean outdoor concentration of $PM_{2.5}$.

Hourly readings were extracted from the London Air Quality Network (LAQN) [33]. LAQN consists of automatic monitoring equipment in fixed cabins, which measures air pollution at breathing height. It provides electronically available data on concentrations of major urban pollutants and has been used in several studies [35,36]. The ratified concentration data from 23 available monitoring stations in GLA were downloaded and added to BenMap-CE. Only the monitors that could provide at least 70% of the data for the whole year were selected. The ambient concentration was considered as representative of the total population exposure.

Tier model 2: Indoor

$$E = C_{in},\tag{2}$$

where C_{in} is the mean indoor (i.e., home-indoor) concentration.

This Tier model utilized the information of the spatially distributed concentration and the total average Indoor/Outdoor (I/O) ratio in GLA to estimate the exposure inside the residence [37].

Tier model 3: Indoor (dwellings)

$$E = \sum C_{out}{}^* F_i{}^* x_i,\tag{3}$$

where F_i is the infiltration rate of each dwelling type (i) and x_i is the frequency (%) of this type in London.

In this study, all the indoor environments were combined into one single ME (i.e., home-indoor) without considering other indoor environments, such as office or commercial buildings, due to the lack of infiltration data. Subsequently, the I/O ratios that we used also represented offices and other indoor places, assuming that the I/O ratios for other indoor MEs had the same values as domestic home buildings [32]. The I/O ratios of London's housing stock were obtained from Taylor et al. [37]. In their study they estimated the Indoor/Outdoor ratio of 15 building archetypes. We grouped these archetypes into five main dwelling types in response to available housing stock data in Middle-Super-Output-Area resolution obtained from the Mayor of London, Datastore [38]: i) flat, ii) bungalow, iii) terraced, iv) semi-detached and v) detached (Table 2). The frequency of each type could be calculated from the number of properties in the GLA, which represented 98.7% of the housing (The average I/O ratio was assigned to the unknown 1.13%). Figure 2 shows the annual average I/O ratios of $PM_{2.5}$ concentration in the GLA. The average ratios, including all dwelling types and their frequency, ranged from less than 0.54 to 0.59. The highest ratios were observed in Outer London, whereas the lowest ratios were observed in Inner and South West London, probably due to the newer building stock and the large number of flats in large buildings (London Datastore), where the available surface for infiltration was considerably smaller.

Table 2. London's dwelling group type descriptions, frequency in stock and average Indoor/Outdoor (I/O) ratios.

Dwelling Type	Frequency %	I/O Ratios	Total Average I/O Ratio (All Dwellings)
Bungalow	1.81	0.63	
Flat	50.4	0.54	
Terraced	28.1	0.56	0.56
Semi-detached	14.5	0.585	
Detached	4.06	0.585	
Unknown	1.13	0.56	

Tier model 4: Outdoor + Indoor + Transportation (aboveground and underground)

$$E = (C_{out}{}^* t_{out}) + (\sum C_{out}{}^* F_i{}^* x_i) {}^* t_{ind} + (\sum C_{out}{}^* F_j) {}^* t_{abg} + (C_{undg}{}^* t_{undg}),\tag{4}$$

where (j) is each aboveground transport-ME (tMEs) and t_{out}, t_{ind}, t_{abg} and t_{undg} is the fractional time spent (%) annually outdoors, indoors, aboveground tME and London Underground (LU) tME, respectively.

This Tier-model includes transportation as an additional microenvironment, where an urban population spends time during the day. This ME was categorized into aboveground and underground transportation. Aboveground transportation refers to car, bus and train, whereas underground to

London subway. By separating transportation into 2 groups we were able to evaluate the influence of a highly polluted ME, like the London Underground (described in the next section), on the total population exposure concentration.

Figure 2. Map of annual average Indoor/Outdoor (I/O) ratios used in our study.

The space–time–activity data for our study were based on the London Travel Transport Agency (LTDS) of Transport for London (TfL) [39] for the period between 2005 and 2010 (Table 3). The data were generated from the interviews of approximately 8000 households per year, providing very useful information about their daily time–activity patterns, including travel modes and trip times. The data were scaled to represent the population of London, excluding children under five years old [32].

Table 3. Summary table of the time–activity data.

Microenvironments (Groups)	Mode/Place	Time Spent (%)
Outdoor	Walking	1.3
	Cycling	0.1
Transportation (public/private)	Bus	0.7
	Car	1.6
	Rail	0.2
Indoor	Underground/DLR	0.4
	Home, office, other indoor	95.7

According to Smith et al. [32], the average daily percentage of time spent indoors was 95.7 %, whereas people spent 2.5%, 0.4% and 1.4% in aboveground transportation, London Underground and outside (walking or cycling), respectively. This proportion of time spent indoors also includes approximately 20% of surveyed people, who did not leave their house. In this study, these percentages were used as annual averages for the whole population over five years old, including the different times spent during weekdays and weekends.

For the in-vehicle exposure of the aboveground sub-microenvironment, we calculated the $PM_{2.5}$ concentration by solving the mass balance equation [30]:

$$dC_{in} / dt = \lambda_{win} * (C_{out} - C_{in}) - \eta\lambda_{HVAC} * C_{in} - V_g * (A' / V) * C_{in} + Q/V, \tag{5}$$

where C_{out} is the outdoor concentration around the vehicle, C_{in} the concentration inside the vehicle, λ_{win} and λ_{HVAC} are the hourly air exchange rates from the windows and mechanical ventilation system, respectively, n is the filter removal efficiency taking values between 0–1, V_g is the deposition velocity in (m/h), A' is the internal surface area, V is the volume of the vehicle and Q is the in-vehicle particle emission rate in µg/h. To solve this equation, the same values with Smith et al. [32] were used except for the concentrations and the commuter's surface was derived from Song et al. [40], in order to calculate A'.

Tier model 5: Outdoor + Indoor + Transportation (aboveground and underground→ deep lines + subsurface lines).

The time-weighted exposure equation associated with this Tier model stage is:

$$E = (C_{out}* t_{out}) + [(\textstyle\sum C_{out}* F_i * x_i) * t_{ind}] + (\textstyle\sum C_{out}* F_j) * t_{abg} + (C_{undg\text{-}hvac} * t_{undg\text{-}hvac}) + (C_{deep\text{-}undg} * t_{deep\text{-}undg}),$$

(6)

In the 5th and most complex Tier model, the same procedure as in Tier 4 was followed but the London underground microenvironment was further divided into subsurface and deep lines to reflect the significant difference in concentration on two types of lines. The use of mechanical ventilation in the subsurface lines results in much lower $PM_{2.5}$ concentrations than the deep lines due to air filtration (explicitly described in the next section). Hence, by dividing the underground into two subgroups we were able to improve the exposure estimates and to examine the contribution of a very highly polluted microenvironment to the total exposure. The proportion of time spent in each of those two subcategories was assumed according to the number of annual journeys completed in each line during 2017, where 77% were made by the deep-line underground and 33% by the subsurface.

$PM_{2.5}$ Concentration in the London Underground

As the London Underground microenvironment was unable to be accurately represented by the outdoor measurements, due to its high concentration of $PM_{2.5}$ and its limited connection to the outside world, a series of air pollution measurements were conducted inside the London Underground. The $PM_{2.5}$ measurements took place on five major London Underground platforms and trains (Bakerloo line, Circle line, Central line, District line and Victoria line) by using the portable DustTrak II Aerosol Monitor 8534, a light scatter laser photometer, which could provide a large number of real-time readings. The current selection of the lines was decided in order for both the deep without mechanical ventilation lines and the subsurface with HVAC lines to be represented by our measurements.

Our original intention was that the measurements would reflect the cold and the warm period of 2017. Hence, the experiment was conducted during the morning and the afternoon for one week in February and one week in July. The average concentration in the London Underground for the whole year was very high, approximately 218 $\mu g/m^3$, albeit when we grouped the lines into deep without HVAC lines (Central, Bakerloo and Victoria) and subsurface lines with HVAC (Circle, District) we noticed a remarkable difference between the two concentrations (70.2 $\mu g/m^3$ for the subsurface lines and 365.6 $\mu g/m^3$ for the deep lines). The $PM_{2.5}$ concentration levels in the unmeasured lines were assumed to be similar to these measured. The classification of the unmeasured lines was made according to their depth and ventilation system.

In the London Underground, Seaton et al. [41] reported higher platform concentrations of 480 $\mu g/m^3$. Recently, Smith et al. [42] assessed day to day variation in LU concentrations and compared them with those above ground. During their campaign, 22 repeat journeys were made on weekday mornings over a period of five months. They found that the subsurface ventilated District line had the lowest $PM_{2.5}$ concentration levels (i.e., mean 32 $\mu g/m^3$) and the deep unventilated Victoria line the highest (i.e., mean 381 $\mu g/m^3$), while the mean concentration in the LU, according to their measurements, was 302 $\mu g/m^3$. Although their monitoring method and equipment were different from those used in this study and the sampling period was longer, their findings do not differ significantly from ours. Even though the station measurements in the UK are limited, most of the studies made so far have measured approximately two times higher concentrations in the London Underground than in other undergrounds worldwide [43,44], probably due to its age and the limited ventilation systems.

2.2. Simulating $PM_{2.5}$ Exposure Concentration and Estimating Health Impact Using BenMap-CE

The environmental Benefits Mapping and Analysis Program—Community Edition (BenMap-CE) is a powerful Geographical Information system (GIS)-based program that estimates the health effects associated with the change in air quality [34,45]. These data consisted of a middle layer super output areas (MSOA) map of GLA, the derived monitoring data and London's population data, in order to estimate the health impact. BenMap-CE provides three interpolation methods: the closest monitor, the fixed radius, and Voronoi Neighbour Averaging (VNA). Among the incorporated methods, VNA was the most suitable for our case, covering the unmonitored areas and giving the best spatial distribution of the concentration.

After uploading the essential data and determining the appropriate Health Impact Function (HIF) for our analysis, we were able to quantify the health impact misclassification (i.e., change in all-cause mortality, either incurred or avoided) resulting from the exposure metric differences. In this study, the following long-term health impact function was used to estimate the change in all-cause mortality [46]:

$$\Delta Y = Y_0 * (1 - e^{-\beta \Delta PM}) * \text{Pop}, \tag{7}$$

where ΔY is the change in health effect, Y_0 is the baseline mortality rate (the mortality rate at minimum risk concentration), β is the unitless beta coefficient, ΔPM is the change in the exposure rates between Tier 1 and the other Tier models (Tier 1 is the base case) and Pop is the exposed population.

One limitation of the aforementioned effort to estimate the health impact of indoor air pollution is the use of the mortality effect estimate (i.e., beta coefficient) that is usually taken directly from the epidemiology literature on the studies conducted for outdoor air pollution. Therefore, to account for that fact, some studies on the health effects of outdoor-generated $PM_{2.5}$ introduced a method for modifying the mortality effect estimate (i.e., beta coefficient) based on the average infiltration factor combined with the mean fraction of time spent in indoor MEs [13,47,48]. However, the application of the adjusted coefficient is solely for the component of indoor $PM_{2.5}$ of outdoor origin and not of indoor $PM_{2.5}$ in total. The way indoor particle sources are treated has a larger impact than the adjustment of the coefficient for the outdoor-generated fine particles and remains an evidence gap of considerable public health importance. In another study, Logue et al. [49] used a central estimate of the beta coefficient for premature mortality related to both indoor- and outdoor-generated $PM_{2.5}$, which was directly derived from the epidemiology literature. In our case, due to the mobile monitoring conducted in the LU and the distinct function of BenMap-CE, a central mortality effect coefficient from Pope et al. [50] was used as an input. The mortality effect coefficient was utilized to generate BenMap's health impact functions in the direction of estimating the change in estimates of mortality (either avoided or incurred) when using different exposure metrics. Furthermore, we estimated the percentage decrease in the predicted avoided cases when moving from the less complex (static) metrics to more dynamic metrics.

3. Results

3.1. Exposure Metrics Summary

The highest annual average exposure concentration was approximately 13.1 µg/m^3 for Tier model 1. Tier model 2 and Tier model 3 indicated that the exposure that occurred indoors was much lower than outdoors due to the infiltration rates of the buildings, resulting in annual average exposure concentrations of 7.18 µg/m^3 and 7.26 µg/m^3, respectively. There was an approximately 45% reduction between Tier 1 and Tier 3. This result clearly suggests that spending long periods of time indoors, reduces the exposure to outdoor-generated air pollution. The incorporation of transportation and predominately the highly polluted London Underground in Tier model 4 resulted in an elevated exposure concentration (8.28 µg/m^3), pinpointing that even though the time spent in transit is only 2.9%, this microenvironment has a significant contribution to the total exposure. By dividing the

London Underground into subsurface with HVAC and deep line without HVAC, we were able to quantify the impact of the most highly polluted ME on the total exposures (the deep-line underground). Tier 5 showed an approximately 0.30 $\mu g/m^3$ higher exposure concentration (8.60 $\mu g/m^3$) than Tier 4, where an average concentration for the whole underground was used (Table 4).

Table 4. Annual exposure calculated in each model stage.

Tier Models	Annual Exposure ($\mu g/m^3$)	Standard Deviation (+/– $\mu g/m^3$)
Tier model 1	13.07	1.2
Tier model 2	7.18	0.66
Tier model 3	7.26	0.66
Tier model 4	8.3	0.67
Tier model 5	8.6	0.67

$PM_{2.5}$ exposure concentration maps for each Tier-model stage were created by BenMap-CE showing how the exposure was distributed across GLA. Figure 3a,b illustrate the spatial distribution of the annual exposures in Tier 1 and 5. The maps of Tiers 2, 3 and 4 are included in the Supplementary Information (Figure A1a–c in the Appendix A.1). The highest exposure concentrations occurred in Inner London for both Tier 1 and Tier 5 (15.4 $\mu g/m^3$ and 10.1 $\mu g/m^3$, respectively), whereas the lowest exposures were observed in Western GLA (less than 10.9 and less than 7.10 $\mu g/m^3$ for Tier 1 and 5, respectively). The incorporation of indoor infiltration along with time-activity data led to an overall mitigation of the exposure concentrations in GLA when Tiers 2, 3, 4 and 5 were used. After the utilization of our most complex model, Tier 5 had the highest difference observed at the centre with approximately 37% (Figure 3c), while average reduction in GLA was approximately 34%. Inner London continued to show the highest values (Figure 3b), although the infiltration factors in Inner London were lower than in the outskirts. This could be due to the much higher outdoor concentrations in Inner GLA than in the Outer. In Inner London, the higher number of sources of anthropogenic and traffic-related pollutants, including $PM_{2.5}$, generate significantly higher ambient pollution levels. Several studies suggest that traffic pollutants are elevated above background concentrations around major roads and highways [13,51]. The percentage of exposure concentration reduction in Tiers 2, 3 and 4 after comparison with our baseline exposure concentration (Tier 1), is illustrated in Figure A2a–c in the Appendix A.2. Apart from proximity to roads, fewer green spaces and the densely constructed city center may also contribute to the higher levels of outdoor particulate pollution [52–54]. Urban populations are subject to daily activity patterns, so that exposure is not a static phenomenon but should be quantified as a function of concentration and time [4]. Therefore, by assigning people's exposure to a single location (e.g., at their residence) and ignoring highly polluted MEs such as the subway, it is unlikely to accurately represent total exposure. Hence, by gradually incorporating time-activity data and indoor MEs, the spatial variability of the exposure concentration across GLA increased. Since we used annual average time-activity data for the entire GLA, time-activity could not change the spatial pattern of the exposure. In our case, the spatial variability of the housing stock and I/O ratios across GLA were the main reasons for any increase in the spatial variability of the exposure concentration.

Figure 4 presents the contribution of each examined microenvironment to the total exposure estimated by Tier-model 5. Indoor exposure concentration is clearly the dominating contributor (approximately 83%) to the total exposure (due to the time that people spent there–95.7%) followed by the deep-line underground ME (14%) albeit people spent on average only 0.31% of their annual time. According to our measurements, the $PM_{2.5}$ concentration in deep underground lines was around 28 times higher than the outdoor levels, which rationalized the high contribution of that ME to total exposure. In contrast, London population spent only 1.4% of its annual time outside and the outdoor ME contributed only 2% to the total exposure concentration. The findings described above indicate that outdoor $PM_{2.5}$ levels are unlikely to accurately represent the total exposure of an urban population like in London.

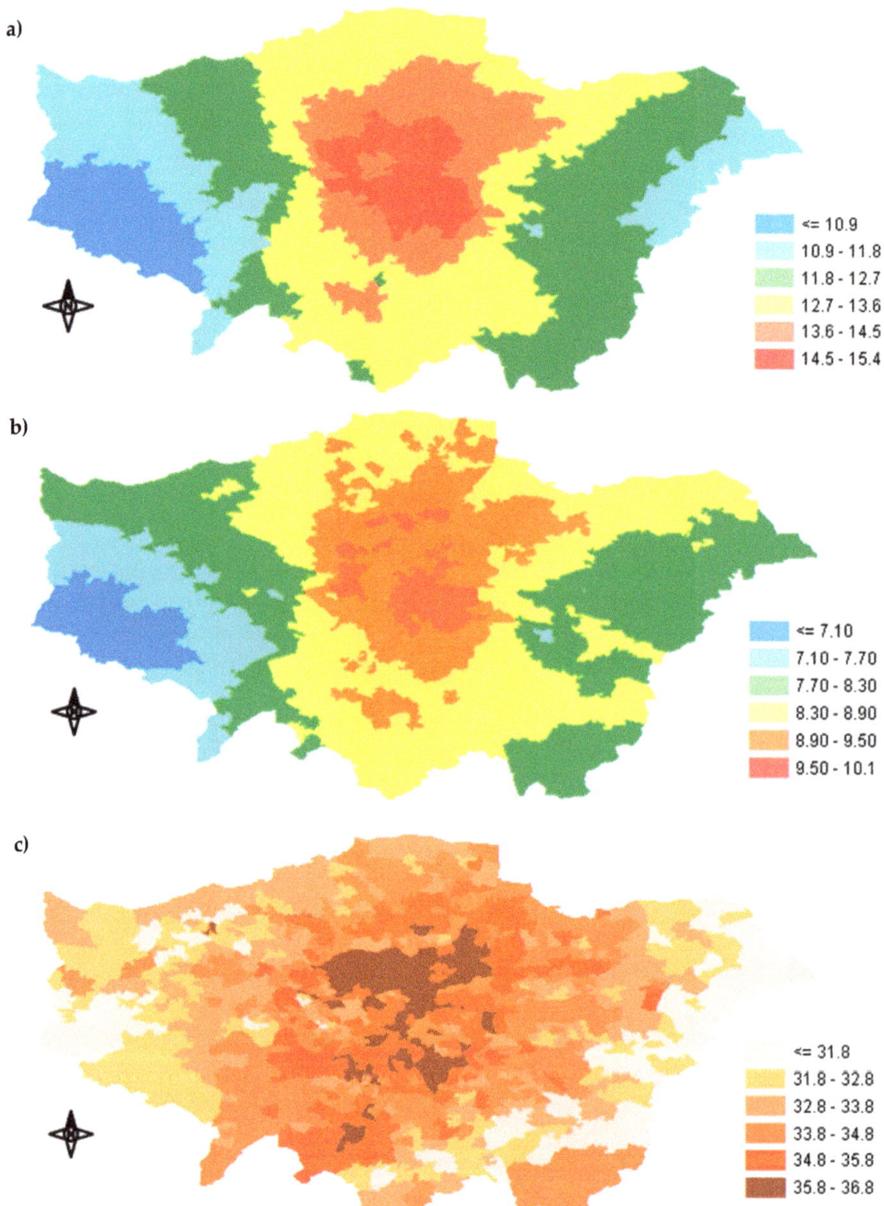

Figure 3. Maps of annual distributions across the Greater London Area (GLA): (**a**) Tier-model 1 annual mean PM$_{2.5}$ exposure concentration (μg/m^3), (**b**) Tier-model 5 annual mean PM$_{2.5}$ exposure concentration (μg/m^3), and (**c**) percentage of the PM$_{2.5}$ exposure concentration difference between Tier-model 1 and Tier-model 5.

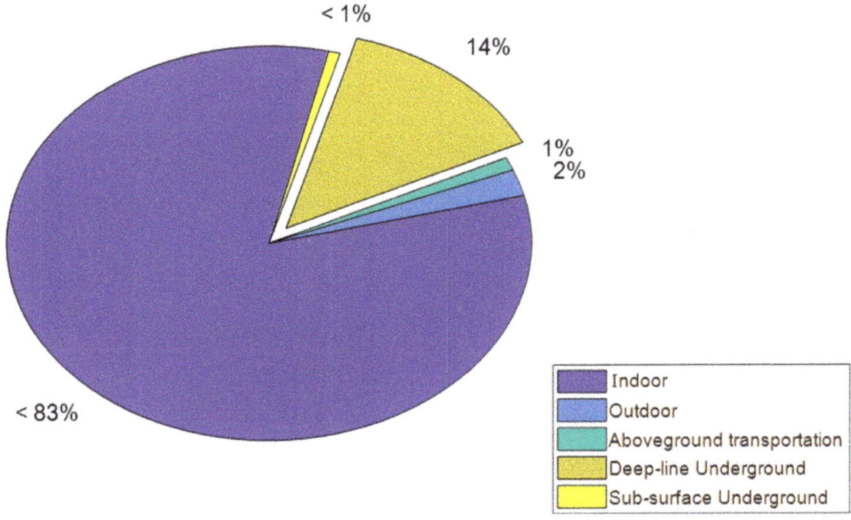

Figure 4. Contribution of each microenvironment (ME) to the total exposure. Indoor exposure shows the greater contribution followed by the deep underground lines.

3.2. Epidemiological Implications and Health Impact Misclassification

Because in epidemiology the concentration from central-site monitors is used as a proxy for the exposure to air pollution, we selected Tier 1 as our reference and compared it with the estimates of Tiers 2, 3, 4 and 5. The mean change in the estimates of all-cause mortality when applying Tier model 2 was predicted to be 1541 (95% CI: (427–2633)) deaths, while when using Tier 3 exposure concentration estimates the death cases were reduced to 1521 (95% CI: (421– 2598)). The impact on mortality when applying the 4th Tier model, which included the transportation microenvironments (tMEs), was estimated to be 1257 (95% CI: (347–2151)) cases. Due to the significance of the deep-line underground, the most complex Tier model 5 presented the lowest number of cases compared with the other 3 metrics (Tiers 2, 3 and 4). Namely, once Tier 5 was applied the prediction for the estimated avoided mortalities were 1174 (95% CI: (324 – 2010)). We can assume that the calculated change in mortality represents the potential health burden misclassification that might occur when changing the exposure metrics to assess the population exposure. Subsequently, we were able to estimate the percentage decrease in predicted mortalities when altering the exposure metric's complexity. The substantial changes in avoided mortality predictions indicate that using a static exposure approach in a study might lead to significant uncertainty in a health burden assessment. As anticipated, the predicted mortality was significantly reduced when increasing the model complexity. The highest changes were observed in Tier-model 2 and 3, due to the time that people spent indoors in urban areas, the big difference between outdoor and indoor exposure and the absence of highly polluted transportation MEs, pinpointing the importance of taking into serious consideration the exposure that occurs inside buildings when estimating health effects. The model predicted most avoided cases when Tier 2 was applied and while increasing complexity the cases showed a decrease of 1.95%, 18.4% and 23.8% for Tier 3, 4 and 5, respectively. As explained above, the London Underground contributes significantly to the total average exposure concentration of the study population by increasing the estimates. Therefore, we can securely presume that this is the main reason for the high decrease in avoided mortalities when Tier 4 and, predominantly, Tier 5 were used in BenMap-CE.

All results are summarized in Table 5.

Table 5. Change in the annual mean estimates of mortality (predicted avoided mortalities) between the different exposure metrics and decrease between the estimated change in mortality predictions.

Tier Models	2.5th percentile	97.5th percentile	Mean	Decrease (%)
Tier models 1–2	427	2633	1541	
Tier models 1–3	421	2598	1521	1.95
Tier models 1–4	347	2151	1257	18.4
Tier models 1–5	324	2010	1174	23.8

Looking at the spatial distribution of the predicted change in mortalities shown in Figure 5 (Tier 1–Tier 5) we can notice that the biggest change in mortality occurred in central GLA. Several factors could explain this result such as the outdoor PM$_{2.5}$ concentration, the housing stock (I/O ratios) and the population. As described above, after the inclusion of the time-activity data there was an overall reduction in exposure concentration because people usually spend most of their time (>95%) in indoor MEs (excluding transportation), where the concentration of outdoor PM$_{2.5}$ is lower than the measured ambient levels. Because the health impact function used by BenMap-CE is a concentration response function, the amount of the reduced exposure concentration determines the fraction of the mortality reduction. In our case, knowing that moving from Tier 1 to Tier 5 would result in a greater reduction of exposure concentration that appeared in central GLA (Figure 3c), we could presume that the high outdoor PM$_{2.5}$ concentration and the building type of that area, were largely responsible for the mortality change. The similar distribution patterns between Figures 3c and 5 also supported this argument. As already shown in Figure 2, the infiltration factors of the buildings there were lower than the rest of the GLA, leading to higher mitigation of the exposure concentration. In the Appendix A.3, Figure A3a–c show the spatial distribution of the predicted change in mortality between Tier 1 and Tiers 2, 3 and 4, respectively.

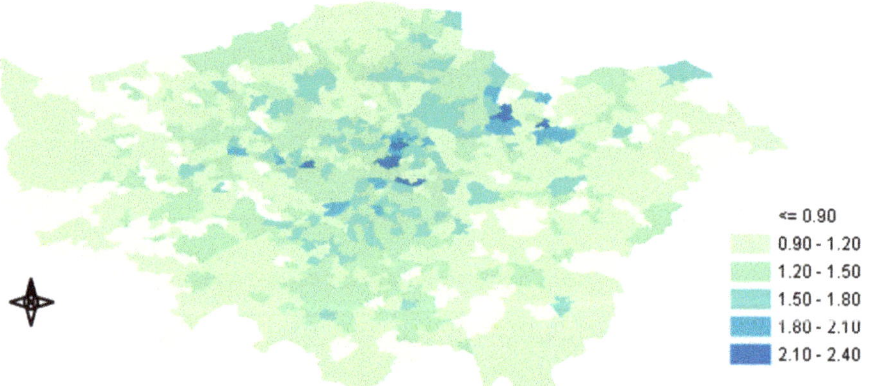

Figure 5. Spatial distribution of the predicted change in the estimates of mortality burden misclassification (in death cases) between Tier model 1 and Tier model 5.

Overall, these outcomes demonstrate the importance of the complexity of an exposure metric when incorporated into an epidemiological study. Here, we proved that indoor MEs such as the home and the subway are governing human exposure to air pollution and any possible absence in a metric is likely to cause considerable misclassification of the magnitude of mortality.

4. Discussion

Due to the limited time most people spend outside, the amount of ambient concentration of PM$_{2.5}$ that people are directly exposed to is likely to be different based on variation in people's behavior and the performance characteristics of the buildings they are occupying [55]. Consequently, spatial variability, time-activity and losses due to outdoor-to-indoor transport are all sources of exposure uncertainty in the epidemiological analysis, when fixed-site monitor concentrations are used as surrogates for exposure to air pollution. In this work we established a more comprehensive understanding of population exposure concentration and the impact that different exposure metrics can make on all-cause mortality predictions. We showed that the I/O ratios and individual's patterns of movement play a key role in estimating exposure to $PM_{2.5}$ and that transportation-MEs, predominately the highly polluted London Underground, are important in accurately establishing exposure. We demonstrated that subway and Indoor MEs make a significant contribution to the exposure misclassification and therefore mortality change predictions. Azimi and Stephens [31] highlighted the importance of including indoor MEs when estimating the total exposure and the need for a better understanding of how the infiltration factors vary by building type in order to improve the exposure estimates and reduce the uncertainty. Based on field measurements, they found that exposure to $PM_{2.5}$ of outdoor origin inside the residence contributed around 67% to the total U.S. mortality burden. In our analysis, we found that the Indoor environment contributed approximately 83% to the total mortality burden in London. The difference in our results may be explained by the different MEs considered in each study. As our aim was to quantify the misclassification and give an insight into how the absence of significant MEs from an exposure assessment could increase the uncertainty, we mainly focused on the different infiltration factors of home types and the LU. Martins et al. [56] determined the $PM_{2.5}$ exposure and estimated the daily $PM_{2.5}$ dose during Barcelona subway commuting. They estimated that the $PM_{2.5}$ dose received by an adult in the subway contributed approximately 46% to the total daily dose in the respiratory tract. In our study, LU contributed approximately 15% to the total health burden. Due to the different methods used and different health endpoints, their results cannot directly be compared to ours. However, their outcomes indicate the non-trivial contribution from subway ME on health effects estimates. Several studies have compared static (home address-based) with more dynamic air pollution methods and proved that there is a reduction in average total exposure levels in urban areas with related characteristics as GLA [32,57]. Tang et al. [57] used a staged modelling approach to evaluate the use of static ambient concentrations as exposure estimates and examined the impact of dynamic components on estimated air pollution exposure. They found that the mean population exposures in Hong Kong for their full dynamic model were approximately 20% lower than the ambient baseline estimates of the static approach. Smith et al. [32] combined a dispersion modelling approach with building infiltration factors and travel behavior in order to create the London Hybrid Exposure Model (LHEM). They found that their model's estimates were around 37% lower for $PM_{2.5}$ than the static approach (residential address-based). Similarly, by adopting a staged modelling approach to evaluate the effect of including dynamic components to our exposure models we found that the absence of mobility and infiltration factors in the static Tier-model 1 led to an overestimation of annual $PM_{2.5}$ population exposure. Overall, the exposure estimates of our most complex model (Tier 5) were around 34% lower than those of the static baseline model (Tier 1). These findings were different from Tang et al.'s [57] study but very similar to the LHEM study, mainly because the study population was the same and similar travel behavior data was used. Recently, Singh et al. [58] quantified the population exposure to $PM_{2.5}$ concentrations in London and assessed the importance of including movement and indoor infiltration to total population exposure. They found that their refined exposure assessment predicted 28% lower total population exposure than the traditional static exposure method. As in this study, the time-activity data were derived from the LTDS [39] and the study area was London. However, the small difference between their results and ours could be explained by the different datasets used for the infiltration factors and the different concept used for the key MEs (e.g., the London Underground). Results from other similar studies are difficult to find as we compare different exposure estimates during the same time period (2017) in an effort to examine the effect on all-cause mortality predictions. We showed that using a static exposure metric instead of a more dynamic approach (based on time-activity data and indoor infiltration) to predict the mortality in the GLA population would lead to an overestimation of 1174–1541 mean predicted estimates of mortality attributed to $PM_{2.5}$. Ebelt et al. [59] found for several health outcomes associated with cardiopulmonary diseases, analyses with ambient exposures resulted in larger effect estimates. These results strongly supported their original hypothesis that the reduced exposure misclassification resulting from the utilization of ambient exposures instead of ambient concentrations provide more precise estimates of effects in epidemiology.

This work provides further understanding as to the impact of an exposure assessment on the mortality predictions and helps to mitigate the uncertainty in health risk assessments of air pollution. As a result, it would be possible to increase the efficiency of regional or local air quality management strategies.

Limitations and Future Work

The current study contains several limitations. Only some of the deep and subsurface underground lines were monitored and only for a small sampling period. In this study, we assumed that these measurements also represented the corresponding lines that were not measured. Moreover, only 23 monitoring stations were available for $PM_{2.5}$ and their locations were not uniformly spread across the study area. Consequently, this may have affected the simulation accuracy and the interpolated ambient concentration estimates in the unmonitored areas that were far from the stations that might have contained higher uncertainty. Furthermore, another limitation was the assumption that the Indoor microenvironment and the average dwelling I/O ratios also represented the office and commercial buildings. The toxicity of $PM_{2.5}$ was not included, but mainly because it was out of the scope of the study to investigate the toxicity of the particles.

The space–time–activity data is based upon the London Travel Demand Survey for the period 2005–2010 and may not be fully accurate locally, spatially and temporally, for the year 2017. Moreover, the annual average of the time-activity data that we used, assuming that people followed the typical daily mobility patterns for the whole year, may have increased the uncertainty in our models, because those data might not have accurately represented a part of the population. Since the main body of our study was based on averages and the population was not divided into different age groups, our health burden predictions may be less accurate for special groups of people that have different behaviours (e.g., ill or elderly that spend most of the day inside their residence).

Parameters that could affect particle infiltration, such as differences in indoor-outdoor air pressure due to the impact of the surrounding micro-environment, and the existence and efficiency of mechanical filtration, were not the focus of the current study and were therefore not investigated.

In the future, this study could be improved by conducting further measurements in the London Underground and for larger periods of time. Simple sensitivity tests could be made in order to check each model's response and how the misclassification affects our estimates. As the next stage of this work we could investigate how this framework applies to other cities with higher ambient $PM_{2.5}$ concentrations and different indoor characteristics (such as interventions-PACs, HVAC). Taking into consideration that each urban area may have different characteristics, it is important to examine how the incorporation of the local urban or building features could make an impact on exposure concentration estimates and health burden predictions.

5. Conclusions

The use of ambient centralized monitoring concentrations as a surrogate for people's exposure may not provide an accurate representation in a population study. In this study we developed a static exposure approach, commonly used in epidemiology, as our baseline metric and by incrementally enhancing the metric we were able to report the potential impact that the application of different metrics would have on a health outcomes assessment. We demonstrated that studies focusing on centralized monitoring ambient concentrations may show reduced ability to detect the true associations between exposure to $PM_{2.5}$ and health effects due to inadequate spatial variability of the concentration and the absence of people's mobility. The magnitude of the misclassification related to the inclusion of indoor MEs and the metric's complexity was large relative to the dynamic nature of human exposure to air pollution.

This analysis illustrates the significance of allowing for population activity and indoor infiltration. The indoor ME showed the highest contribution to the total population exposure (i.e., 83%), while the LU contributed approximately 15%, although people spend only 0.4% of their time there. Consequently, all our models showed lower total exposures than the traditional exposure approach that assumes that the $PM_{2.5}$ concentrations outside the residence are representative of the total population exposure. Particularly, our most complex and accurate Tier-model estimated an approximately 34% lower mean exposure concentration compared with using simply an outdoor concentration.

The exposure misclassification due to home infiltration and underground ME is likely important in assessing the health burden in an urban area because people in cities spend the majority of their time inside the residence or workplace and the pollution concentrations that occur underground are remarkably high. The misclassification between the traditional exposure approach to estimate health outcomes and our most dynamic metric was found to be 1174 mean predicted mortalities in GLA, with the highest numbers observed in Inner London.

Overall, by quantifying the health burden misclassification we managed to pinpoint the importance of developing a metric that can adequately represent the study population concerned and showed that the use of more dynamic data in epidemiology could significantly increase the accuracy of health impact assessments.

Author Contributions: Z.L.: Conceptualization; V.K. and Z.L. Methodology; V.K.: Data analysis; Z.L. and I.E.: Supervision; V.K.: Visualization; V.K.: Original draft preparation; V.K., Z.L. and I.E.: Review and editing. All authors have read and agreed to the published version of the manuscript.

Funding: This research received no external funding.

Conflicts of Interest: The authors declare no conflict of interest.

Appendix A

Appendix A.1. Spatial Distribution of Exposure Concentration

Figure A1 shows the spatial distribution of exposure concentrations estimated by Tier-model 2, Tier-model 3 and Tier-model 4.

a)

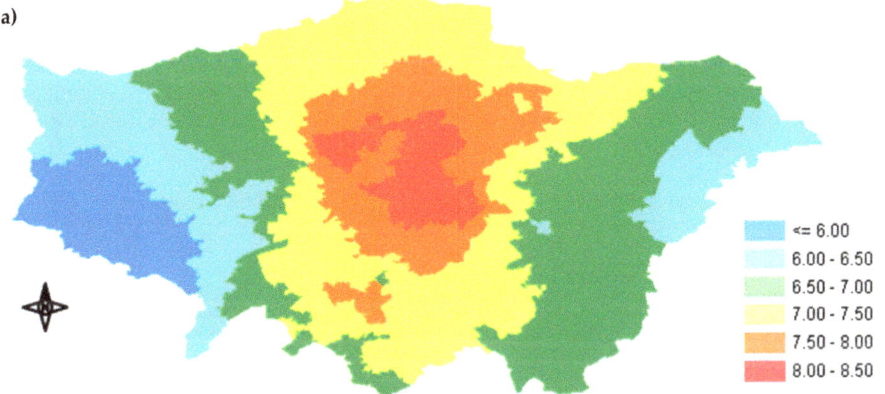

Figure A1. *Cont.*

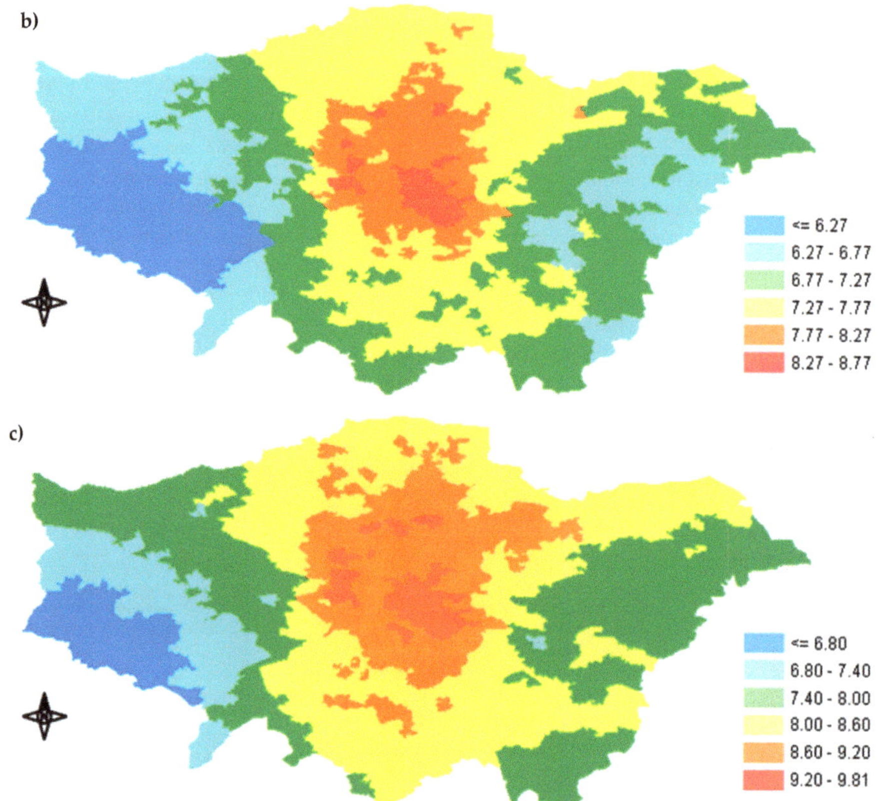

Figure A1. Map of GLA showing the spatial distribution of the annual mean exposure concentration ($\mu g/m^3$): (**a**) Tier 2, (**b**) Tier 3 and (**c**) Tier 4.

In Tiers 2 and 3, although the exposures range similarly, we can see different distribution patterns across GLA. The inclusion of the I/O ratios of the different dwelling types increased the spatial resolution of the exposure concentration, while in Tier 2 (Figure A1a) the spatial distribution was similar to our baseline Tier-model 1 (Figure 3a) due to the average infiltration factor that was used. Figure A1c is presenting higher exposure concentrations than Figure A1a,b due to the incorporation of the highly polluted transportation MEs such as LU. In all three maps, central GLA showed the highest exposure concentrations because of the high ambient $PM_{2.5}$ concentrations in that area.

Appendix A.2. Percentage of Exposure Concentration Reduction across GLA

The percentage of reduction in exposure concentration after using Tier 3 and Tier 4 are shown in Figure A2a,b.

After including the I/O ratios of the different dwelling types (Tier 3), Central and South-eastern GLA had the highest reduction (between 45.5% and 47.6%), while in Figure A2b, Central and parts of the Southern GLA showed the highest percentage of mitigation (up to approximately 39.4%). As shown in Figure 1, Central and South-eastern GLA had the lowest I/O ratios. Because in Tier 3 transportation and outdoor MEs were not considered and people were assumed to spend 100% of their time indoors, the percentage of the reduction was mainly driven by the infiltration factors. However, after increasing the complexity and including outdoor and transportation MEs, the percentage of reduction was also strongly influenced by the ambient $PM_{2.5}$ concentration in each area, in addition to the infiltration factors.

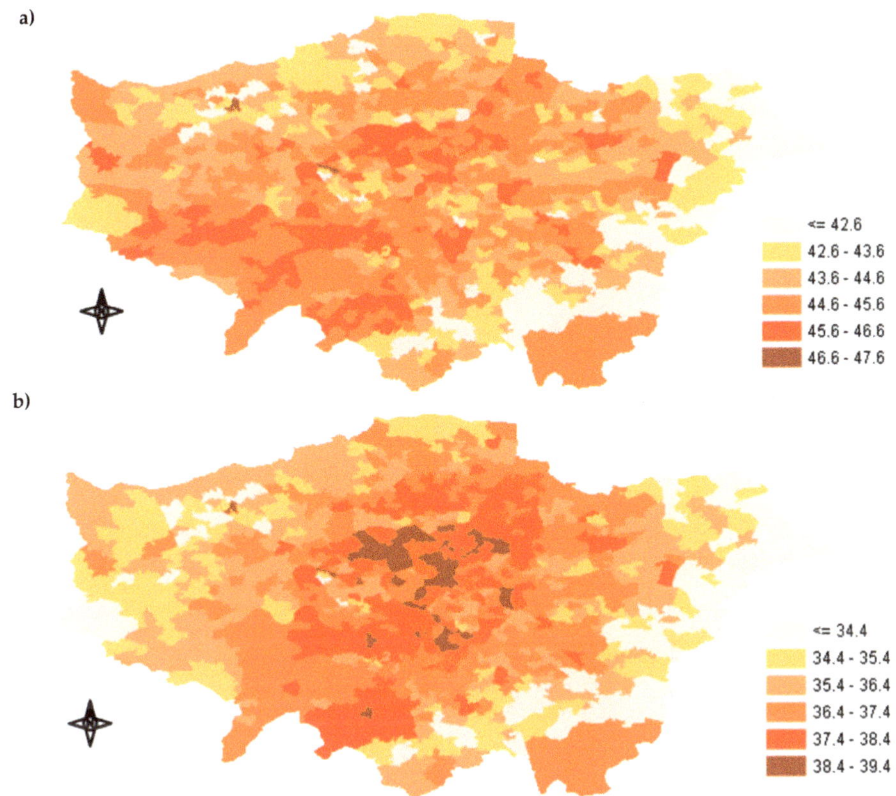

Figure A2. Percentage (%) of exposure concentration reduction between Tier 1 and: (**a**) Tier 3 and (**b**) Tier 4.

Appendix A.3. Spatial Distribution of the Predicted Avoided Mortality

The annual mean predicted avoided mortality due to the utilization of Tiers 2, 3 and 4 is illustrated in Figure A3a–c.

Figure A3a–c show similar spatial distribution of the predicted avoided mortality. As anticipated, the number of predicted mortalities after using Tiers 2 and 3 was higher, because the concentration-response function used by BenMap-CE to calculate the mortality was affected by the reduction (or increase) of the exposure concentration. The absence of the time-activity data in Tier 2 and Tier 3 led to an underestimation of the total exposure concentration and as a result the exposure difference between those models and our baseline model (Tier 1) was higher.

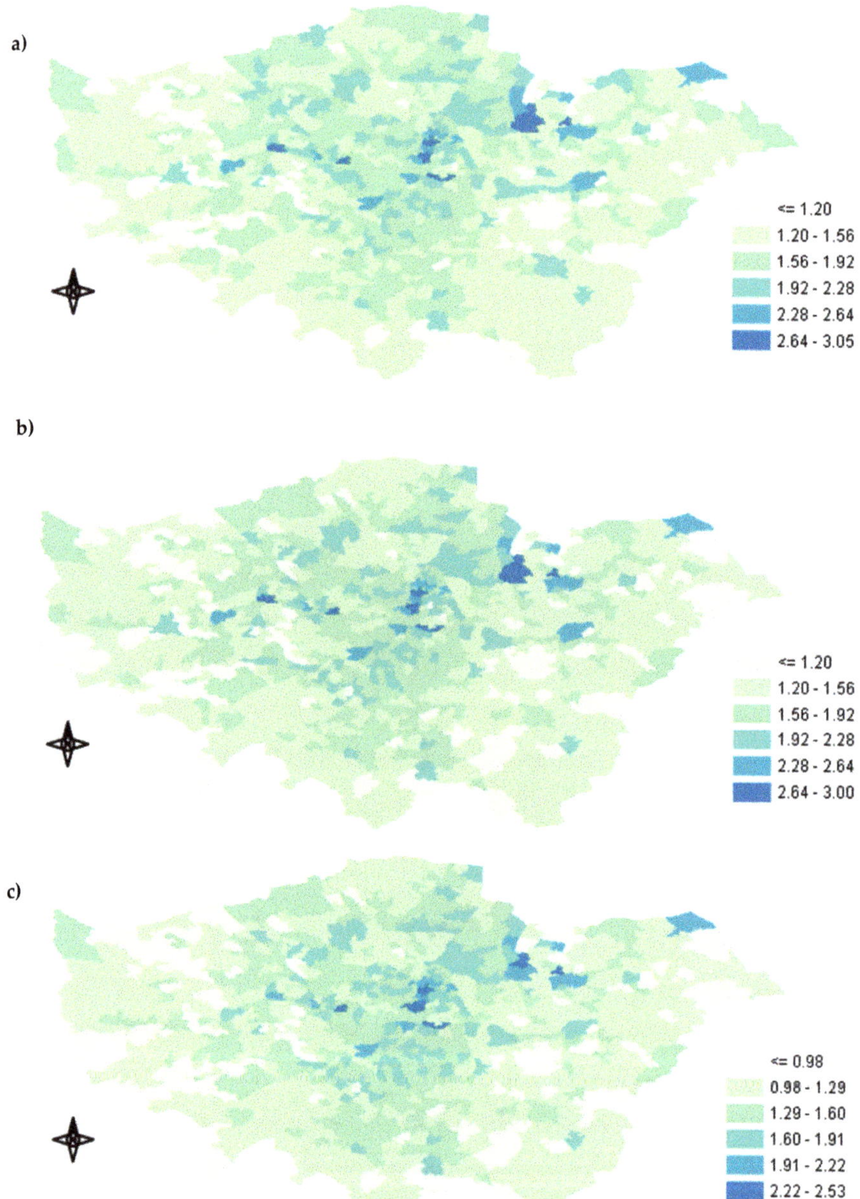

Figure A3. Map of the Greater London Area showing the spatial distribution of the mean avoided mortality (in death cases): (**a**) Tier 1–2, (**b**) Tier 1–3, (**c**) Tier 1–4.

References

1. Public Health England, COMEAP (2009): Long-term exposure to air pollution: Effect on mortality. GOV.UK Website. Available online: https://www.gov.uk/government/publications/comeap-london-term-exposure-to-air-pollution-effect-on-mortality (accessed on 9 March 2019).

2. Cohen, A.J.; Brauer, M.; Burnett, R.; Anderson, H.R.; Frostad, J.; Estep, K.; Balakrishnan, K.; Brunekreef, B.; Dandona, L.; Dandona, R.; et al. Estimates and 25-year trends of the global burden of disease attributable to ambient air pollution: An analysis of data from the Global Burden of Diseases Study 2015. *Lancet* **2017**, *389*, 1907–1918. [CrossRef]

3. Forouzanfar, M.H.; Alexander, L.; Anderson, H.R.; Bachman, V.F.; Biryukov, S.; Brauer, M.; Burnett, R.; Casey, D.; Coates, M.M.; Cohen, A.; et al. Global, regional, and national comparative risk assessment of 79 behavioural, environmental and occupational, and metabolic risks or clusters of risks in 188 countries, 1990–2013: A systematic analysis for the Global Burden of Disease Study 2013. *Lancet* **2015**, *386*, 2287–2323. [CrossRef]

4. Dias, D.; Tchepel, O. Spatial and Temporal Dynamics in Air Pollution Exposure Assessment. *Int. J. Environ. Res. Public. Health* **2018**, *15*, 558. [CrossRef] [PubMed]

5. Batty, M. *Cities as Complex Systems: Scaling, Interactions, Networks, Dynamics and Urban Morphologies. Working Paper. CASA Working Papers (131)*; Centre for Advanced Spatial Analysis (UCL): London, UK, 2012.

6. Laden, F.; Schwartz, J.; Speizer, F.E.; Dockery, D.W. Reduction in fine particulate air pollution and mortality—Extended follow-up of the Harvard six cities study. *Am. J. Respir. Crit. Care* **2006**, *173*, 667–672. [CrossRef] [PubMed]

7. Zeger, S.L.; Dominici, F.; McDermott, A.; Samet, J.M. Mortality in the Medicare Population and Chronic Exposure to Fine Particulate Air Pollution in Urban Centers (2000–2005). *Environ. Health Perspect.* **2008**, *116*, 1614–1619. [CrossRef] [PubMed]

8. Belleudi, V.; Faustini, A.; Stafoggia, M.; Cattani, G.; Marconi, A.; Perucci, C.A.; Forastiere, F. Impact of Fine and Ultrafine Particles on Emergency Hospital Admissions for Cardiac and Respiratory Diseases. *Epidemiology* **2010**, *21*, 414–423. [CrossRef]

9. Gao, Y.; Chan, E.Y.Y.; Li, L.P.; Lau, P.W.C.; Wong, T.W. Chronic effects of ambient air pollution on respiratory morbidities among Chinese children: A cross-sectional study in Hong Kong. *BMC Public Health* **2014**, *14*, 105. [CrossRef]

10. Collart, P.; Coppieters, Y.; Mercier, G.; Kubuta, V.M.; Leveque, A. Comparison of four case-crossover study designs to analyze the association between air pollution exposure and acute myocardial infarction. *Int. J. Environ. Health Res.* **2015**, *25*, 601–613. [CrossRef]

11. Nikolova, I.; Janssen, S.; Vrancken, K.; Vos, P.; Mishra, V.; Berghmans, P. Size resolved ultrafine particles emission model—A continues size distribution approach. *Sci. Total Environ.* **2011**, *409*, 3492–3499. [CrossRef]

12. Dionisio, K.L.; Isakov, V.; Baxter, L.K.; Sarnat, J.A.; Sarnat, S.E.; Burke, J.; Rosenbaum, A.; Graham, S.E.; Cook, R.; Mulholland, J.; et al. Development and evaluation of alternative approaches for exposure assessment of multiple air pollutants in Atlanta, Georgia. *J. Expo. Sci. Environ. Epidemiol.* **2013**, *23*, 581–592. [CrossRef]

13. Jerrett, M.; Arain, A.; Kanaroglou, P.; Beckerman, B.; Potoglou, D.; Sahsuvaroglu, T.; Morrison, J.; Giovis, C. A review and evaluation of intraurban air pollution exposure models. *J. Expo. Anal. Environ. Epidemiol.* **2005**, *15*, 185–204. [CrossRef] [PubMed]

14. Willers, S.M.; Eriksson, C.; Gidhagen, L.; Nilsson, M.E.; Pershagen, G.; Bellander, T. Fine and coarse particulate air pollution in relation to respiratory health in Sweden. *Eur. Respir. J.* **2013**, *42*, 924–934. [CrossRef] [PubMed]

15. Liu, L.J.S.; Tsai, M.Y.; Keidel, D.; Gemperli, A.; Ineichen, A.; Hazenkamp-von Arx, M.; Bayer-Oglesby, L.; Rochat, T.; Kunzli, N.; Ackermann-Liebrich, U.; et al. Long-term exposure models for traffic related NO2 across geographically diverse areas over separate years. *Atmos. Environ.* **2012**, *46*, 460–471.

16. Emaus, M.J.; Bakker, M.F.; Beelen, R.M.J.; Veldhuis, W.B.; Peeters, P.H.M.; van Gils, C.H. Degree of urbanization and mammographic density in Dutch breast cancer screening participants: Results from the EPIC-NL cohort. *Breast Cancer Res. Treat.* **2014**, *148*, 655–663. [CrossRef] [PubMed]

17. Batterman, S.; Burke, J.; Isakov, V.; Lewis, T.; Mukherjee, B.; Robins, T. A Comparison of Exposure Metrics for Traffic-Related Air Pollutants: Application to Epidemiology Studies in Detroit, Michigan. *Int. J. Environ. Res. Public Health* **2014**, *11*, 9553–9577. [CrossRef] [PubMed]

18. Korek, M.J.; Bellander, T.D.; Lind, T.; Bottai, M.; Eneroth, K.M.; Caracciolos, B.; de Faire, U.H.; Fratiglioni, L.; Hilding, A.; Leander, K.; et al. Traffic-related air pollution exposure and incidence of stroke in four cohorts from Stockholm. *J. Expo. Sci. Environ. Epidemiol.* **2015**, *25*, 517–523. [CrossRef]

19. Punger, E.M.; West, J.J. The effect of grid resolution on estimates of the burden of ozone and fine particulate matter on premature mortality in the USA. *Air Qual. Atmos. Health* **2013**, *6*, 563–573. [CrossRef]

20. Li, Y.; Henze, D.K.; Jack, D.; Kinney, P.L.J. The influence of air quality model resolution on health impact assessment for fine particulate matter and its components. *Air Qual. Atmos. Health* **2016**, *9*, 51–68. [CrossRef]

21. Fenech, S.; Doherty, R.M.; Heaviside, C.; Vardoulakis, S.; Macintyre, H.L.; O'Connor, F.M. The influence of model spatial resolution on simulated ozone and fine particulate matter for Europe: Implications for health impact assessments. *Atmos. Chem. Phys.* **2018**, *18*, 5765–5784. [CrossRef]

22. Korhonen, A.; Lehtomäki, H.; Rumrich, I.; Karvosenoja, N.; Paunu, V.-V.; Kupiainen, K.; Sofiev, M.; Palamarchuk, Y.; Kukkonen, J.; Kangas, L.; et al. Influence of spatial resolution on population $PM_{2.5}$ exposure and health impacts. *Air Qual. Atmos. Health* **2019**, *12*, 705–718. [CrossRef]

23. Klepeis, N.E.; Nelson, W.C.; Ott, W.R.; Robinson, J.P.; Tsang, A.M.; Switzer, P.; Behar, J.V.; Hern, S.C.; Engelmann, W.H. The National Human Activity Pattern Survey (NHAPS): A resource for assessing exposure to environmental pollutants. *J. Expo. Anal. Environ. Epidemiol.* **2001**, *11*, 231–252. [CrossRef] [PubMed]

24. Bhangar, S.; Mullen, N.A.; Hering, S.V.; Kreisberg, N.M.; Nazaroff, W.W. Ultrafine particle concentrations and exposures in seven residences in northern California. *Indoor Air* **2011**, *21*, 132–144. [CrossRef] [PubMed]

25. Kearney, J.; Wallace, L.; MacNeill, M.; Xu, X.; VanRyswyk, K.; You, H.; Kulka, R.; Wheeler, A.J. Residential indoor and outdoor ultrafine particles in Windsor, Ontario. *Atmos. Environ.* **2011**, *45*, 7583–7593. [CrossRef]

26. Meng, Q.Y.; Spector, D.; Colome, S.; Turpin, B. Determinants of indoor and personal exposure to $PM_{2.5}$ of indoor and outdoor origin during the RIOPA study. *Atmos. Environ.* **2009**, *43*, 5750–5758. [CrossRef] [PubMed]

27. Baxter, L.K.; Burke, J.; Lunden, M.; Turpin, B.J.; Rich, D.Q.; Thevenet-Morrison, K.; Hodas, N.; Ozkaynak, H. Influence of human activity patterns, particle composition, and residential air exchange rates on modeled distributions of $PM_{2.5}$ exposure compared with central-site monitoring data. *J. Expo. Sci. Environ. Epidemiol.* **2013**, *23*, 241–247. [CrossRef] [PubMed]

28. Ma, J.; Tao, Y.; Kwan, M.-P.; Chai, Y. Assessing Mobility-Based Real-Time Air Pollution Exposure in Space and Time Using Smart Sensors and GPS Trajectories in Beijing. *Ann. Am. Assoc. Geogr.* **2019**, 1–15. [CrossRef]

29. Ji, W.J.; Zhao, B. Estimating Mortality Derived from Indoor Exposure to Particles of Outdoor Origin. *PLoS ONE* **2015**, *10*, e0124238. [CrossRef]

30. Azimi, P.; Stephens, B. A framework for estimating the US mortality burden of fine particulate matter exposure attributable to indoor and outdoor microenvironments. *J. Expo. Sci. Environ. Epidemiol.* **2018**. [CrossRef]

31. Fenech, S.; Aquilina, N.J. Trends in ambient ozone, nitrogen dioxide, and particulate matter concentrations over the Maltese Islands and the corresponding health impacts. *Sci. Total Environ.* **2020**, *700*, 134527. [CrossRef]

32. Smith, J.D.; Mitsakou, C.; Kitwiroon, N.; Barratt, B.M.; Walton, H.A.; Taylor, J.G.; Anderson, H.R.; Kelly, F.J.; Beevers, S.D. London Hybrid Exposure Model: Improving Human Exposure Estimates to NO_2 and $PM_{2.5}$ in an Urban Setting. *Environ. Sci. Technol.* **2016**, *50*, 11760–11768. [CrossRef]

33. King's College London. London Air Quality Network. Available online: https://www.londonair.org.uk/LondonAir/Default.aspx (accessed on 10 December 2018).

34. Sacks, J.D.; Lloyd, J.M.; Zhu, Y.; Anderton, J.; Jang, C.J.; Hubbell, B.; Fann, N. The Environmental Benefits Mapping and Analysis Program—Community Edition (BenMAP-CE): A tool to estimate the health and economic benefits of reducing air pollution. *Environ. Model. Softw.* **2018**, *104*, 118–129. [CrossRef] [PubMed]

35. Tang, R.; Blangiardo, M.; Gulliver, J. Using Building Heights and Street Configuration to Enhance Intraurban PM_{10}, NOX, and NO2 Land Use Regression Models. *Environ. Sci. Technol.* **2013**, *47*, 11643–11650. [CrossRef] [PubMed]

36. Yu, Y.; Sokhi, R.S.; Kitwiroon, N.; Middleton, D.R.; Fisher, B. Performance characteristics of MM5–SMOKE–CMAQ for a summer photochemical episode in southeast England, United Kingdom. *Atmos. Environ.* **2008**, *42*, 4870–4883. [CrossRef]

37. Taylor, J.; Shrubsole, C.; Davies, M.; Biddulph, P.; Das, P.; Hamilton, I.; Vardoulakis, S.; Mavrogianni, A.; Jones, B.; Oikonomou, E. The modifying effect of the building envelope on population exposure to $PM_{2.5}$ from outdoor sources. *Indoor Air* **2014**, *24*, 639–651. [CrossRef] [PubMed]

38. Mayor of London, London assembley, London Datastore. Available online: https://www.data.london.gov.uk (accessed on 25 January 2019).

39. Mayor of London, Transport for London, London Travel Demand Survey (LTDS). Available online: https://www.tfl.gov.uk (accessed on 9 February 2019).

40. Song, W.W.; Ashmore, M.R.; Terry, A.C. The influence of passenger activities on exposure to particles inside buses. *Atmos. Environ.* **2009**, *43*, 6271–6278. [CrossRef]
41. Seaton, A.; Cherrie, J.; Dennekamp, M.; Donaldson, K.; Hurley, J.F.; Tran, C.L. The London Underground: Dust and hazards to health. *Occup. Environ. Med.* **2005**, *62*, 355–362. [CrossRef]
42. Smith, J.D.; Barratt, B.M.; Fuller, G.W.; Kelly, F.J.; Loxham, M.; Nicolosi, E.; Priestman, M.; Tremper, A.H.; Green, D.C. $PM_{2.5}$ on the London Underground. *Environ. Int.* **2019**, *134*, 105–188. [CrossRef]
43. Querol, X.; Alastuey, A.; Rodriguez, S.; Plana, F.; Ruiz, C.R.; Cots, N.; Massague, G.; Puig, O. PM 10 and $PM_{2.5}$ source apportionment in the Barcelona Metropolitan area, Catalonia, Spain. *Atmos. Environ.* **2001**, *35*, 6407–6419. [CrossRef]
44. Johansson, C.; Johansson, P.A. Particulate matter in the underground of Stockholm. *Atmos. Environ.* **2003**, *37*, 3–9. [CrossRef]
45. Altieri, K.E.; Keen, S.L. Public health benefits of reducing exposure to ambient fine particulate matter in South Africa. *Sci. Total Environ.* **2019**, *684*, 610–620. [CrossRef]
46. U.S. EPA Environmental Benefits Mapping and Analysis Program-Community edition: User's Manual. 2017. Available online: https://www.epa.gov/sites/production/files/2015-04/documents/benmap-ce_user_manual_march_2015.pdf (accessed on 21 January 2019).
47. Zhao, D.; Azimi, P.; Stephens, B. Evaluating the Long-Term Health and Economic Impacts of Central Residential Air Filtration for Reducing Premature Mortality Associated with Indoor Fine Particulate Matter ($PM_{2.5}$) of Outdoor Origin. *Int. J. Environ. Res. Public Health* **2015**, *12*, 8448–8479. [CrossRef] [PubMed]
48. Milner, J.; Armstrong, B.; Davies, M.; Ridley, I.; Chalabi, Z.; Shrubsole, C.; Vardoulakis, S.; Wilkinson, P. An Exposure-Mortality Relationship for Residential Indoor $PM_{2.5}$ Exposure from Outdoor Sources. *Climate* **2017**, *5*, 66. [CrossRef]
49. Logue, J.M.; Price, P.N.; Sherman, M.H.; Singer, B.C. A Method to Estimate the Chronic Health Impact of Air Pollutants in US Residences. *Environ. Health Perspect.* **2012**, *120*, 216–222. [CrossRef] [PubMed]
50. Pope, C.A.; Burnett, R.T.; Thun, M.J.; Calle, E.E.; Krewski, D.; Ito, K.; Thurston, G.D. Lung cancer, cardiopulmonary mortality, and long-term exposure to fine particulate air pollution. *JAMA J. Am. Med. Assoc.* **2002**, *287*, 1132–1141. [CrossRef] [PubMed]
51. Zhu, Y.; Hinds, W.C.; Kim, S.; Shen, S.; Sioutas, C. Study of ultrafine particles near a major highway with heavy-duty diesel traffic. *Atmos. Environ.* **2002**, *36*, 4323–4335. [CrossRef]
52. Selmi, W.; Weber, C.; Riviere, E.; Blond, N.; Mehdi, L.; Nowak, D. Air pollution removal by trees in public green spaces in Strasbourg city, France. *Urban For. Urban Green.* **2016**, *17*, 192–201. [CrossRef]
53. Yli-Pelkonen, V.; Scott, A.A.; Viippola, V.; Setala, H. Trees in urban parks and forests reduce O3, but not NO2 concentrations in Baltimore, MD, USA. *Atmos. Environ.* **2017**, *167*, 73–80. [CrossRef]
54. Abhijith, K.V.; Kumar, P.; Gallagher, J.; McNabola, A.; Baldauf, R.; Pilla, F.; Broderick, B.; Di Sabatino, S.; Pulvirenti, B. Air pollution abatement performances of green infrastructure in open road and built-up street canyon environments—A review. *Atmos. Environ.* **2017**, *162*, 71–86. [CrossRef]
55. Strand, M.; Hopke, P.K.; Zhao, W.; Vedal, S.; Gelfand, E.; Rabinovitch, N. A study of health effect estimates using competing methods to model personal exposures to ambient $PM_{2.5}$. *J. Expo. Sci. Environ. Epidemiol.* **2007**, *17*, 549. [CrossRef]
56. Martins, V.; Moreno, T.; Minguillon, M.C.; Amato, F.; de Miguel, E.; Capdevila, M.; Querol, X. Exposure to airborne particulate matter in the subway system. *Sci. Total Environ.* **2015**, *511*, 711–722. [CrossRef]
57. Tang, R.; Tian, L.W.; Thach, T.Q.; Tsui, T.H.; Brauer, M.; Lee, M.; Allen, R.; Yuchi, W.; Lai, P.C.; Wong, P.; et al. Integrating travel behavior with land use regression to estimate dynamic air pollution exposure in Hong Kong. *Environ. Int.* **2018**, *113*, 100–108. [CrossRef] [PubMed]
58. Singh, V.; Sokhi, R.S.; Kukkonen, J. An approach to predict population exposure to ambient air $PM_{2.5}$ concentrations and its dependence on population activity for the megacity London. *Environ. Pollut.* **2020**, *257*, 113623. [CrossRef] [PubMed]
59. Ebelt, S.T.; Wilson, W.E.; Brauer, M. Exposure to Ambient and Nonambient Components of Particulate Matter: A Comparison of Health Effects. *Epidemiology* **2005**, *16*, 396–405. [CrossRef] [PubMed]

Article

Short-Term Effect of Air Pollution on Tuberculosis Based on Kriged Data: A Time-Series Analysis

Shuqiong Huang [1], Hao Xiang [2,3], Wenwen Yang [1], Zhongmin Zhu [4,5], Liqiao Tian [5], Shiquan Deng [5], Tianhao Zhang [5], Yuanan Lu [6], Feifei Liu [2,3], Xiangyu Li [2,3] and Suyang Liu [2,3,*]

[1] Hubei Provincial Center for Disease Control and Prevention, Wuhan 430079, China; hsq7513@163.com (S.H.); wenwenyanglinyi@163.com (W.Y.)

[2] Department of Global Health, School of Health Sciences, Wuhan University, 115 Donghu Road, Wuhan 430071, China; xianghao@whu.edu.cn (H.X.); 2018103050011@whu.edu.cn (F.L.); lxy329880@163.com (X.L.)

[3] Global Health Institute, Wuhan University, 115 Donghu Road, Wuhan 430071, China

[4] College of Information Science and Engineering, Wuchang Shouyi University, Wuhan 430064, China; zhongmin.zhu@whu.edu.cn

[5] State Key Laboratory of Information Engineering in Surveying, Mapping and Remote Sensing, Wuhan University, Wuhan 430079, China; tianliqiao@whu.edu.cn (L.T.); dyv587@foxmail.com (S.D.); tianhaozhang@whu.edu.cn (T.Z.)

[6] Environmental Health Laboratory, Department of Public Health Sciences, University of Hawaii at Manoa, 1960 East-West Road, Biomed Building, D105, Honolulu, HI 96822, USA; yuanan@hawaii.edu

* Correspondence: dayangwater@hotmail.com or dywater@whu.edu.cn; Tel.: +86-027-68759118; Fax: +86-027-68758648

Received: 10 January 2020; Accepted: 19 February 2020; Published: 27 February 2020

Abstract: Tuberculosis (TB) has a very high mortality rate worldwide. However, only a few studies have examined the associations between short-term exposure to air pollution and TB incidence. Our objectives were to estimate associations between short-term exposure to air pollutants and TB incidence in Wuhan city, China, during the 2015–2016 period. We applied a generalized additive model to access the short-term association of air pollution with TB. Daily exposure to each air pollutant in Wuhan was determined using ordinary kriging. The air pollutants included in the analysis were particulate matter (PM) with an aerodynamic diameter less than or equal to 2.5 micrometers ($PM_{2.5}$), PM with an aerodynamic diameter less than or equal to 10 micrometers (PM_{10}), sulfur dioxide (SO_2), nitrogen dioxide (NO_2), carbon monoxide (CO), and ground-level ozone (O_3). Daily incident cases of TB were obtained from the Hubei Provincial Center for Disease Control and Prevention (Hubei CDC). Both single- and multiple-pollutant models were used to examine the associations between air pollution and TB. Seasonal variation was assessed by splitting the all-year data into warm (May–October) and cold (November–April) seasons. In the single-pollutant model, for a 10 μg/m^3 increase in $PM_{2.5}$, PM_{10}, and O_3 at lag 7, the associated TB risk increased by 17.03% (95% CI: 6.39, 28.74), 11.08% (95% CI: 6.39, 28.74), and 16.15% (95% CI: 1.88, 32.42), respectively. In the multi-pollutant model, the effect of $PM_{2.5}$ on TB remained statistically significant, while the effects of other pollutants were attenuated. The seasonal analysis showed that there was not much difference regarding the impact of air pollution on TB between the warm season and the cold season. Our study reveals that the mechanism linking air pollution and TB is still complex. Further research is warranted to explore the interaction of air pollution and TB.

Keywords: tuberculosis; infectious disease; air pollution; time-series; Poisson regression; kriging

1. Introduction

Tuberculosis (TB) is one of the top ten deadly diseases in the world [1]. In 2016, there were about 10.4 million people diagnosed with TB worldwide, and 1.7 million people died from the disease [1]. Seven developing countries, China, India, Indonesia, Nigeria, the Philippines, Pakistan, and South Africa, account for 64% of the total cases [1]. As one of the countries with the highest number of TB cases, China has been working hard to lower the incidence of TB and has successfully reduced the TB-related mortality by 80% from 1990 to 2010 [2]. Despite this progress, China still has an estimated 1 million new cases each year [3] and faces serious drug-resistant TB epidemics [4]. China's estimated TB incidence rate was 63/100,000 in 2017, ranking the 28th globally [3]. In Hubei Province where Wuhan is located, the reported incidence of TB in 2017 was 68.3/100,000, ranking ninth in the nation [5]. As the capital of Hubei Province and the largest city in central China, Wuhan also has a high incidence of TB. In 2017, 5952 new cases of tuberculosis were reported in Wuhan, with an estimated incidence of about 54.64/100,000 [5].

Previous studies have identified air pollution as one of the possible risk factors for TB. Most of these studies have focused on indoor exposure, which has been well summarized in several review articles [6–9]. These articles reveal that the use of biomass fuels (e.g., scrap lumber, crops, and manure) and fossil fuels (e.g., coal and diesel) are two environmental risk factors for TB infection. Tremblay (2007) studied the relationship between coal consumption and TB disease using historical statistics and proposed a hypothesis that the combustion of coal and possibly town gas aggravated the TB epidemics. Fullerton et al. (2008) claimed that the aggregated evidence from different studies across the world supports a causal relationship between biomass smoke exposure and the development of TB. Besides biomass and fossil fuels, tobacco smoking is also recognized as a non-communicable environmental risk factor for TB [10,11]. Compared with studies that have linked air pollution to TB in an indoor environment, fewer studies have focused on outdoor environments where the pollution level is relatively lower.

In epidemiologic studies that specifically investigated the associations between TB and ambient air pollution, the air pollutants associated with tuberculosis were not the same, and their effects varied across studies. Jassal et al. (2013) and You et al. (2016) found a positive association between exposure to ambient particulate matter (PM) with aerodynamic diameter of ≤ 2.5 μm ($PM_{2.5}$) and the risk of TB. Zhu et al. (2018) observed a positive association of PM with an aerodynamic diameter of ≤ 10 μm (PM_{10}) with the incidence of TB. Zhu et al. (2018) and Xu et al. (2019) found that gaseous pollutants sulfur dioxide (SO_2) and nitrogen dioxide (NO_2) were positively associated with the risk of TB, though the effects were modified by age and gender in the former study. In addition to those reported positive associations [12–15], other studies found no or weak associations [16,17], and one study even reported a protective effect of ambient SO_2 on TB [18]. This inconsistency warrants further epidemiological studies on the topic. There are some possible mechanisms to explain the link between air pollution and the increased risk of TB. These mechanisms include (1) both gaseous pollutants and PM may trigger the disease by weakening the immune system (e.g., affecting T cells or impairing macrophage function); (2) oxidative stress and inflammatory reactions induced by air pollution may result in damage to the respiratory tract or lung epithelial cells; and (3) the direct transport of bacteria and those attached to particles infect the healthy population [19,20]. These mechanisms are discussed in more detail in the discussion section.

Researchers studying the health effects of ambient air pollution often rely on monitoring data that are routinely collected at ground environmental monitoring stations [21]. Typically, the collected data are averaged across the study area as a proxy measure to represent the exposure of the study population, resulting in measurement bias. Alternatively, spatial interpolation methods such as inverse distance weighting (IDW) and kriging are used to estimate ambient air pollution. Kriging can utilize the data collected at given positions (e.g., air monitoring stations) to predict the concentrations at unmeasured locations. In the past decades, kriging has been broadly used for mapping air pollution levels [22,23] and to estimate exposures in epidemiologic studies [24–26]. In the current study, we estimated air pollution levels using kriging and applied generalized additive models to assess associations between short-term exposure to ambient air pollutants and TB incidence in Wuhan city, China, during the 2015–2016 period.

2. Methods

2.1. Study Location

Wuhan, with a population of 10.6 million and an area of 8594 km^2, is the largest city in central China and the capital city of Hubei Province. Wuhan has 13 administrative districts. As of 2014, Wuhan's gross domestic product (GDP) per capita reached 15.6 thousand U.S. dollars, making it one of the fastest growing cities in China [27]. However, its urbanization and industrialization have exposed local residents to a high level of air pollution. For example, the annual average concentration of PM$_{2.5}$ in Wuhan was 106.5 μg/m^3 as of 2017, which was significantly higher than that of other cities with similar economic development levels [27]. Figure 1 shows the geographical location of Wuhan, provincially-designated air sampling stations, and population by administrative districts.

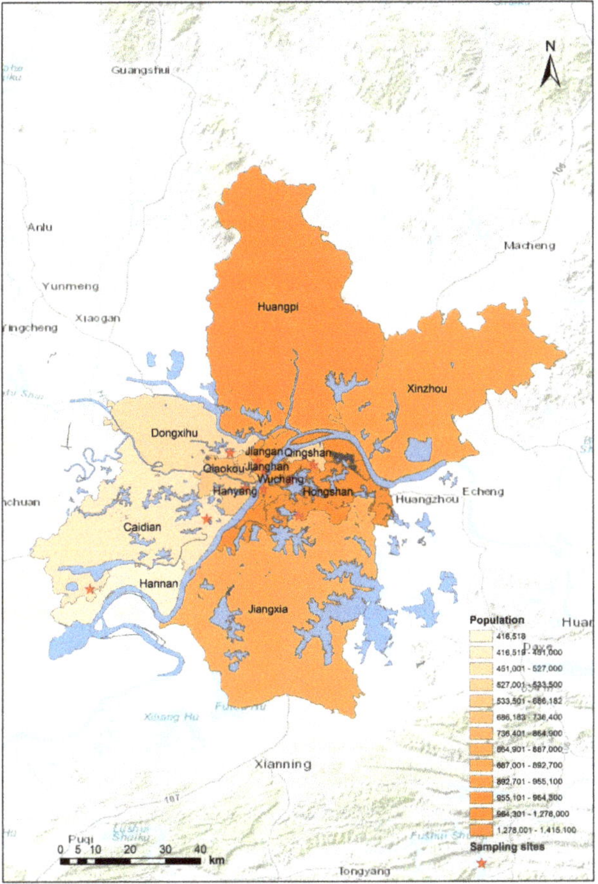

Figure 1. Map of Wuhan. The red pentagram represents monitoring stations. Population density by district is represented by gradient color.

2.2. Air Pollution and Meteorological Data

The monitoring data used in the present study were collected by the Hubei Provincial Environmental Quality Supervision and Administration Bureau (http://www.hbemc.com.cn/). All air pollutants were measured based on a 24-h sampling schedule, except for ozone, which was collected using an 8-h schedule. In the current study, rather than calculating daily average concentrations for each air pollutant across multiple monitors as many studies usually do, we used kriging, a spatial interpolation method for exposure assessment, to estimate daily levels of ambient air pollutants including $PM_{2.5}$, PM_{10}, SO_2, NO_2, carbon monoxide (CO), and tropospheric ozone (O_3) in Wuhan city. Although our interest was the Wuhan metropolitan area, we did include monitoring data from all sampling sites in Hubei Province to increase the statistical stability of our analysis. This led to a total of 51 monitoring sites included in the analysis, with 10 sites in Wuhan and an additional 41 sites across Hubei Province. We used ordinary kriging to determine the concentration of each air pollutant on each day for the Wuhan metropolitan area throughout the entire study period from 2015 to 2016. Specifically, kriging values generated from concentrations of air pollutants collected at the monitoring sites were first exported to a raster. The centroid of a given county or city was then used as the point data (represented by the geographical coordinates) to extract the interpolation results corresponding to the output raster. The analysis was repeated for each air pollutant on each day throughout the study period to obtain the complete exposure data set. It must be emphasized that kriging was used for predicting exposures, not TB incidence. The analysis was achieved using ArcMap 10.3 (ArcGIS. Version 10.3. Environmental Systems Research Institute, Redlands, CA, USA).

Daily meteorological data including temperature and relative humidity were collected at Wuhan Tianhe International Airport and obtained from https://rp5.ru/Weather_in_the_world.

2.3. TB Data

Daily TB incidence data were obtained from the Hubei Provincial Center for Disease Control and Prevention (Hubei CDC) for the 2015–2016 period. Hubei CDC collects information on a total of 39 infectious diseases that are legally regulated in China. Information on infectious diseases is collected at multiple levels, including health care institutions, individual practitioners, and rural doctors. Upon the collection of infectious disease information, designated epidemic reporters will diagnose infectious disease patients or suspected patients according to the latest diagnostic criteria for infectious diseases (Chinese Center for Disease Control and Prevention, 2008). In short, a chest X-ray is a common method for detecting tuberculosis, but the following tests are needed to confirm the diagnosis. First, check the patient's sputum for *Mycobacterium tuberculosis* to confirm the diagnosis. Second, use a lung X-ray examination to diagnose the location, scope, and nature of the disease. Third, conduct a tuberculin test. A positive test indicates infection. County-level CDCs will review the information collected to ensure the quality of the data and then file it online in the Infectious Disease Reporting Information Management System. We extracted data from this online reporting system and used the International Classification of Disease, 10th revision (WHO, 2007), to identify TB cases (coded as A15.0–A15.3; A16.0–A16.2).

Ethics approval and consent to participate: The study was approved by the Ethics Committee of Wuhan University School of Medicine.

2.4. Statistical Analysis

We used ageneralized additive model (GAM) to relax the assumption of linearity between predictors and response variables. GAM is the most widely used method in air pollution epidemiology because it allows for nonparametric adjustment of nonlinear confounding effects of seasonality, trends, and weather variables. Variables such as temperature and humidity can be controlled in the model by specifying a spline, a function to smooth the parameters for predicting the optimizing results [28]. We used GAMs to examine the associations between daily incidence of TB and daily concentrations of air pollutants (PM$_{2.5}$, PM$_{10}$, SO$_2$, NO$_2$, CO, and O$_3$) in Wuhan during the 2015–2016 period. We also examined the potential seasonal variation by stratifying the data set during warm (May–October) and cold (November–April) seasons. Typically, time-series data are "overly dispersed", which means the variance in a data set exceeds the mean. To account for overdispersion, the daily TB counts were assigned by quasi-Poisson distribution, a common way to deal with over-dispersed data. We used natural cubic spline functions to adjust for the time-varying variables. Specifically, the daily average temperature and relative humidity were adjusted with 3 degrees of freedom (df). The long-term and seasonal effects were adjusted with 8 df per year of data. Also included in the model is day of week (DOW), which is a set of indicator variables shown on each day of the week. The selection of covariates and df were based on a previously published study [25]. To explore the lag structures between incident TB and each air pollutant, a single-day lag model was examined from lag 0 through lag 14. The "lag" is defined as the time between the reporting of a TB case and days of exposure prior to this reporting. To report the results, we calculated the relative risk (RR) of TB associated with a 10 μg/m^3 increase in each pollutant concentration, with corresponding 95% confidence intervals (CIs).

To test the robustness of the results, we performed two sensitivity analyses. In the first sensitivity analysis, we used different df values for long-term effects (e.g., 4, 6, 10, and 12 df per year). In the second sensitivity analysis, we adopted different temperature and humidity values in our model (e.g., values at t-1 and the cumulative effect calculated by three-day moving averages) to test whether the results were different from those calculated using daily averages in the current model.

We used SAS (version 9.3; SAS Institute Inc., Cary, NC, USA) to combine TB data, meteorological data, and air pollution data, as well as generate descriptive statistics. Our time-series analysis used the R (version 3.0.2; http://R-project.org) statistical package "mgcv" to establish the model and calculate the results.

3. Results

Overall, there were 12,648 incident cases of TB included in our analysis during this two-year study. The cumulative incidence rate of TB was 1.3% in Wuhan during the 2015–2016 period. Table 1 shows the descriptive statistics for TB incident cases, air pollutants, and meteorological data. On average, there were 17 incident cases of TB per day. The average daily number of TB was the highest during spring season (March, April, and May), whereas the lowest was observed in winter months (December, January, and February). Air pollutant data were close to completeness, with only 9 days missing. Daily average concentrations of air pollutants were all higher during the cold season than in the warm season. The respective daily mean and standard deviation of PM$_{2.5}$ concentration were 58.94 μg/m^3 and 35.41 μg/m^3, lower than those of PM$_{10}$ (89.34 ± 42.16 μg/m^3). The highest level of PM$_{2.5}$ was observed in January (113.4 μg/m^3) while the lowest was observed in July (29.7 μg/m^3). For gaseous pollutants SO$_2$, NO$_2$, CO and O$_3$, the daily mean concentrations were 15.32, 38.61, 1.03 and 119.23 μg/m^3, respectively.

We fit a separate model to the relative risk (RR) of TB incident cases for each air pollutant at lag 0 through 14 and present the results in Figure 2. The single-day lag models show that the effect of air pollutants on TB was greatest at lag 7. Among them, PM$_{2.5}$, PM$_{10}$, and O$_3$ were statistically significantly ($p < 0.05$) associated with TB. We therefore report our time-series analysis results based on exposure at lag 7 because of this lag structure. The effect of air pollutants on TB at lag 7 is presented in Table 2. For a 10 μg/m^3 increase in PM$_{2.5}$, PM$_{10}$, and O$_3$, the TB risk increased by 17.03% (95% CI: 6.39, 28.74), 11.08% (95% CI: 6.39, 28.74), and 16.15% (95% CI: 1.88, 32.42), respectively. We also fit a multi-pollutant model including all pollutants simultaneously to assess whether associations found in single-pollutant models were confounded by the presence of co-pollutants. After controlling for co-pollutants, the association of PM$_{2.5}$ with TB was still robust and remained statistically significant. The effect estimates were slightly attenuated for all remaining pollutants in the multi-pollutant model.

Table 1. Descriptive statistics of TB incidence counts, air pollutants, and meteorological variables in Wuhan, China, during the 2015–2016 period.

Variable	Days	Mean ± SD			Percentiles [a]				
		All-year	Warm [b]	Cold	Min	25%	50%	75%	Max
Daily TB [c] cases	732	17.28 ± 13.49	17.67 ± 12.87	16.89 ± 14.10	2	11	14	19	134
Pollutant ($\mu g/m^3$)									
$PM_{2.5}$ [d]	723	58.94 ± 35.41	40.93 ± 20.21	77.71 ± 38.04	8.31	33.60	49.03	76.84	227.20
PM_{10}	723	89.34 ± 42.16	74.86 ± 33.47	104.42 ± 44.95	11.36	57.30	83.71	115.23	282.05
SO_2	723	15.32 ± 7.80	11.66 ± 4.81	19.13 ± 8.47	3.28	9.74	13.50	18.97	53.53
NO_2	723	38.61 ± 15.01	32.43 ± 11.65	45.06 ± 15.42	11.53	27.20	35.78	46.89	94.72
CO	723	1.03 ± 0.30	0.88 ± 0.20	1.18 ± 0.30	0.42	0.80	0.99	1.22	2.11
O_3	731	119.23 ± 53.98	147.95 ± 51.49	89.96 ± 38.56	9.30	74.40	114.29	159.04	272.01
Weather variables									
Temperature (°F)	732	63.02 ± 16.02	75.92 ± 8.31	49.92 ± 10.29	26.08	48.29	65.56	76.79	91.36
Relative Humidity (%)	732	79.23 ± 10.80	79.55 ± 9.79	78.9 ± 11.74	41	72.25	80	87	100

[a] Min = minimum, Max = Maximum; [b] Warm = May–October; Cold = November–April; [c] TB = Tuberculosis; [d] $PM_{2.5}$ = particulate matter (PM) with an aerodynamic diameter less than or equal to 2.5 µm, PM_{10} = PM with an aerodynamic diameter less than or equal to 10 µm, SO_2 = sulfur dioxide, NO_2 = nitrogen dioxide, CO = carbon monoxide, O_3 = ground-level ozone.

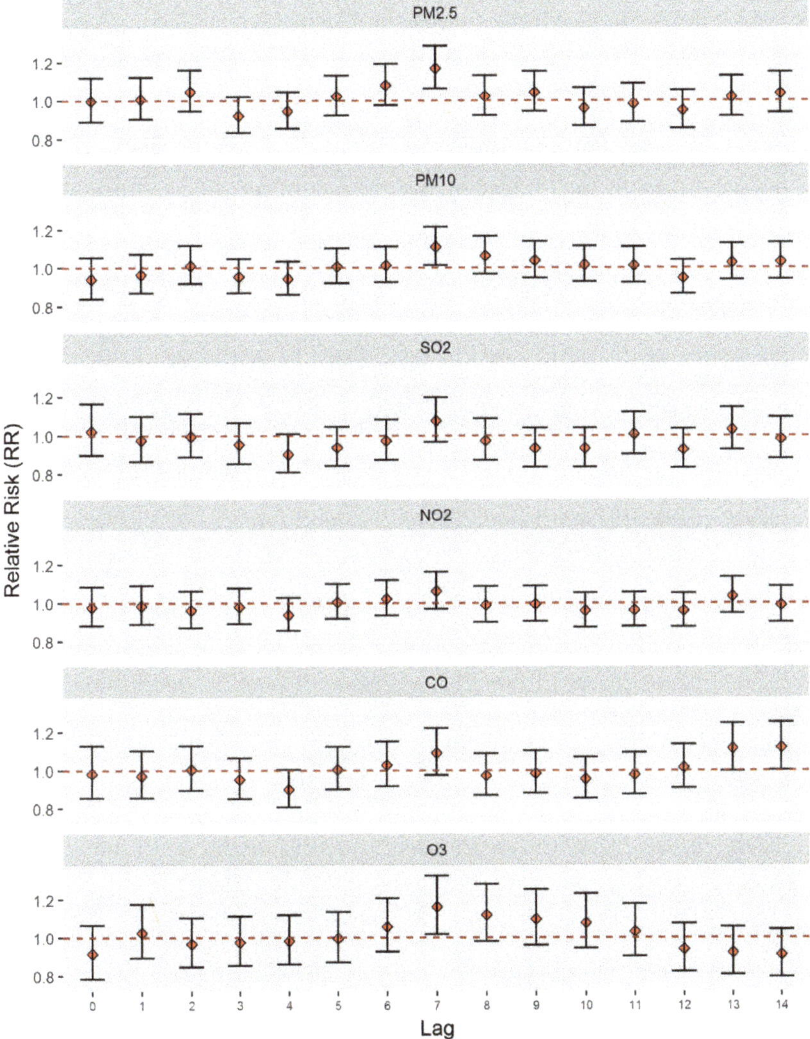

Figure 2. Relative risks (and 95% CIs) of TB incidence associated with an increase of 10 µg/m^3 in air pollutant concentrations at lag 0–14 days in Wuhan, China, during the 2015–2016 period.

The season-specific effects of each air pollutant on TB are presented in Table 2 and further illustrated in Figure 3. We did not observe an obvious trend in seasonality between the warm season (May–October) and the cold season (November–April). The associations between PM$_{2.5}$ and TB were statistically significant ($p < 0.05$) for both warm and cold seasons. The effect of NO$_2$ on TB during the cold season was also statistically significant ($p < 0.05$). It is noteworthy that dividing the data into warm and cold seasons reduced the sample size and led to an expanded confidence interval.

Table 2. Increased risk of TB and corresponding 95% CIs associated with 10 µg/m^3 increase in air pollutants [a] on the 7th lag day for single- and multi-pollutant models.

Pollutant	Single Pollutant Model			Multi-Pollutant Model [b]
	All-Year	Warm Season [c]	Cold Season	All-Year
PM$_{2.5}$	**1.17 (1.06, 1.28)** [d]	**1.13 (1.01, 1.26)**	**1.22 (1.06, 1.40)**	**1.20 (1.02, 1.41)**
PM$_{10}$	**1.11 (1.01, 1.22)**	1.12 (0.99, 1.26)	1.12 (0.99, 1.27)	0.98 (0.84, 1.15)
SO$_2$	1.08 (0.97, 1.20)	1.11 (0.96, 1.29)	1.08 (0.94, 1.24)	1.00 (0.86, 1.16)
NO$_2$	1.06 (0.97, 1.16)	**1.13 (1.02, 1.24)**	1.03 (0.91, 1.16)	0.97 (0.84, 1.12)
CO	1.09 (0.97, 1.21)	1.07 (0.94, 1.22)	1.10 (0.97, 1.24)	0.96 (0.82, 1.12)
O$_3$	**1.16 (1.02, 1.32)**	1.11 (0.95, 1.30)	1.16 (0.99, 1.35)	1.11 (0.95, 1.30)

[a] PM$_{2.5}$ = particulate matter (PM) with an aerodynamic diameter less than or equal to 2.5 µm, PM$_{10}$ = PM with an aerodynamic diameter less than or equal to 10 µm, SO$_2$ = sulfur dioxide, NO$_2$ = nitrogen dioxide, CO = carbon monoxide, O$_3$ = ground-level ozone; [b] The effect of each pollutant was adjusted for all the remaining pollutants; [c] Warm season = May–October; Cold season = November–April; [d] Statistically significant associations are shown in bold.

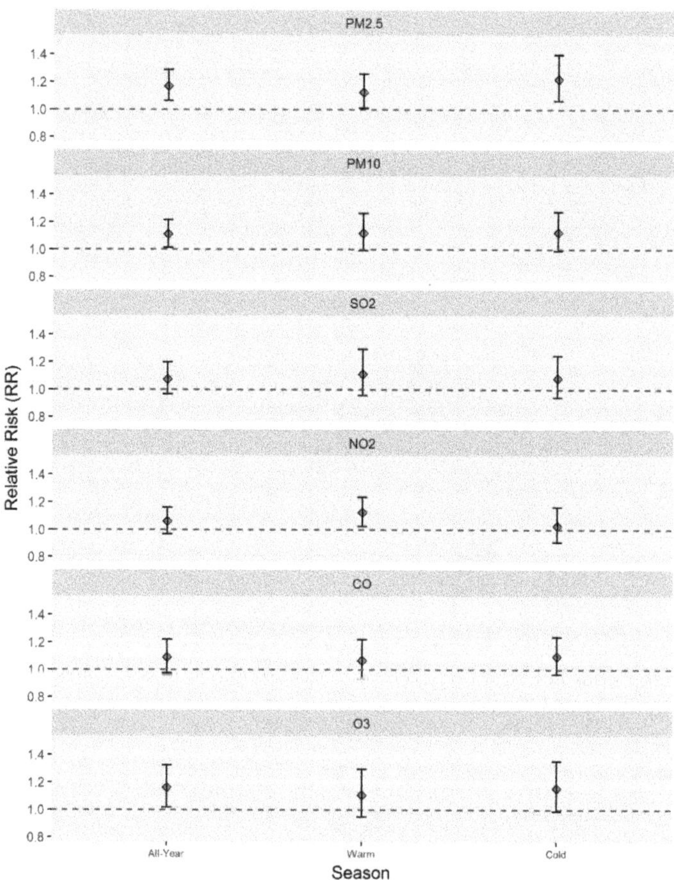

Figure 3. Relative risks (and 95% CIs) of TB incidence associated with an increase of 10 µg/m^3 in air pollutant concentrations by season (all-year, warm season [May–October] and cold season [May–October]) at lag 7 in Wuhan, China, during the 2015–2016 period.

4. Discussion

In this study, we conducted a time-series analysis to estimate short-term effects of air pollution on TB incidence in Wuhan city during the 2015–2016 period. We found that the effect of air pollutants on TB was greatest at lag 7. $PM_{2.5}$ and PM_{10}, and O_3 were statistically significantly associated with the increased risk of TB at the 7th day's exposure. We fitted a multi-pollutant model to adjust for the possible confounding from the presence of co-pollutants. Our results showed that the effect of $PM_{2.5}$ on TB remained robust after the adjustment while the effects for the rest of the pollutants were reduced or remained almost unchanged (O_3). The seasonal analysis showed that the impact of air pollution on TB was not that different between the warm season and the cold season.

One noteworthy finding of this research is that the effect of air pollution on TB was greatest at lag 7. The surprising consistency observed in lag 7 may be due to the ubiquitous collinearity among the pollutants, that is, the effect of any pollutant on TB actually reflects the effect of another or a combination of several pollutants on TB. Our results are contrary to many previous studies in which the effect was larger in early lags (typically at lag 0–3). This is likely due to the incubation period of the TB bacterium *Mycobacterium tuberculosis*. Our finding agrees with a previously published article in which significant associations between respiratory infection and $PM_{10}/PM_{2.5}$ were observed at lag 9–10 days and 7–8 days, respectively [29]. Chen et al. (2017) also reported a lagged effect of $PM_{2.5}$ on influenza and attributed this to the incubation of the influenza virus [30]. Considering that the incubation period of the TB bacterium varies from weeks to decades [18], it is possible that exposure to ambient air pollution shortened the incubation period and accelerated the progress of the disease. Many previous studies have provided a theoretical basis for this hypothesis. Animal studies and cellular-level studies suggest that biomass smoke impairs alveolar macrophage function [31–33]. $PM_{2.5}$ contains heavy metals that may damage macrophages, resulting in weakened phagocytosis [33]. It has been suggested that air pollutants $PM_{2.5}$, CO, NO_2, and SO_2 can modify the regulation of tumor necrosis factor-alpha (TNF-α) and interferon-gamma (IFN-γ) [9,33–35], which functions to induce macrophage activation [33,34]. Consequently, the damaged system may lead to an increased incidence of TB [6]. Additionally, studies have found that bacteria and viruses can attach to particles in the air and be inhaled into the respiratory system [36–38]. This makes the mechanism even more complex as one cannot judge if infection occurred before or during exposure to air pollution.

We observed a statistically significant association between $PM_{2.5}$ and PM_{10} and TB in our analysis. We noted that another study in Wuhan also examined the relationship between short-term exposure to air pollution and the risk of TB [15]. However, our research differs from theirs in three ways. First, the two studies used different statistical models. We used GAM, and they used distributed lag non-linear models (DLNMs). Second, exposure to air pollution was measured differently. We used ordinary kriging to estimate the exposure, whereas they used data collected from air monitoring stations directly. Third, our study conducted seasonal analysis, and they did not. Possibly due to these differences, the results of the two studies are also different. They observed the effects of NO_2, SO_2, CO, and $PM_{2.5}$ on the risk of TB, while we only observed the effects of PM on the risk of TB. In addition, their study had more years of data and was, therefore, easier to detect significant statistical associations, which could explain the differences in results between the two articles. Because no consensus has been reached on which combination of co-pollutants should be included in the multi-pollutant model [39], we included all air pollutants in the model. After the adjustment, the effect of $PM_{2.5}$ on TB remained robust, while the effects of other pollutants were attenuated. However, the attenuated effects of other pollutants in the multi-contaminant model may be due to the collinearity between the pollutants, which does not necessarily mean that they have no effect on TB. This finding is consistent with previous epidemiologic studies on TB and ambient $PM_{2.5}$ [12,13]. Nevertheless, the interpretation should be regarded with caution, as the presence of collinearity cannot be ignored to influence the association. Our results showed certain degrees of positive correlations between pairs of air pollutants (r: 0.59 to 0.80). Therefore, a single-pollutant model might be more reasonable for these types of data. We also observed positive but insignificant associations between gaseous air pollutants and TB. As mentioned earlier, CO, NO$_2$, and SO$_2$ all have roles in impairing the normal function of alveolar macrophages and result in an increased number of TB cases. Additionally, NO$_2$ and SO$_2$ can form ammonium sulfate and ammonium nitrate by naturally occurring chemical processes and can form part of PM and then further threaten human health [39].

Weather, especially temperature and humidity (or dew point temperature), are factors that can confound the relationship between health effects and air pollution at both the long- and short-term timescales. Temperature and humidity can be correlated with both air pollutants and health effects such as infectious diseases. We used GAM, a type of semiparametric model, to control for temperature and humidity. The semiparametric model combines the advantages of both parametric and nonparametric models, allowing for explicit parameter items that include exposures of interest and smooth nonparametric items such as temperature and humidity. The adjustment was achieved by using the smooth functions of temperature and humidity in the model. The smoothness of a model fit is controlled by the number of knots used, which was determined by the number of degrees of freedom assigned to each function. It has been reported that the potential confounding effect of temperature on short-term time scales may be less pronounced than long-term time scales, and the association between air pollution and health effects is relatively insensitive to the choice of the temperature model [40].

This article showed that TB incidence was highest in the spring. Despite that the underlying mechanism remains unknown, several factors may explain this phenomenon. First, people will reduce outdoor activities during the cold season. Consequently, the overcrowded, humid, and low airflow indoor environment makes TB easier to spread [41]. Meanwhile, more indoor activities reduce ultraviolet light exposure that can kill microorganisms [41]. Second, the decrease in human serum vitamin D in winter will increase the infection and re-activation of TB [42]. In addition, delays in seeking health care after the onset and diagnosis may also result in delayed reporting from the winter to the following spring [41]. This also led to misclassification of the actual exposure date. Our seasonal analysis suggested that the effects of air pollution on TB during the cold and warm seasons are not significantly different, which may mean that the increase in TB during spring is related to the indoor environment and is less associated with outdoor air pollution.

We examined the seasonal variation by splitting the all-year data into warm (May–October) and cold (November–April) seasons. There was no clear trend that the effect of air pollutants on TB was influenced by season. The effect estimates of SO$_2$ and NO$_2$ during the warm season were slightly higher than during the cold season. Certain factors may explain this phenomenon. First, during the warm season, people tend to go outdoors more often than during the cold season, so they are exposed to more ambient air pollutants as compared to the opposite. Second, more outdoor activities increase the possibility of TB infection, as elevated levels of airborne *Mycobacterium tuberculosis* may be present in the air. We also observed that the effect estimates of PM$_{2.5}$, PM$_{10}$, CO, and O$_3$ during the warm season were slightly lower than during the cold season. This was consistent with findings of another study that the effect of ambient PM$_{2.5}$ on TB was most significant during winter [13]. The study explained that staying indoors during high ambient air pollution episodes would lead to an increase in human contact and thus the increased risk of TB transmission. The same hypothesis was given elsewhere as well [43]. Therefore, the real mechanism is not yet clear and warrants more related research.

A feature of this study is the use of kriging to estimate the concentration of air pollution. Kriging has been broadly used to estimate ambient air pollution levels in epidemiological studies [22,24,25,44,45]. However, only a small part of the literature used kriging to study TB. Many such studies used kriging for the detection of TB clusters for animals and humans [46–48].

To the best of our knowledge, no studies have used the exposures estimated by kriging to conduct time-series research on TB. We attempted to use kriging to overcome a long-standing limitation existing in the epidemiology of air pollution, that is, to average the pollution level across the study area as the proxy for the population exposure level. By using kriging, we hope to reduce the misclassification raised by the simple averaging method. Nevertheless, it must be pointed out that kriging itself also has limitations. The accuracy of prediction using kriging largely depends on the number of monitoring stations. Due to a low ratio of monitoring stations to total land area, prediction error is quite possible. Furthermore, kriging also assumes isotropy, that is, it assumes uniformity in all directions. This obviously ignores the impact of the real environment (such as the built environment) on exposure [49]. A previous study compared the values estimated by kriging with the values estimated by the Assessment System for Population Exposure Nationwide (ASPEN) model and found that the former was generally smaller than the latter [21]. As estimates from the ASPEN model yielded a good agreement with monitoring data in multiple studies [21], it is considered more reliable than kriged values. In China, we do not have an air pollution prediction model similar to the ASPEN for comparison; thus, the accuracy of the prediction results is still uncertain.

Limitations and Implications

Our study has several limitations. First, our ecological study design cannot capture exposure levels at an individual level, resulting in exposure misclassifications. Second, two years of data assume a small sample size and may reduce the statistical power for the detection of significance. Compared to studies with many years of data, this led to overly wide confidence intervals. Previous studies have reported that many air pollutants are statistically significantly associated with TB. However, we only found that there was a statistically significant association of $PM_{2.5}$, PM_{10}, and O_3 with TB. Third, our O_3 data were collected 8 h per day, while other pollutants were collected based on a 24-h sampling schedule. This will cause bias in our results. Fourth, due to the lack of data, some factors that can lead to confounding or effect modification, such as gender, age, socioeconomic status, and so on, have not been included in the analysis, which may result in bias in results. Fifth, because TB patients may not be able to seek health care immediately after they become ill, and it takes time from the diagnosis to the reporting of a case, the results of this article are affected by this unavoidable measurement error. Lastly, the existence of collinearity among air pollutants may create bias and affect the magnitude of the estimated associations. Considering the above limitations, caution is needed in interpreting the results of the article.

5. Conclusion

In summary, this study explored associations between short-term exposure to ambient air pollutants and TB incidence in Wuhan city, China, during the 2015–2016 period. We found that the impact of air pollutants on tuberculosis was greatest at the 7th day's lag, and $PM_{2.5}$, PM_{10}, and O_3 were significantly associated with an increased risk of TB. Considering that the health impacts of air pollution and infectious disease are typically studied separately, the findings of this study not only add to the literature regarding the short-term effect of air pollution on TB but also help us to understand the mechanism of the interaction of air pollution and infectious disease. Future preventive efforts against TB should consider the reduction of exposure to air pollution at the community or personal level.

Author Contributions: Data curation, L.T.; Formal analysis, S.D., T.Z., F.L., and X.L.; Resources, S.H., W.Y., and Z.Z.; Supervision, H.X.; Writing—original draft, S.L.; Writing—review & editing, Y.L. All authors have read and agreed to the published version of the manuscript.

Funding: This work was supported by the National Undergraduate Innovation and Entrepreneurship Grant of Wuhan University (Grant No.201810486113) and Research Fund of Hubei Health and Family Planning Commission (Grant No. WJ2017M139).

Conflicts of Interest: The authors declare that they have no competing interests.

References

1. Tuberculosis—World Health Organization. Available online: http://www.whoint/mediacentre/factsheets/fs104/en/ (accessed on 9 March 2018).
2. Wang, L.; Zhang, H.; Ruan, Y.; Chin, D.P.; Xia, Y.; Cheng, S.; Chen, M.; Zhao, Y.; Jiang, S.-W.; Du, X.; et al. Tuberculosis prevalence in China, 1990–2010; a longitudinal analysis of national survey data. *Lancet* **2014**, *383*, 2057–2064. [CrossRef]

3. Tuberculosis in China. Available online: http://www.wprowhoint/china/mediacentre/factsheets/tuberculosis/en/ (accessed on 3 January 2020).

4. Zhao, Y.; Xu, S.; Wang, L.; Chin, D.P.; Wang, S.; Jiang, G.; Xia, H.; Zhou, Y.; Li, Q.; Ou, X.; et al. National Survey of drug-resistant tuberculosis in China. *N. Engl. J. Med.* **2012**, *366*, 2161–2170. [CrossRef] [PubMed]

5. Reported Incidence of Tuberculosis in Hubei. Available online: http://www.hbcdc.cn/ (accessed on 3 January 2019).

6. Fullerton, D.; Bruce, N.; Gordon, S.B. Indoor air pollution from biomass fuel smoke is a major health concern in the developing world. *Trans. R. Soc. Trop. Med. Hyg.* **2008**, *102*, 843–851. [CrossRef] [PubMed]

7. Kim, K.-H.; Jahan, S.A.; Kabir, E. A review of diseases associated with household air pollution due to the use of biomass fuels. *J. Hazard. Mater.* **2011**, *192*, 425–431. [CrossRef]

8. Sumpter, C.; Chandramohan, D. Systematic review and meta-analysis of the associations between indoor air pollution and tuberculosis. *Trop. Med. Int. Heal.* **2012**, *18*, 101–108. [CrossRef]

9. Tremblay, G. Historical statistics support a hypothesis linking tuberculosis and air pollution caused by coal. *Int. J. Tuberc. Lung Dis.* **2007**, *11*, 722–732.

10. Ferrara, G.; Murray, M.; Winthrop, K.; Centis, R.; Sotgiu, G.; Migliori, G.B.; Maeurer, M.; Zumla, A. Risk factors associated with pulmonary tuberculosis. *Curr. Opin. Pulm. Med.* **2012**, *18*, 233–240. [CrossRef]

11. Lin, H.-H.; Ezzati, M.; Murray, M.B. Tobacco smoke, indoor air pollution and tuberculosis: A systematic review and meta-analysis. *PLoS Med.* **2007**, *4*, e20. [CrossRef]

12. Jassal, M.; Bakman, I.; Jones, B. Correlation of ambient pollution levels and heavily-trafficked roadway proximity on the prevalence of smear-positive tuberculosis. *Public Heal.* **2013**, *127*, 268–274. [CrossRef]

13. You, S.; Tong, Y.W.; Neoh, K.G.; Dai, Y.; Wang, C.-H. On the association between outdoor $PM_{2.5}$ concentration and the seasonality of tuberculosis for Beijing and Hong Kong. *Environ. Pollut.* **2016**, *218*, 1170–1179. [CrossRef]

14. Zhu, S.; Xia, L.; Wu, J.; Chen, S.; Chen, F.; Zeng, F.; Chen, X.; Chen, C.; Xia, Y.; Zhao, X.; et al. Ambient air pollutants are associated with newly diagnosed tuberculosis: A time-series study in Chengdu, China. *Sci. Total Environ.* **2018**, *631*, 47–55. [CrossRef] [PubMed]

15. Xu, M.; Liao, J.; Yin, P.; Hou, J.; Zhou, Y.; Huang, J.; Liu, B.; Chen, R.; Ke, L.; Chen, H.; et al. Association of air pollution with the risk of initial outpatient visits for tuberculosis in Wuhan, China. *Occup. Environ. Med.* **2019**, *76*, 560–566. [CrossRef] [PubMed]

16. Lai, T.-C.; Chiang, C.-Y.; Wu, C.-F.; Yang, S.-L.; Liu, D.-P.; Chan, C.-C.; Lin, H.-H. Ambient air pollution and risk of tuberculosis: A cohort study. *Occup. Environ. Med.* **2015**, *73*, 56–61. [CrossRef] [PubMed]

17. Smith, G.S.; Schoenbach, V.J.; Richardson, D.B.; Gammon, M.D. Particulate air pollution and susceptibility to the development of pulmonary tuberculosis disease in North Carolina: An ecological study. *Int. J. Environ. Heal. Res.* **2013**, *24*, 103–112. [CrossRef]

18. Ge, E.; Fan, M.; Qiu, H.; Hu, H.; Tian, L.; Wang, X.; Xu, G.; Wei, X. Ambient sulfur dioxide levels associated with reduced risk of initial outpatient visits for tuberculosis: A population based time series analysis. *Environ. Pollut.* **2017**, *228*, 408–415. [CrossRef]

19. Popovic, I.; Magalhaes, R.J.S.; Ge, E.; Marks, G.B.; Dong, G.-H.; Wei, X.; Knibbs, L.D. A systematic literature review and critical appraisal of epidemiological studies on outdoor air pollution and tuberculosis outcomes. *Environ. Res.* **2019**, *170*, 33–45. [CrossRef]

20. Torres, M.; Carranza, C.; Sarkar, S.; Gonzalez, Y.; Osornio-Vargas, A.R.; Black, K.; Meng, Q.; Quintana-Belmares, R.; Hernandez, M.; Garcia, J.J.F.A.; et al. Urban airborne particle exposure impairs human lung and blood Mycobacterium tuberculosis immunity. *Thorax* **2019**, *74*, 675–683. [CrossRef]

21. Whitworth, K.; Symanski, E.; Lai, D.; Coker, A.L. Kriged and modeled ambient air levels of benzene in an urban environment: An exposure assessment study. *Environ. Heal.* **2011**, *10*, 21. [CrossRef]

22. Liao, D.; Peuquet, N.J.; Duan, Y.; Whitsel, E.A.; Dou, J.; Smith, R.L.; Lin, H.-M.; Chen, J.-C.; Heiss, G. GIS approaches for the estimation of residential-level ambient PM concentrations. *Environ. Heal. Perspect.* **2006**, *114*, 1374–1380. [CrossRef]

23. Rohde, R.A.; Muller, R.A. Air Pollution in China: Mapping of Concentrations and Sources. *PLoS ONE* **2015**, *10*, e0135749. [CrossRef]

24. Jerrett, M.; Burnett, R.T.; Ma, R.; Pope, C.A.; Krewski, D.; Newbold, K.B.; Thurston, G.; Shi, Y.; Finkelstein, N.; Calle, E.E.; et al. Spatial analysis of air pollution and mortality in Los Angeles. *Epidemiology* **2005**, *16*, 727–736. [CrossRef] [PubMed]

25. Jerrett, M.; Buzzelli, M.; Burnett, R.T.; DeLuca, P.F. Particulate air pollution, social confounders, and mortality in small areas of an industrial city. *Soc. Sci. Med.* **2005**, *60*, 2845–2863. [CrossRef] [PubMed]

26. Roberts, J.D.; Voss, J.; Knight, B. The association of ambient air pollution and physical inactivity in the United States. *PLoS ONE* **2014**, *9*, e90143. [CrossRef] [PubMed]

27. Chu, Y.; Liu, Y.; Lu, Y.; Yu, L.; Lu, H.; Guo, Y.; Xiang, H.; Liu, F.; Wu, Y.; Mao, Z.; et al. Propensity to Migrate and Willingness to Pay Related to Air Pollution among Different Populations in Wuhan, China. *Aerosol Air Qual. Res.* **2017**, *17*, 752–760. [CrossRef]

28. Zanobetti, A.; Franklin, M.; Koutrakis, P.; Schwartz, J. Fine particulate air pollution and its components in association with cause-specific emergency admissions. *Environ. Heal.* **2009**, *8*, 58. [CrossRef]

29. Xia, X.; Zhang, A.; Liang, S.; Qi, Q.; Jiang, L.; Ye, Y. The Association between Air Pollution and Population Health Risk for Respiratory Infection: A Case Study of Shenzhen, China. *Int. J. Environ. Res. Public Heal.* **2017**, *14*, 950. [CrossRef]

30. Chen, G.; Zhang, W.; Li, S.; Zhang, Y.; Williams, G.; Huxley, R.R.; Ren, H.; Cao, W.; Guo, Y. The impact of ambient fine particles on influenza transmission and the modification effects of temperature in China: A multi-city study. *Environ. Int.* **2017**, *98*, 82–88. [CrossRef]

31. Aam, B.B.; Fonnum, F. Carbon black particles increase reactive oxygen species formation in rat alveolar macrophages in vitro. *Arch. Toxicol.* **2006**, *81*, 441–446. [CrossRef]

32. Arredouani, M.S.; Yang, Z.; Imrich, A.; Ning, Y.; Qin, G.; Kobzik, L. The Macrophage Scavenger Receptor SR-AI/II and Lung Defense against Pneumococci and Particles. *Am. J. Respir. Cell Mol. Boil.* **2006**, *35*, 474–478. [CrossRef]

33. Zhou, H.; Kobzik, L. Effect of concentrated ambient particles on macrophage phagocytosis and killing of Streptococcus pneumoniae. *Am. J. Respir. Cell Mol. Boil.* **2006**, *36*, 460–465. [CrossRef]

34. Drumm, K.; Buhl, R.; Kienast, K. Additional NO_2 exposure induces a decrease in cytokine specific mRNA expression and cytokine release of particle and fibre exposed human alveolar macrophages. *Eur. J. Med. Res.* **1999**, *4*, 59–66.

35. Knorst, M.M.; Kienast, K.; Müller-Quernheim, J.; Ferlinz, R. Effect of sulfur dioxide on cytokine production of human alveolar macrophages in vitro. *Arch. Environ. Heal. Int. J.* **1996**, *51*, 150–156. [CrossRef]

36. Chen, P.-S.; Tsai, F.T.; Lin, C.K.; Yang, C.-Y.; Chan, C.-C.; Young, C.-Y.; Lee, C.-H. Ambient influenza and avian influenza virus during dust storm days and background days. *Environ. Heal. Perspect.* **2010**, *118*, 1211–1216. [CrossRef]

37. Jaspers, I.; Ciencewicki, J.M.; Zhang, W.; Brighton, L.E.; Carson, J.L.; Beck, M.A.; Madden, M.C. Diesel exhaust enhances influenza virus infections in respiratory epithelial cells. *Toxicol. Sci.* **2005**, *85*, 990–1002. [CrossRef]

38. Schafer, M.P. Detection and characterization of airborne Mycobacterium tuberculosis H37Ra particles, A surrogate for airborne pathogenic M. tuberculosis. *Aerosol Sci. Technol.* **1999**, *30*, 161–173. [CrossRef]

39. Liu, S.; Zhang, K. Fine particulate matter components and mortality in Greater Houston: Did the risk reduce from 2000 to 2011? *Sci. Total. Environ.* **2015**, *538*, 162–168. [CrossRef]

40. Welty, L.J.; Zeger, S.L. Are the acute effects of PM_{10} on mortality in NMMAPS the result of inadequate control for weather and season? A sensitivity analysis using flexible distributed lag models. *Am. J. Epidemiol.* **2005**, *162*, 80–88. [CrossRef]

41. Fares, A. Seasonality of tuberculosis. *J. Glob. Infect. Dis.* **2011**, *3*, 46–55. [CrossRef]

42. Nnoaham, K.E.; Clarke, A. Low serum vitamin D levels and tuberculosis: A systematic review and meta-analysis. *Int. J. Epidemiol.* **2008**, *37*, 113–119. [CrossRef]

43. Chen, S.-C.; Liao, C.-M.; Li, S.-S.; You, S.-H. A probabilistic transmission model to assess infection risk from Mycobacterium tuberculosis in commercial passenger trains. *Risk Anal. Int. J.* **2010**, *31*, 930–939. [CrossRef]

44. Iñiguez, C.; Ballester, F.; Estarlich, M.; Llop, S.; Fernández-Patier, R.; Aguirre-Alfaro, A.; Esplugues, A. Estimation of personal NO_2 exposure in a cohort of pregnant women. *Sci. Total Environ.* **2009**, *407*, 6093–6099. [CrossRef] [PubMed]

45. Leem, J.-H.; Kaplan, B.M.; Shim, Y.K.; Pohl, H.R.; Gotway, C.A.; Bullard, S.M.; Rogers, J.F.; Smith, M.M.; Tylenda, C.A. Exposures to air pollutants during pregnancy and preterm delivery. *Environ. Heal. Perspect.* **2006**, *114*, 905–910. [CrossRef] [PubMed]

46. Liu, Y.; Jiang, S.-W.; Liu, Y.; Wang, R.; Li, X.; Yuan, Z.; Wang, L.; Xue, F. Spatial epidemiology and spatial ecology study of worldwide drug-resistant tuberculosis. *Int. J. Heal. Geogr.* **2011**, *10*, 50. [CrossRef]

47. Martínez, H.Z.; Suazo, F.M.; Gil, J.Q.C.; Bello, G.C.; Escalera, A.M.A.; Márquez, G.H.; Casanova, L.G. Spatial epidemiology of bovine tuberculosis in Mexico. *Vet. Ital.* **2010**, *43*, 629–634.
48. Nunes, C. Tuberculosis incidence in Portugal: Spatiotemporal clustering. *Int. J. Heal. Geogr.* **2007**, *6*, 30. [CrossRef]
49. Remy, N.; Boucher, A.; Wu, J. *Applied Geostatistics with SgeMS: A User's Guide*; Cambridge University Press: Cambridge, UK, 2009.

 International Journal of
*Environmental Research
and Public Health*

Article

The Effects of Indoor Pollutants Exposure on Allergy and Lung Inflammation: An Activation State of Neutrophils and Eosinophils in Sputum

Khairul Nizam Mohd Isa [1,2], Zailina Hashim [1,*], Juliana Jalaludin [1], Leslie Thian Lung Than [3] and Jamal Hisham Hashim [4]

[1] Department of Environmental and Occupational Health, Faculty of Medicine and Health Sciences, Universiti Putra Malaysia, UPM, Serdang 43400, Selangor, Malaysia; khairulnizamm@unikl.edu.my (K.N.M.I.); juliana@upm.edu.my (J.J.)

[2] Environmental Health Research Cluster (EHRc), Environmental Healthcare Section, Institute of Medical Science Technology, Universiti Kuala Lumpur, Kajang 43000, Selangor, Malaysia

[3] Department of Microbiology and Parasitology, Faculty of Medicine and Health Sciences, Universiti Putra Malaysia, UPM, Serdang 43400, Selangor, Malaysia; leslie@upm.edu.my

[4] IIGH United Nations University, UKM Medical Centre, Cheras 56000, Kuala Lumpur, Malaysia; jamalhas@hotmail.com

* Correspondence: drzhashim@gmail.com; Tel.: +601-7636-1367

Received: 28 April 2020; Accepted: 22 June 2020; Published: 28 July 2020

Abstract: Background: To explore the inflammation phenotypes following indoor pollutants exposure based on marker expression on eosinophils and neutrophils with the application of chemometric analysis approaches. Methods: A cross-sectional study was undertaken among secondary school students in eight suburban and urban schools in the district of Hulu Langat, Selangor, Malaysia. The survey was completed by 96 students at the age of 14 by using the International Study of Asthma and Allergies in Children (ISAAC) and European Community Respiratory Health Survey (ECRHS) questionnaires. The fractional exhaled nitric oxide (FeNO) was measured, and an allergic skin prick test and sputum induction were performed for all students. Induced sputum samples were analysed for the expression of CD11b, CD35, CD63, and CD66b on eosinophils and neutrophils by flow cytometry. The particulate matter ($PM_{2.5}$ and PM_{10}), NO_2, CO_2, and formaldehyde were measured inside the classrooms. Results: Chemometric and regression results have clustered the expression of CD63 with $PM_{2.5}$, CD11b with NO_2, CD66b with FeNO levels, and CO_2 with eosinophils, with the prediction accuracy of the models being 71.88%, 76.04%, and 76.04%, respectively. Meanwhile, for neutrophils, the CD63 and CD66b clustering with $PM_{2.5}$ and CD11b with FeNO levels showed a model prediction accuracy of 72.92% and 71.88%, respectively. Conclusion: The findings indicated that the exposure to $PM_{2.5}$ and NO_2 was likely associated with the degranulation of eosinophils and neutrophils, following the activation mechanisms that led to the inflammatory reactions.

Keywords: lung inflammation; allergy; indoor pollutants; biomarkers; FeNO; eosinophil; neutrophil

---

1. Introduction

Exposure to indoor pollutants are strongly associated with increased morbidity and mortality, mainly among school children who spend most of their time in the classrooms [1]. Reports have shown that exposure to high concentration of particles, nitrogen dioxide (NO_2), carbon dioxide (CO_2), ozone (O_3), volatile organic compounds (VOCs), and fibres induce persistent airway inflammation, which is mediated by the immune system [2,3]. Biomarkers, such as fractional exhaled nitric oxide (FeNO), cytokines, chemokines, lipid mediators, enzymes, adhesion molecules, and other growth factors, have been considered as the indicators of allergic airway inflammation [4]. The indicators that underlie the complex molecular pathways that regulate inflammation have not been fully elucidated; however, numerous studies have reported that the process certainly involves the activation of eosinophils and neutrophils [5,6]. They play roles in innate host defence via effector mechanisms, including degranulation, DNA traps, and cytolysis, following the activation cascaded by diverse mediators [7].

Extensive investigation of the mediators that are implicated in allergy, lung inflammation, and asthma has been documented but there was insufficient evidence available for the multi-dimensional characters of the activation and degranulation markers expression of CD11b, CD35, CD63, and CD66b on eosinophils and neutrophils. Activation markers, such as integrin Mac-1 (CD11b/CD18), L-selectin (CD62L), CBRM1/5, ICAM-1 (CD54), PD-L1 (CD274), PSGL-1 (CD162), FcγRII (CD32), CD16, CD44, and CD69, were shown to be upregulated on circulating or sputum granulocytes from asthma patients [8,9]. Response to allergens and environmental stimuli among asthma patients had a detrimental outcome and was characterized by the infiltration of different granule contents from eosinophils and neutrophils following activation [10]. Eosinophils and neutrophils have four distinct granules, namely azurophilic, secondary, tertiary, and secretary, which are formed throughout development in the bone marrow [11]. Several models have demonstrated the link between degranulation of these granules and pollutants exposure at different levels of stimuli with the expression of CD11c, CD63, and CD66b [12,13]. At present, biomarkers help tailor the management of respiratory illness and studies have highlighted precision therapy approaches based on disease mechanisms by targeting the cytokines and chemokines [14,15]. Recently, advanced statistical analysis approaches, such as hierarchical clustering and latent class analyses, have renewed the interest of researchers in the investigation of airway translational inflammation phenotypes [16,17]. This new vision helps the researchers to better understand and precisely describe the inflammatory phenotypes [18,19].

This study aimed to investigate the concentrations and sources of indoor pollutants in classrooms located in the urban and suburban areas of Hulu Langat, Selangor, Malaysia. Moreover, it aimed to predict the toxicodynamic effects of indoor pollutants in the classrooms towards marker expression on the eosinophils and neutrophils in sputum samples by using chemometric analysis techniques.

2. Materials and Methods

2.1. Study Population

The study population was randomly sampled from eight secondary schools in Hulu Langat, Selangor, Malaysia. The researchers of this study targeted school children at the age of 14 of which they were randomly selected from four classrooms in each school. The total number of students who received their guardian's consent and was thus recruited in the study was 470. Among them, only 50 (10.6%) students were diagnosed with asthma by a doctor based on the survey questionnaire. Another 46 students out of the remaining 420 students were randomly selected as a potential control group. The control group was selected among students who produced an adequate sputum cell count from the same class and school. In total, 96 students were included in the final study group. Students with a history of smoking in the last 12 months and students who received antibiotic treatments in the past four weeks were excluded from this study. The school areas were classified as urban and suburban by the Ministry of Education, Malaysia, based on the locale classification of the ecological measures. The data collection from the clinical assessments and indoor air monitoring were carried out at the same time in August until November 2018 and February 2019.

The researchers used the questionnaire adopted from the European Community Respiratory Health Survey (ECRHS) and International Study of Asthma and Allergies in Childhood (ISAAC) that are inclusive of questions on doctor-diagnosed asthma, allergies, and respiratory symptoms. The self-administered questionnaire was distributed to the selected students in the same week of the technical measurements. Subsequently, the researchers went through the questionnaires during a face-to-face interview with the students to clarify any uncertainty. The doctor-diagnosed asthma in this study is defined as having asthma medication, asthma attacks, and wheezes with breathlessness in the last 12 months, which were diagnosed by the physician [20,21]. This information was verified during face-to-face interviews and telephone calls with the students' respective guardians.

2.2. FeNO Assessment and Allergy Skin Test

FeNO measurement was performed by a chemiluminescence analyser (NIOX VERO, Circassia, Sweden) as recommended by the American Thoracic Society/European Respiratory Society (ATS/ERS) [22]. The detection limits and accuracy for this device are 5–300 ppb and ±5 ppb, respectively. The students were asked to inhale deeply through the mouthpiece attached to the patient filter and then slowly exhale for about six to ten seconds at a constant flow rate (50 mL/s)—the single-breath technique. This process was repeated at least twice to get an average value. Students were instructed to refrain from eating and drinking for one hour before the FeNO assessment.

An allergy skin prick test was performed on the volar side of the forearm alongside *Dermatophagoides pteronyssinus* (Derp1) (house dust mite), *Dermatophagoides farina* (Derf1) (house dust mite), *Cladosporium herbarium* (fungi), *Alternaria alternate* (fungi), and *Felis domesticus* (cat) allergens in liquid form (Prick-Test Diagnostic, ALK-Abelló, Madrid, Spain). The same amount of histamine (10 mg/mL) and normal saline were used as the positive and negative controls, respectively. The reaction was measured after 15 min of which the wheal diameter was recorded. The allergen's wheal diameter of 3 mm was considered as a positive control. Atopy was defined as a significant positive skin test reaction to at least one of the applied allergens [23].

2.3. Sputum Induction and Processing

Sputum induction was conducted by the inhalation of a nebulised, sterile mixture of 4.5% sodium chloride (hypertonic) and salbutamol 200 µg, followed by the coughing and expectoration of airway secretions. For nebulisation, an ultrasound nebuliser (Model CUN60 Citizen System Japan Co. Ltd., Tokyo, Japan) was was used as recommended [24] with a mouthpiece that fitted an output of ~1 mL·min^{-1} to achieve successful sampling. The induced sputum samples collected from the respondents were kept in an icebox and further processed within two hours by using flow cytometry. The method of processing sputum samples has been previously described [25]. In short, the sputum sample was diluted with freshly prepared phosphate buffer saline and gently vortexed at room temperature for homogenisation. These steps were repeated thrice. Subsequently, the sputum samples were centrifuged at 800× *g* for 10 min. Next, cytospin slides of sputum cells were stained with May–Grunwald–Giemsa for the cell differentiation count. Samples with >80% of squamous epithelial cells were excluded for the flow cytometry analysis.

2.4. Flow Cytometry

The processed sputum sample at a concentration of 1×10^6 cells in 100 µL was resuspended in 1 mL of stain buffer (FBS) (BD Pharmingen^{TM}, San Diego, CA, USA) and washed twice in cold stain buffer by centrifugation at 200× *g* for 5 min. Subsequently, the supernatant was removed, and the cell pellet was resuspended in 300 µL of stain buffer. A volume of 100 µL aliquots of the cell suspension were transferred into three different sterile polypropylene round-bottom tubes, and monoclonal antibodies and isotype controls were added to the cells according to the manufacturer protocol. The cells were immunostained with antibodies for neutrophil and eosinophil surface markers of PE-Cy 7 Mouse Anti-Human CD11b/Mac-1, BB515 Mouse Anti-Human CD35, PE Mouse Anti-Human CD63, APC-H7 Mouse Anti-Human CD16, PerCP-Cy 5.5 CD41a, and Alexa Fluor 647 Mouse Anti-Human CD66b purchased from BD Biosciences, US. The isotype controls of PE-Cy 7 Mouse IgG1, BB515 Mouse IgG1 K, PE Mouse IgG1 K, APC-H7 Mouse IgG1 K, PerCP-Cy 5.5 Mouse IgG1 K, and Alexa Fluor 647 Mouse IgM K (BD Biosciences, US) were added to the cells. The cell samples were incubated for 15 min at room temperature in the dark. Next, the cell samples were washed twice in 1 mL of stain buffer and centrifuged at 300× *g* for 5 min. The supernatant was removed, and the cells were loosened up by tapping the tube. Subsequently, the cells were carefully resuspended in 500 µL of the stain buffer.

Samples without antibodies and isotype were used as the controls. Hereafter, the cells were analysed by using the BD FACSCanto II flow cytometry instrument (BD Bioscience, US) equipped with blue, red, and violet lasers. Compensation was set to account for spectral overlap between the four fluorescent channels. The gating region was set so that less than 1% of the samples were stained with negative controls. Data were processed by using FlowJo software (Version 10.1r1) [26,27].

The activation of eosinophils was measured by using CD11b (integrin αM), CD35 (CR1), and CD66b (CEACAM-8). Meanwhile, CD11b was used as an activation marker for neutrophils. The degranulation activity was measured by the expression of CD11b and CD63 for eosinophils as a marker for cyctalloid (specific/secondary) and secretory (sombrero) vesicles, respectively. For neutrophils, the CD11b, CD35, CD63, and CD66b markers were used as tertiary, secretory, azurophilic, and specific granule expression markers in the degranulation activities [28]. Isotype controls were used as the positive control and to address the background produced by non-specific antibody binding, whereas the Anti-Mouse Ig K Negative Control Compensation Particle Set was used to optimise the fluorescence compensation settings [29] (Figure 1).

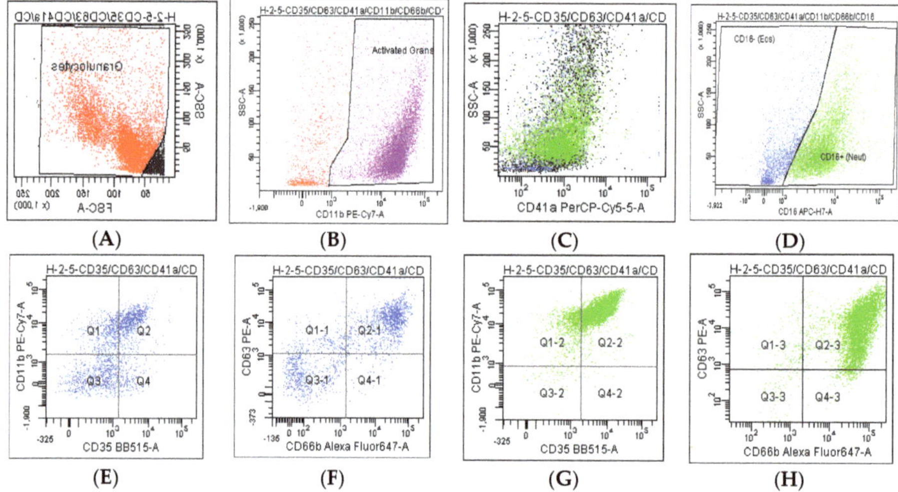

Figure 1. Gating strategy to identify activated sputum granulocytes. (**A**) A forward-/side-scatter (FSC/SSC) gate to identify the granulocytes. (**B**) The activated granulocytes were based on CD11b+. (**C**) CD41a was gated with SSC to confirm that the activated granulocytes were neutrophils and eosinophils. (**D**) Subsequently, the eosinophils (blue) and neutrophils (green) were separated based on the expression of CD16− and CD16+ on SSC, respectively. The activation markers of CD11b, CD35, CD63, and CD66b for eosinophils (**E,F**) and neutrophils (**G,H**) were based on the negative gates of the isotype control for all antibodies.

2.5. Assessment of Indoor Air Quality in Classrooms and Building Inspection

Indoor pollutants and physical parameters, including temperature (°C), relative humidity (%), and carbon dioxide (ppm), were measured in the classrooms during learning session within an hour by using a Q-TrakTM IAQ monitor (Model 7565 TSI Incorporated, Shoreview, MN, USA) with the average log interval values over one minute. The accuracy of this device on temperature, relative humidity, and CO_2 are ±0.6 °C, ±3%, and ±50 ppm, respectively. The sampling for the particles was measured by using a Dust-Trak monitor (Model 8532 TSI Incorporated, Shoreview, MN, USA) at a sampling rate of 1.7 L/min with a resolution of 0.001 mg/m^3 and detection limit of 0.001–150 mg/m^3. The PPM Formaldemeter^{TM} htV-M (PPM Technology Ltd, Wales, UK) with accuracy of 10% at 2 ppm was used to measure the concentration of formaldehyde. In each school, a total of four hours of measurements were taken from four randomly selected classrooms for a period of an hour each during the learning session, as has been previously described [30–32]. The instruments used were placed one metre from the ground in the centre of the classrooms. All instruments used were calibrated regularly. The NO_2 ($\mu g/m^3$) concentration was measured by using a diffusion sampler (IVL, Goteborg, Sweden) for a period of a week. The sampler was place at height of approximately 2–3 m above the ground and returned to the IVL Swedish Environmental Research Institute Laboratory (Goteborg, Sweden), an accredited laboratory for further analysis. This measurement technique provides an average concentration of NO_2 in the air during a week, with a limit of detection (LOD) of 0.5 $\mu g/m^3$ and a 10% (at the 95% confidence level) measurement uncertainty [33]. A building inspection was carried out before the indoor air quality assessments were obtained. Details on the building information, floor furnish, furniture, and type of ventilation system were noted [34].

2.6. Ethical Statement

The Ethics Committee for Research Involving Human Subjects Universiti Putra Malaysia (JKEUPM) has approved this study (JKEUPM-2018-189) and each of the students was given a written consent form for their guardian's approval.

2.7. Data Analysis

The descriptive analysis was carried out by Mann–Whitney tests using the Statistical Package for the Social Sciences (SPSS) 25.0. The differences in biomarker expression between the doctor-diagnosed asthmatic children and healthy children were made by using GraphPad Prism 8 for Windows. Subsequently, a principle component analysis (PCA) and agglomerative hierarchical clustering (AHC) were applied to explore the association and pattern recognition between the biomarker expression and the concentrations of indoor pollutants. The final prediction models were generated by logistic regression analysis in which the models' performance was based on the coefficient of determinant, overall accuracy, sensitivity, specificity, and the area under the curve (AUC) of the receiver operator characteristics (ROC) [35]. The researchers used the standardized data of the indoor pollutants and biomarkers in the chemometric and regression analyses. The multivariate analysis was carried out by using the Statistical Package XLSTAT Evaluation 2019.2.3 (Addinsoft, New York, US).

3. Results

3.1. Data Analysis for the Personal and Clinical Characteristics of School Children

The study population was well-balanced between the doctor-diagnosed asthmatic children (52%) and healthy children (48%), in which 60% of them were from urban schools. The majority of asthmatic children tested positive for at least one of the allergens, with 74.0% of them sensitised towards house dust mites (Derp1 and Derf1), followed by cat dander (30.0%) (Supplementary Table S1). The FeNO levels were statistically higher among the doctor-diagnosed asthmatic children than healthy children ($p < 0.001$). The researchers observed that the total percentage of eosinophil count in the sputum samples was slightly higher in the doctor-diagnosed asthmatic children but not statistically different between the two groups ($p > 0.05$) (Table 1).

Table 1. Students' characteristic and inflammation status for the doctor-diagnosed asthma and control group.

Characteristics	Doctor Diagnosed Asthma ($n = 50$)	Healthy ($n = 46$)	p-Value
Female (%)	44.2	55.8	0.094
Male (%)	61.4	38.6	
Atopic (%)	65.0	35.0	<0.001 **
Non atopic (%)	30.6	69.4	
Parental with asthma/allergy (%)	65.9	34.1	0.020 *
Parental without asthma/allergy (%)	41.8	58.2	
School area—urban (%)	60.4	39.6	0.102
School area—suburban (%)	43.8	56.3	
Clinical characteristics			
FeNO levels (ppb)	56 (66.5)	23 (32.3)	0.002 *
Eosinophil count (%)	11.6 (11.3)	10.2 (9.2)	0.259
Neutrophil count (%)	11.6 (5.0)	13.4 (14.0)	0.130

Values are the median (IQR) for clinical characteristics. IQR = Interquartile range. * $p < 0.05$; ** $p < 0.001$.

Both the eosinophils and neutrophils expressed an activated phenotype in both groups of induced sputum samples. The expression of CD11b was significantly upregulated in both their cell surfaces among the doctor-diagnosed asthmatic children ($p < 0.05$). The expression of CD63 on the neutrophils was also significantly higher in the sputum samples of the doctor-diagnosed asthmatic children as compared to healthy children ($p < 0.001$). Samples from the doctor-diagnosed asthmatic children with a high expression of CD11b (tertiary) and CD63 (azurophilic/crystalloid) on the eosinophil and neutrophil surfaces indicated a moderately degranulated state of sputum granulocytes (Figure 2).

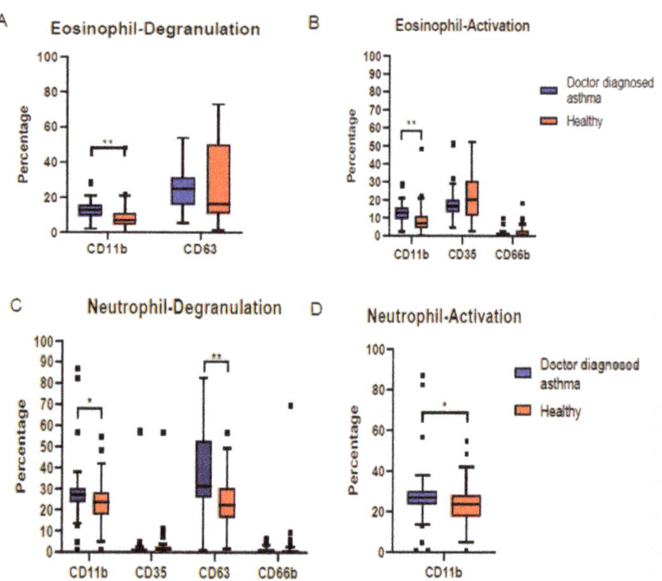

Figure 2. Expression profile of sputum granulocytes compared between the doctor-diagnosed asthmatic and healthy school children. Expression of degranulation and activation markers on eosinophil (**A**,**B**) and neutrophil (**C**,**D**). The expression of CD11b as the degranulation and classical activation marker is displayed twice in this graph. * $p < 0.05$, ** $p < 0.001$.

3.2. Levels of Indoor Pollutants and Building Inspection Data

All the classrooms were designed with natural ventilation, which is equipped with glass windowpanes on both sides of the wall. The classrooms were equipped with an average of three ceiling fans in each classroom. The schools were painted, and the floor surface was furnished with concrete. There were bookshelves, whiteboard, and soft boards in every classroom. Some of the classrooms had window curtains fixed on both sides of the class. The statistical analysis showed a significant difference in all of the indoor environmental parameters between the urban and suburban schools (Table 2).

Table 2. Comparison of the environmental parameters between schools located in urban and suburban areas.

Parameter	Urban $n = 16$			Suburban $n = 16$			p-Value
	Median (IQR)	Min	Max	Median (IQR)	Min	Max	
Temperature (°C)	29.0 (2.0)	28.0	32.0	27.5 (1.0)	27.0	28.0	<0.001 **
Relative Humidity (%)	74.7 (9.5)	63.6	88.1	81.4 (7.5)	74.8	88.1	<0.001 **
Formaldehyde (mg/m^3)	13.2 (9.3)	5.2	19.5	3.1 (5.2)	2.1	15.9	<0.001 **
CO_2 (ppm)	453.0 (34.5)	417.0	468.0	455.5 (25.5)	402.0	471.0	0.462
NO_2 ($\mu g/m^3$)	32.0 (7.0)	15.0	45.0	19.0 (22.5)	16.0	48.0	<0.001 **
$PM_{2.5}$ ($\mu g/m^3$)	24.6 (2.4)	17.9	26.5	22.0 (1.9)	16.8	26.9	<0.001 **
PM_{10} ($\mu g/m^3$)	41.6 (7.5)	34.7	48.0	37.0 (4.9)	32.3	44.9	<0.001 **

$n = 32$; IQR = Interquartile range, Min = Minimum, Max = Maximum. ** $p < 0.001$.

3.3. Chemometrics Analysis of the Biomarkers and Indoor Air Pollutants

A PCA was performed to explain the variance observed between the biomarkers and indoor pollutants in a more efficient way. Prior to the PCA, Bartlett's sphericity and Kaiser–Meyer–Olkin (KMO) tests were conducted to determine the correlation difference and sampling adequacy, with both measurements achieving the required levels. The PCA was applied with a normalization procedure and the coefficient factor loadings produced also expressed the correlation between the variables [36]. The factor loadings of the four factors with eigenvalues >1 were extracted from the eosinophil and neutrophil expression markers. A total of 39.0% of the variation for both Factor 1 and Factor 2 was observed in the expression of markers on eosinophils tested with indoor pollutants. Moderate factor loadings were identified for expression of CD11b and CD35, together with a strong factor loading for the concentrations of NO_2 and formaldehyde in Factor 1. In other words, the upregulation of CD11b and CD35 expression in the eosinophils were associated with exposure to NO_2 and formaldehyde; in turn, Factor 2 had moderate factor loadings of FeNO levels, PM_{10}, and $PM_{2.5}$. Moderate factor loadings were observed for FeNO levels, CD66b, and PM_{10} in Factor 3, with a total variation of 14.9%. High factor loading of CD63 was observed in Factor 4, with a total variation of 13.0%. For neutrophils, a strong factor loading of CD66b expression together with moderate factor loadings for FeNO levels, CD11b, and CD63 were recorded in Factor 1, with a variation of 24.2%. Factor 2 showed a strong factor loading for the concentration of formaldehyde together with a moderate factor loading for the concentrations of CO_2, PM_{10}, and $PM_{2.5}$, with 19.3% of the total variation. There were moderate loading factors for expression of CD35 and the concentrations of NO_2 and PM_{10} in Factor 3, with 14.3% of the total variation. Factor 4 showed moderate factor loadings of CD11b expression and concentration of $PM_{2.5}$, with 13.2% of the total variation (Table 3).

Table 3. Factor loadings using PCA for eosinophils and eosinophils. The moderate (0.5–0.75) and strong (>0.75) factor loadings are highlighted in bold.

Biomarkers and Indoor Pollutants	Eosinophils				Neutrophils			
	F1	F2	F3	F4	F1	F2	F3	F4
FeNO levels	0.194	**−0.575**	**0.501**	0.235	**0.653**	−0.306	−0.108	−0.120
CD11b	**0.528**	0.083	−0.398	0.059	**0.571**	−0.053	0.175	**0.685**
CD35	**0.559**	0.129	0.432	0.348	0.337	0.213	**0.612**	−0.445
CD63	−0.193	0.379	0.070	**0.732**	**−0.665**	0.394	0.117	−0.288
CD66b	0.285	−0.465	**0.561**	−0.419	**0.727**	0.067	−0.235	−0.039
CO_2	−0.346	−0.329	−0.456	0.084	−0.287	**−0.506**	0.123	−0.360
NO_2	**0.771**	−0.116	−0.118	0.375	0.470	0.301	**0.657**	−0.412
PM_{10}	0.102	**0.582**	**0.584**	−0.035	0.074	**0.684**	**−0.523**	−0.047
$PM_{2.5}$	−0.053	**0.699**	−0.061	−0.490	−0.409	**0.503**	−0.099	**0.502**
Formaldehyde	**0.788**	0.380	0.100	0.123	0.322	**0.746**	0.467	−0.089
Eigenvalue	2.10	1.81	1.49	1.30	2.42	1.93	1.43	1.32
Variability (%)	21.0	18.1	14.9	13.0	24.2	19.3	14.3	13.2
Cumulative %	21.0	39.0	54.0	67.0	24.2	43.4	57.8	71.0

We further analysed the biomarkers and indoor pollutants to categorize them based on their homogeneity levels using agglomerative hierarchical clustering (AHC). Our data showed the cluster of markers and indoor pollutants from the PCA and AHC analyses were relatively identical. Three clusters were generated for eosinophils, which consisted of CD66b, FeNO levels, and concentrations of CO_2 in Cluster 1, with 98.4% of variance within-class; whereas Cluster 2 presented 56.5% of the variation within-class for CD11b expression and concentrations of NO_2 and formaldehyde. High (82.9%) variance within-class was observed for CD35 and CD63 expression and concentrations of PM_{10} and $PM_{2.5}$ in Cluster 3; three clusters were also generated for neutrophils, which consisted of CD35, CD63, CD66b, and concentrations of CO_2, PM_{10}, and $PM_{2.5}$, with 95.7% of the variation within-class for Cluster 1. The concentrations of NO_2 and formaldehyde were grouped in Cluster 2 with 26.0% of the variation within-class. Cluster 3 showed 56.7% of the variation within-class for CD11b expression and FeNO levels. The clusters generated through this process confirmed the contributing factors in the PCA earlier. This showed that exposure to CO_2, PM_{10}, and $PM_{2.5}$ have resulted in degranulation of both eosinophils and neutrophils, with upregulation of the markers for tertiary (CD11b), specific (CD66b), and azurophilic/crystalloid (CD63). In this study, exposure to NO_2 and formaldehyde also triggered the activation of tertiary eosinophil surface markers (Figure 3A).

3.4. Binary Logistic Regression (LR)

Finally, models were built to predict the toxicodynamic effects of indoor pollutants towards marker expression on the eosinophil and neutrophil in the sputum samples among doctor diagnosed asthmatic children. For this objective, the researchers next used the binary logistic regression to model the prediction with the potential confounders of gender, atopy, parental asthma/allergy status, and area of schools. Cluster 1 showed an overall accuracy of 76.0% in predicting asthmatic children by using CD66b expression markers on the eosinophil and FeNO levels in relation to CO_2 exposure. Meanwhile, the model generated for Cluster 2 showed a 76.0% accuracy and 68.0% sensitivity in predicting asthmatic children by using the upregulation of CD11b expression on eosinophils in relation to the NO_2 exposure. In the model generated for Cluster 3, the upregulation of CD63 expression on eosinophils and $PM_{2.5}$ concentration was significantly associated ($p < 0.05$), with a 71.9% accuracy and 60.0% sensitivity. Similarly, the upregulation of the neutrophil expression markers, CD63 and CD66b, was significant ($p < 0.05$), and could be predicted from the $PM_{2.5}$ exposure with a 72.9% accuracy and 72.0% sensitivity. Overall, children with a status of atopy, parental asthma/allergy, and from urban school were more likely to develop asthma ($p < 0.05$) (Table 4 and Figure 4).

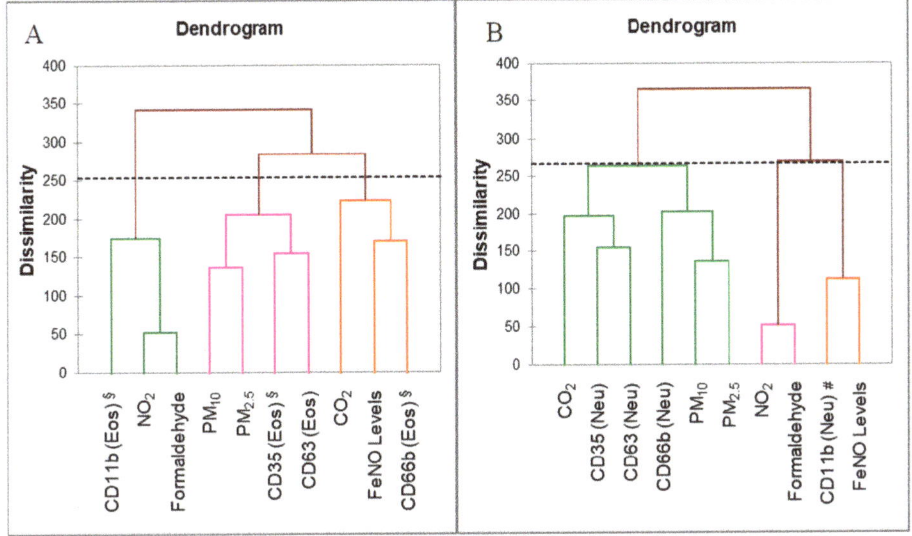

Figure 3. Agglomerative hierarchical clustering (AHC) analysis using the Ward linkage method and using Euclidean distances to generate the clustering of degranulation and the activation of markers and environmental pollutants measured inside classrooms for (**A**) eosinophils and (**B**) neutrophils. The dotted line represents the pruning level to generate distinct clusters. § Activation marker for eosinophils; # Activation marker for neutrophils.

Table 4. Summary of the binary logistic regression models based on the clusters generated in the AHC analysis.

Variable	B	SE	*p*-Value	R^2
Eosinophils				
Cluster 1				
Constant	−0.860	0.409	0.035 *	0.494
FeNO Levels	0.615	0.222	0.006 *	
CD66b	−0.850	0.250	<0.001 **	
CO_2	0.457	0.191	0.016 *	
Atopy	0.893	0.377	0.018 *	
Parental Asthma/Allergy	0.643	0.319	0.044 *	
Area—Urban	0.747	0.354	0.035 *	
Cluster 2				
Constant	−0.109	0.536	0.840	0.504
CD11b	0.454	0.260	0.018 *	
NO_2	1.305	0.375	0.002 *	
Formaldehyde	−0.684	0.440	0.120	
Atopy	0.993	0.378	0.009 *	
Parental Asthma/Allergy	0.916	0.424	0.031 *	
Cluster 3				
Constant	−1.387	0.462	0.003 *	0.377
CD35	−0.097	0.189	0.608	
CD63	−0.080	0.166	0.042 *	
PM_{10}	−0.379	0.212	0.074	
$PM_{2.5}$	−0.378	0.181	0.037 *	
Atopy	0.909	0.377	0.015 *	
Parental Asthma/Allergy	0.842	0.350	0.016 *	

Table 4. *Cont.*

Variable	B	SE	*p*-Value	R^2
Neutrophils				
Cluster 1				
Constant	−1.039	0.490	0.034 *	0.510
CD35	−0.054	0.169	0.751	
CD63	−0.632	0.256	0.014 *	
CD66b	−1.794	0.858	0.036 *	
CO_2	0.230	0.200	0.250	
PM_{10}	−0.253	0.218	0.244	
$PM_{2.5}$	−0.582	0.234	0.013 *	
Atopy	1.134	0.439	0.010 *	
Parental Asthma/Allergy	0.944	0.404	0.021 *	
Area—Urban	1.348	0.514	0.009 *	
Cluster 3				
Constant	−0.876	0.432	0.042 *	0.298
FeNO Levels	0.063	0.171	0.712	
CD11b	0.390	0.197	0.047 *	

* $p < 0.05$, ** $p < 0.001$.

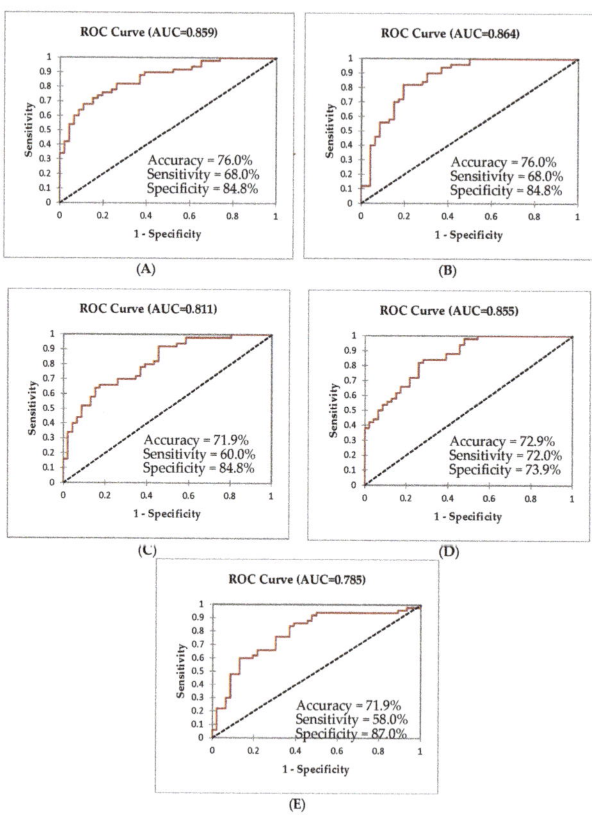

Figure 4. Receiver operating curve (ROC) for the models predicting an asthmatic or healthy child classification based on the clusters generated in the AHC analysis. The corresponding predictors for each curve are presented in Table 4. For eosinophils: (**A**) ROC for Cluster 1, (**B**) Cluster 2, and (**C**) for Cluster 3. For neutrophils: (**D**) ROC for Cluster 1 and (**E**) for Cluster 3.

4. Discussion

The role of biomarkers in airways is complex and specific, which is helpful in evaluating the aetiology, characterisation of phenotyping, and treatment of allergy and lung inflammation [37]. In this study, the FeNO levels were significantly higher among asthmatic school children, which are similar with the studies conducted in China [38], Terengganu, Malaysia [39], and Penang, Malaysia [40]. The result showed that there was inflammation in the airways and the average value was above the threshold of 50 ppb, which could reflect a high degree of inflammation. Liu et al. [41] and Carlsen et al. [42] reported that there was a significantly positive relationship between the FeNO levels and almost all pollutants, namely PM_{10}, $PM_{2.5}$, SO_2, NO_2, CO, and VOCs. This advocates a relationship between the high levels of all pollutants measured inside the classroom of urban schools and the high levels of FeNO among school children in this study. Some researchers estimated that FeNO is positively correlated up to five-fold and two-fold when exposed to NO_2 [43] and finer particles, such as $PM_{2.5}$ [44], respectively, which could be modulated by DNA methylation in the arginase–nitric oxide synthase pathway [45,46].

The CO_2 concentration in both school areas was below the recommended limit of 1000 ppm [47]. Similarly, the PM_{10} and $PM_{2.5}$ concentrations were below the 24 h mean of the World Health Organisation (WHO) guideline (PM_{10} = 50 $\mu g/m^3$, $PM_{2.5}$ = 25 $\mu g/m^3$), the National Ambient Air Quality Standard by USEPA (PM_{10} = 150 $\mu g/m^3$, $PM_{2.5}$ = 35 $\mu g/m^3$), and the new Malaysian Ambient Air Quality Standard 2018 Interim Target-2 (PM_{10} = 120 $\mu g/m^3$, $PM_{2.5}$ = 50 $\mu g/m^3$) [48]. A few classrooms recorded a concentration of $PM_{2.5}$ that exceeded the value of 25 $\mu g/m^3$, especially in the urban areas (37.5%) compared to the suburban (12.5%) areas. The median level of NO_2 for the urban and suburban areas was also below the WHO guideline of 40 $\mu g/m^3$ (annual mean), with only 18.8% and 25.0% of the classrooms in the urban and suburban areas, respectively, exceeding the limit. Overall, the levels of indoor air pollutants were below the guideline limits. This was due to the sufficient natural ventilation system and a wider window design on both sides of the classroom, together with the adequately equipped ceiling fans. The classroom design has a well-balanced ventilation that suits the temperature of the equatorial region and is able to reduce the particles, NO_2, and CO_2 concentrations [49]. Additionally, Silvestre et al. [50] reported that an opening of 56% of the classroom windows under natural ventilation conditions was able to keep the CO_2 concentration below 1000 ppm.

The schools, classrooms, and children were randomly selected from all secondary schools in the Hulu Langat area, Malaysia. Thus, we concluded that this study was not seriously influenced by selection bias. Moreover, Malaysia has a similar climate all year around; therefore, with the natural ventilation flowing through the windows in the classrooms, the indoor and outdoor levels of pollutants would be expected to be constant throughout the year. This is supported by several studies that have determined equal indoor to outdoor (I/O) ratios for PM_{10}, $PM_{2.5}$, NO_2, CO, and VOCs measured in schools across Peninsular Malaysia [51–53]. Be that as it may, the cross-sectional study design utilized here preludes making conclusions on causality.

In contrast to earlier findings, the total percentage of eosinophil and neutrophil counts in the sputum sample was not statistically different between the doctor-diagnosed asthmatic children and healthy children groups. Previous studies reported that the percentage of eosinophils and neutrophils for asthmatic children was significantly different and in the range of 2.5–13.0% and 15–47%, respectively [54–56]. Meanwhile, for healthy subjects, the percentages were in the range of 0.5–4.0% and 24.1–37.0%, respectively [57,58].

There have been few recent studies on activation and degranulation marker expression in the sputum samples of asthmatic children. The finding of this study confirmed that the sputum granulocytes of the asthmatic children increased the expression of the classical activation markers, CD11 and CD63, in both eosinophil and neutrophil cells, as reported by Tak et al. [28]. The upregulation of these tertiary and azurophilic/crystalloid granules is associated with the circulating cytokines that occur sequentially in response to the stimulus [59]. The mitogen-activated protein kinase (MAPK) pathway is believed to be central to the degranulation process [60]. The present study failed to show the upregulation of CD35 expression on eosinophils. In accordance with the study by Berends et al. [61], the downregulation of CD35 in the sputum of asthmatic children could be partly explained by the absence of intracellular stores for CD35 on eosinophils and neutrophils. Another possible explanation for this discrepancy was that CD35 is highly expressed on blood eosinophils or circulating granulocytes and was only directly associated with antigen inhalation [62] or the lower threshold stimulus required for cell activation [63].

The CD66b (CEACAM8) is a single-chain GPI-anchored glycoprotein and was recognised as an exclusive degranulation marker for neutrophils [64]. CD66b is upregulated when neutrophils are activated. The researchers of this study observed that the expression of CD66b was slightly upregulated on the surface of eosinophils and neutrophils collected from the airways of asthmatic children as compared to healthy children. It represents a normal activation pattern of neutrophils in relation to the migration from the circulating blood in vessels [65]. The late-phase response of the neutrophils could also possibly increase the CD11b, CD11b/18, CD35, CD64, and CD66b expressions [66]. The researchers found one study that identified that CD11b, CD16, and CD66b were consistently expressed on the neutrophils surface and were independent of their location and level of activation [67]. This could be contributed by the increased levels of intracellular cyclic GMP that yielded upregulation in the CD63 and CD66b expression on neutrophils [68].

PCA and AHC are very helpful approaches for dimensionality reduction in proteomics data. In this current study, these analytical approaches depicted similar group factors of biomarkers and air pollutants. The final regression analysis generated relatively moderate prediction models. The researchers noted that the upregulation of CD63 expression on both leukocytes and CD66b expression on neutrophils was related to particle exposure. The likely risks were observed among children under atopic and parental asthmatic/allergic conditions and children from schools located in urban areas. This finding reinforces the previous in vitro study conducted by Jin et al. [69]. They suggested that particulate allergens potentiated the mast cells to modulate the recruitment of eosinophils in the airways by internalising the particulate allergens into the $CD63^+$ intracellular compartments through an endolytic pathway. Another in vitro study reported that CD66b only activated the neutrophils in the peptidoglycan challenge but did not upregulate the surface activation of eosinophils [70]. This is the possible explanation of why CD66b expression is clustered and associated with PM_{10} and $PM_{2.5}$ in neutrophils but not with eosinophils in the AHC and regression analyses. Likewise, the study conducted by Banerje et al. [71] using flow cytometry analysis reported that there were increased CD35, CD16, and CD11b/CD18 expression on circulating neutrophils and a high percentage of eosinophils in the sputum of adults who have been exposed to PM_{10} and $PM_{2.5}$.

To the researchers' knowledge, this study was the first study that explored the interrelation of CD11b, CD63, CD35, and CD66b marker expression on eosinophils and neutrophils with different parameters of air pollutants. This study revealed that CD11b was not clustered together with PM_{10} and $PM_{2.5}$. This result was coherent with the study conducted by Ishii et al. [72] using immunocytochemistry, which showed that the expression of CD11b on alveolar macrophages was unaffected after two hours of stimulation with PM_{10}. They suggested that the adhesive interaction between CD11/CD18 on alveolar macrophages with CD54 on the bronchial epithelial cells contributed to the amplification of cytokine production from the alveolar macrophages. The chemometrics analysis in this current study was also clustered and showed a significant relationship between the CD11b expression on eosinophils with NO_2, especially among asthmatic children under atopic and parental asthmatic/allergic conditions. This finding was consistent with the review article by Hiraiwa and Eeden [73] and an in vitro study reported by Hodgkins et al. [13]. They found that dendritic cells expressed an upregulation of CD11b at 48 h during NO_2-promoted allergic sensitisation. A study has shown that CD11b is directly involved in cellular adhesion, which is expressed in many leukocytes, including neutrophil, monocytes, natural killer cells, and macrophages. The migration of these leukocytes to the inflammation site will only take place if the CD18 subunit is present [74]. The other possible roles of CD11b were reported by Medoff et al. [75] in their experiment in which the CD11b$^+$ had critical roles in mediating the Th2 cell and eosinophil recruitment in the airways via STAT6-dependent chemokine production.

This study also showed that the FeNO levels were positively correlated with the expression of the activation (CD66b, CD11b) and degranulation (CD66b, CD11b) markers for both leukocytes. This result is consistent with the results in the previous studies conducted by Guo et al. [76] and Kobayashi et al. [77], who also indicated that FeNO levels were reflected by eosinophilic airway inflammation [78]. In fact, the activated neutrophils can recruit the Th17/IL-17 and Th1 cells via chemokine release [79] and cause neutrophil infiltration within the airways [80]. This finding also reinforces that eosinophils and eosinophils are the binary indicators for the phenotyping of asthma. It was found in previous studies that exposure to PM_{10} was significantly associated with the increased levels of FeNO in healthy children tested on robust multi-pollutant models [41,81–83]. In line with this report, the chemometric analysis results in this study provided further evidence on the positive effects of PM_{10} on bronchial inflammation and resulted in the increase of FeNO levels among children.

The cluster approach used in this study, which is aimed at improving the interpretability of the data, interestingly revealed that the formaldehyde and NO_2 concentrations were in the same factor, with a total variance of 26.0%. This finding confirmed that the formation of formaldehyde was through the photochemical reactivity of NO_2 in the air with VOCs generating different aldehydes [84]. Formaldehyde also originates from furniture made out of wood and plastics, plywood, textile, table laminate, and consumer products, which is commonly available in the classroom [85]. As reported by Hua [86], NO_2 potentially originates from the process of fossil-fuel combustion, biomass burning, and agricultural activities. Hereafter, all the schools in this current study are located very close to the main road and industrial area, which was considered the probable sources of NO_2 in the classrooms. The researchers found that PM_{10} and $PM_{2.5}$ were grouped together in the dimensionality analysis of the PCA and AHC, and this indicates that both particles originated from the same source. The primary sources of particles in the urban and suburban areas were industrial emissions, transportation, and traffic emissions [87,88]. It was possible that the indoor particles also originated from the occupant's activities or re-suspension of deposited particles, soil materials from the school children's shoes, skin flakes, furniture fragments, and less frequent cleaning [89]. Children are at risk of day-long exposure to the same indoor pollutants, not only at school, but also at home—which was reflected in the time spent during non-school days. This possibility merits further investigation.

5. Conclusions

In conclusion, the chemometric analysis methods produced robust and immunologically meaningful results, which clustered the degranulation markers (CD11b and CD63) expressed on eosinophils with the concentration of NO_2 and $PM_{2.5}$. Besides that, the degranulation markers expressed on the neutrophils, CD63 and CD66b, were clustered together with the $PM_{2.5}$ concentration. A further prospective study is now obligatory to validate the models generated from this current study.

Supplementary Materials: The following are available online at http://www.mdpi.com/1660-4601/17/15/5413/s1, Table S1: The results of allergy skin test.

Author Contributions: Conceptualization, K.N.M.I., Z.H. and J.H.H.; methodology, K.N.M.I., Z.H., J.J., J.H.H. and L.T.L.T.; software, K.N.M.I.; validation, K.N.M.I., Z.H., J.J. and L.T.L.T.; formal analysis, K.N.M.I.; investigation, K.N.M.I. and Z.H.; resources, Z.H. and J.H.H.; data curation, K.N.M.I. and Z.H.; writing—original draft preparation, K.N.M.I.; writing—review and editing, Z.H. and J.J.; visualization, K.N.M.I.; supervision, Z.H.; project administration, K.N.M.I and Z.H.; funding acquisition, Z.H. All authors have read and agreed to the published version of the manuscript.

Funding: This study was funded by High Impact Putra Grant from Universiti Putra Malaysia (UPM) (VOT 9598000).

Acknowledgments: The researchers are grateful to all students who had participated in this study and the great support and hospitality from the school teachers are much appreciated.

Conflicts of Interest: The authors declare no conflict of interest.

References

1. Naja, A.S.; Permaul, P.; Phipatanakul, W. Taming Asthma in School-Aged Children: A Comprehensive Review. *J. Allergy Clin. Immunol. Pract.* **2018**, *6*, 726–735. [CrossRef] [PubMed]
2. Ahmed, F.; Hossain, S.; Hossain, S.; Naieum, A.; Fakhruddin, M. Impact of Household Air Pollution on Human Health: Source Identification and Systematic Management Approach. *SN Appl. Sci.* **2019**, *1*, 1–19. [CrossRef]
3. Gaikwad, A.; Shivhare, N. Indoor Air Pollution—A Threat. *Int. J. Sci. Res. Rev.* **2019**, *7*, 1463–1467.
4. Tiotiu, A. Biomarkers in Asthma: State of the Art. *Asthma Res. Pract.* **2018**, *4*, 1–10. [CrossRef]
5. Carr, F. Use of Biomarkers to Identify Phenotypes and Endotypes of Severe Asthma. *Ann. Allergy Asthma Immunol.* **2018**, *121*, 414–420. [CrossRef]
6. Licari, A.; Castagnoli, R. Asthma Endotyping and Biomarkers in Childhood Asthma. *Pediatr. Allergy Immunol. Pulmonol.* **2018**, *31*, 44–56. [CrossRef]
7. Germic, N.; Frangez, Z.; Yousefi, S.; Simon, H.-U. Regulation of the Innate Immune System by Autophagy: Neutrophils, Eosinophils, Mast Cells, NK Cells. *Cell Death Differ.* **2019**, *26*, 703–714. [CrossRef]
8. Grunwell, J.R.; Stephenson, S.T.; Tirouvanziam, R.; Brown, L.A.S. Children with Neutrophil-Predominant Severe Asthma Have Proinflammatory Neutrophils With Enhanced Survival and Impaired Clearance. *J. Allergy Clin. Immunol. Pract.* **2019**, *7*, 516–525.e6. [CrossRef]
9. Johansson, M.W. Activation States of Blood Eosinophils in Asthma. *Clin. Exp. Allergy* **2014**, *44*, 482–498. [CrossRef]
10. Kämpe, M.; Stolt, I.; Lampinen, M.; Janson, C.; Stålenheim, G.; Carlson, M. Patients with Allergic Rhinitis and Allergic Asthma Share the Same Pattern of Eosinophil and Neutrophil Degranulation after Allergen Challenge. *Clin. Mol. Allergy* **2011**, *9*, 1–10. [CrossRef]
11. Cowland, J.B.; Borregaard, N. Granulopoiesis and Granules of Human Neutrophils. *Immunol. Rev.* **2016**, *273*, 11–28. [CrossRef]
12. Zhao, Y.; Zhang, H.; Yang, X.; Zhang, Y.; Feng, S.; Yan, X. Fine Particulate Matter ($PM_{2.5}$) Enhances Airway Hyperresponsiveness (AHR) by Inducing Necroptosis in BALB/c Mice. *Environ. Toxicol. Pharmacol.* **2019**, *68*, 155–163. [CrossRef]
13. Hodgkins, S.R.; Ather, J.L.; Paveglio, S.A.; Allard, J.L.; Leclair, L.A.W.; Suratt, B.T.; Boyson, J.E.; Poynter, M.E. NO2 Inhalation Induces Maturation of Pulmonary CD11c+ Cells That Promote Antigen-Specific CD4+ T Cell Polarization. *Respir. Res.* **2010**, *11*, 1–18. [CrossRef] [PubMed]
14. Agache, I.; Akdis, C.A.; Agache, I.; Akdis, C.A. Precision Medicine and Phenotypes, Endotypes, Genotypes, Regiotypes, and Theratypes. *J. Clin. Investig.* **2019**, *129*, 1493–1503. [CrossRef] [PubMed]
15. Barnes, P.J. Targeting Cytokines to Treat Asthma and Chronic Obstructive Pulmonary Disease. *Nat. Rev. Immunol.* **2018**, *18*, 454–466. [CrossRef] [PubMed]
16. Hanif, T.; Laulajainen-Hongisto, A.; Luukkainen, A.; Numminen, J.; Kääriäinen, J.; Myller, J.; Kalogjera, L.; Huhtala, H.; Kankainen, M.; Renkonen, R.; et al. Hierarchical Clustering in Evaluating Inflammatory Upper Airway Phenotypes; Increased Symptoms in Adults with Allergic Multimorbidity. *Asian Pac. J. Allergy Immunol.* **2019**, 1–10. [CrossRef]
17. Tinnevelt, G.H.; Jansen, J.J. Resolving Complex Hierarchies in Chemical Mixtures: How Chemometrics May Serve in Understanding the Immune System. *Faraday Discuss.* **2019**, *218*, 317–338. [CrossRef]
18. Kelly, R.S.; Mcgeachie, M.J.; Lee-sarwar, K.A.; Kachroo, P.; Chu, S.H.; Virkud, Y.V.; Huang, M.; Litonjua, A.A.; Weiss, S.T.; Lasky-Su, J. Partial Least Squares Discriminant Analysis and Bayesian Networks for Metabolomic Prediction of Childhood Asthma. *Metabolites* **2018**, *8*, 68. [CrossRef]
19. Hilvering, B.; Vijverberg, S.J.H.; Jansen, J.J.; Houben, L.; Schweizer, R.C.; Go, S.; Xue, L.; Pavord, I.D.; Lammers, J.-W.J.; Koenderman, L. Diagnosing Eosinophilic Asthma Using a Multivariate Prediction Model Based on Blood Granulocyte Responsiveness. *Allergy* **2017**, *72*, 1202–1211. [CrossRef]
20. Norback, D.; Hashim, J.H.; Hashim, Z.; Cai, G.-H.; Sooria, V.; Ismail, S.A.; Wieslander, G. Respiratory Symptoms and Fractional Exhaled Nitric Oxide (FeNO) among Students in Penang, Malaysia in Relation to Sign of Dampness at School and Funfal DNA in School Dust. *Sci. Total Environ.* **2017**, *577*, 148–154. [CrossRef]

21. Zhao, Z.; Sebastian, A.; Larsson, L.; Wang, Z.; Zhang, Z.; Norback, D. Asthmatic Symptoms among Pupils in Relation to Microbial Dust Exposure in Schools in Taiyuan, China. *Pediatr. Allergy Immunol.* **2008**, *19*, 455–465. [CrossRef] [PubMed]

22. Dweik, R.A.; Boggs, P.B.; Erzurum, S.C.; Irvin, C.G.; Leigh, M.W.; Lundberg, J.O.; Olin, A.; Plummer, A.L.; Taylor, D.R. American Thoracic Society Documents An Official ATS Clinical Practice Guideline: Interpretation of Exhaled Nitric Oxide Levels (FENO) for Clinical Applications. *Am. J. Respir. Crit. Care Med.* **2011**, *184*, 602–615. [CrossRef]

23. Australasian Society of Clinical Immunology and Allergy (ASCIA). Skin Prick Testing for the Diagnosis of Allergic Disease—A Manual for Practitioners. Available online: https://www.allergy.org.au/images/stories/pospapers/ASCIA_SPT_Manual_March_2016.pdf (accessed on 14 January 2018).

24. Weiszhar, Z.; Horvath, I. Induced Sputum Analysis: Step by Step. *Breathe* **2013**, *9*, 300–306. [CrossRef]

25. Lay, J.C.; Peden, D.B.; Alexis, N.E. Flow Cytometry of Sputum: Assessing Inflammation and Immune Response Elements in the Bronchial Airways. *Inhal. Toxicol.* **2011**, *23*, 392–406. [CrossRef] [PubMed]

26. De Ruiter, K.; Van Staveren, S.; Hilvering, B.; Knol, E.; Vrisekoop, N.; Koenderman, L.; Yazdanbakhsh, M. A Field-Applicable Method for Flow Cytometric Analysis of Granulocyte Activation: Cryopreservation of Fixed Granulocytes. *Cytom. Part A J. Quant. Cell Sci.* **2018**, *93*, 540–547. [CrossRef] [PubMed]

27. Duan, M.; Steinfort, D.P.; Smallwood, D.; Hew, M.; Chen, W.; Ernst, M.; Irving, L.B.; Anderson, G.P. CD11b Immunophenotyping Identifies Inflammatory Profiles in the Mouse and Human Lungs. *Mucosal Immunol.* **2016**, *9*, 550–563. [CrossRef]

28. Tak, T.; Hilvering, B.; Tesselaar, K. Similar Activation State of Neutrophils in Sputum of Asthma Patients Irrespective of Sputum Eosinophilia. *Clin. Exp. Immunol.* **2015**, *182*, 204–212. [CrossRef]

29. Maestre-Batlle, D.; Pena, O.M.; Hirota, J.A.; Gunawan, E.; Rider, C.F.; Sutherland, D.; Alexis, N.E.; Carlsten, C. Novel Flow Cytometry Approach to Identify Bronchial Epithelial Cells from Healthy Human Airways. *Sci. Rep.* **2017**, *7*, 1–9. [CrossRef]

30. Suhaimi, N.F.; Jalaludin, J.; Bakar, S.A. Cysteinyl Leukotrienes as Biomarkers of Effect in Linking Exposure to Air Pollutants and Respiratory Inflammation among School Children. *Ann. Trop. Med. Public Health* **2017**, *10*, 423–431.

31. Norback, D.; Markowicz, P.; Cai, G.; Hashim, Z.; Ali, F.; Zheng, Y.; Lai, X.; Spangfort, M.D.; Larsson, L.; Hashim, J.H. Endotoxin, Ergosterol, Fungal DNA and Allergens in Dust from Schools in Johor Bahru, Malaysia—Associations with Asthma and Respiratory Infections in Pupils. *PLoS ONE* **2014**, *9*, e88303. [CrossRef]

32. Kamaruddin, A.S.; Jalaludin, J.; Choo, C.P. Indoor Air Quality and Its Association with Respiratory Health among Malay Preschool Children in Shah Alam and Hulu Langat, Selangor. *Adv. Environ. Biol.* **2015**, *9*, 17–26.

33. Foldvary, V.; Beko, G.; Langer, S.; Arrhenius, K.; Petras, D. Effect of Energy Renovation on Indoor Air Quality in Multifamily Residential Buildings in Slovakia. *Build. Environ.* **2017**, *122*, 363–372. [CrossRef]

34. Cai, G.; Hashim, J.H.; Hashim, Z.; Ali, F.; Bloom, E.; Larsson, L.; Lampa, E.; Norback, D. Fungal DNA, Allergens, Mycotoxins and Associations with Asthmatic Symptoms among Pupils in Schools from Johor Bahru, Malaysia. *Pediatr. Allergy Immunol.* **2011**, *22*, 290–297. [CrossRef] [PubMed]

35. Szymanska, E.; Saccenti, E.; Smilde, A.K.; Westerhuis, J.A. Double-Check: Validation of Diagnostic Statistics for PLS-DA Models in Metabolomics Studies. *Metabolomics* **2012**, *8*, S3–S16. [CrossRef]

36. Jolliffe, I.T.; Cadima, J. Principal Component Analysis: A Review and Recent Developments. *Philos. Trans. R. Soc. A* **2016**, *374*, 1–16. [CrossRef]

37. Kim, H.; Eckel, S.P.; Kim, J.H.; Gilliland, F.D. Exhaled NO: Determinants and Clinical Application in Children With Allergic Airway Disease. *Allergy Asthma Immunol. Res.* **2016**, *8*, 12–21. [CrossRef]

38. Xu, F.; Zou, Z.; Yan, S.; Li, F.; Kan, H.; Dan, N.; Gunilla, W.; Xu, J.; Zhao, Z. Fractional Exhaled Nitric Xide in Relation to Asthma, Allergic Rhinitis, and Atopic Dermatitis in Chinese Children. *J. Asthma* **2011**, *48*, 1001–1006. [CrossRef]

39. Ma'pol, A.; Hashim, J.H.; Norbäck, D.; Weislander, G.; Hashim, Z.; Isa, Z.M. FeNO Level and Allergy Status among School Children in Terengganu, Malaysia. *J. Asthma* **2019**, *3*, 1–18. [CrossRef]

40. Norbäck, D.; Hisham, J.; Hashim, Z.; Sooria, V.; Aizat, S.; Wieslander, G. Ocular Symptoms and Tear Film Break up Time (BUT) among Junior High School Students in Penang, Malaysia—Associations with Fungal DNA in School Dust. *Int. J. Hyg. Environ. Health* **2017**, *220*, 697–703. [CrossRef] [PubMed]

41. Liu, D.; Huang, Z.; Huang, Y. Clinical Analysis of Fractional Exhaled and Nasal Nitric Oxide in Allergic Rhinitis Children. *J. Allergy Ther.* **2015**, *6*, 1–4. [CrossRef]

42. Carlsen, H.K.; Boman, P.; Bjor, B.; Olin, A.-C.; Forsberg, B. Coarse Fraction Particle Matter and Exhaled Nitric Oxide in Non-Asthmatic Children. *Int. J. Environ. Res. Public Health* **2016**, *13*, 621. [CrossRef] [PubMed]

43. Kamaruddin, A.S.; Jalaludin, J.; Hamedon, T.R.; Hisamuddin, N.H. FeNO as a Biomarker for Airway Inflammation Due to Exposure to Air Pollutants among School Children Nearby Industrial Areas in Terengganu. *Pertanika J. Sci. Technol.* **2019**, *27*, 589–600.

44. Gong, J.; Zhu, T.; Hu, M.; Wu, Z.; Zhang, J.J. Different Metrics (Number, Surface Area, and Volume Concentration) of Urban Particles with Varying Sizes in Relation to Fractional Exhaled Nitric Oxide (FeNO). *J. Thorac. Dis.* **2019**, *11*, 1714–1726. [CrossRef] [PubMed]

45. Zhang, Q.; Wang, W.; Niu, Y.; Xia, Y.; Lei, X.; Huo, J. The Effects of Fine Particulate Matter Constituents on Exhaled Nitric Oxide and DNA Methylation in the Arginase—Nitric Oxide Synthase Pathway. *Environ. Int.* **2019**, *131*, 105019. [CrossRef] [PubMed]

46. Salam, M.T.; Byun, H.; Lurmann, F.; Breton, C.V.; Wang, X.; Eckel, S.P.; Gilliland, F.D.; Angeles, L. Genetic and Epigenetic Variations in Inducible Nitric Oxide Synthase Promoter, Particulate Pollution, and Exhaled Nitric Oxide Levels in Children. *J. Allergy Clin. Immunol.* **2012**, *129*, 232–239.e7. [CrossRef]

47. ASHRAE. *Ventilation for Acceptable Indoor Air Quality*; American Society of Heating, Refrigerating and Air-Conditioning Engineers, Inc.: Atlanta, GA, USA, 2001.

48. World Health Organization (WHO). *Air Quality Guidelines—Global Update 2005*; World Health Organzantion Regional Office for Europe: Copenhagen, Denmark, 2006.

49. Schibuola, L.; Scarpa, M.; Tambani, C. Natural Ventilation Level Assessment in a School Building by CO_2 Concentration Measures. *Energy Procedia* **2016**, *101*, 257–264. [CrossRef]

50. Silvestre, C.M.; Andre, P.; Michel, T. Air Temperature and CO_2 Variations in a Naturally Ventilated Classroom under a Nordic Climate. In Proceedings of the PLEA2009 26th Conference on Passive and Low Energy Architecture, Quebec, ON, Canada, 22–24 June 2009; pp. 1–6.

51. Norbäck, D.; Hisham, J.; Hashim, Z.; Ali, F. Volatile Organic Compounds (VOC), Formaldehyde and Nitrogen Dioxide (NO_2) in Schools in Johor Bahru, Malaysia: Associations with Rhinitis, Ocular, Throat and Dermal Symptoms, Headache and Fatigue. *Sci. Total Environ.* **2017**, *592*, 153–160. [CrossRef]

52. Abidin, E.Z.; Semple, S.; Rasdi, I.; Ismail, S.N.S.; Ayres, J.G. The Relationship between Air Pollution and Asthma in Malaysian Schoolchildren. *Air Qual. Atmos. Health* **2014**, *7*, 421–432. [CrossRef]

53. Mohamed, N.; Sulaiman, L.H.; Zakaria, T.A.; Kamarudin, A.S.; Rahim, D.A. Health Risk Assessment of PM10 Exposure among School Children and the Proposed API Level for Closing the School during Haze in Malaysia. *Int. J. Public Health Res.* **2016**, *6*, 685–694.

54. Maestrelli, P.; De Fina, O.; Bertin, T.; Papiris, S.; Ruggieri, M.P.; Saetta, M.; Mapp, C.E.; Fabbri, L.M. Integrin Expression on Neutrophils and Mononuclear Cells in Blood and Induced Sputum in Stable Asthma. *Allergy* **1999**, *54*, 1303–1308. [CrossRef]

55. Wark, P.; Gibson, P.; Fakes, K. Induced Sputum Eosinophils in the Assessment of Asthma and Chronic Cough. *Respirology* **2000**, *5*, 51–57. [CrossRef] [PubMed]

56. Belda, J.; Leigh, R.; Parameswaran, K.; Byrne, P.M.O.; Sears, M.R.; Hargreave, F.E. Induced Sputum Cell Counts in Healthy Adults. *Am. J. Respir. Crit. Care Med.* **2000**, *161*, 475–478. [CrossRef] [PubMed]

57. Giudice, M.M.; Pedulla, M.; Brunese, F.P.; Capristo, A.F.; Capristo, C.; Tosca, M.A.; Leonardf, S.; Ciprandp, G. Neutrophilic Cells in Sputum of Allergic Asthmatic Children. *Eur. J. Inflamm.* **2010**, *8*, 151–156. [CrossRef]

58. Hargreave, F.E.; Leigh, R. Induced Sputum, Eosinophilic Bronchitis, and Chronic Obstructive Pulmonary Disease. *Am. J. Respir. Crit. Care Med.* **1999**, *160*, S53–S57. [CrossRef]

59. Kim, K.; Hwang, S.M.; Kim, S.M.; Park, S.W.; Jung, Y. Terminally Differentiating Eosinophils Express Neutrophil Primary Granule Proteins as Well as Eosinophil-Specific Granule Proteins in a Temporal Manner. *Immune Netw.* **2017**, *17*, 410–423. [CrossRef]

60. Carr, T.F.; Berdnikovs, S.; Simon, H.-U.; Bochner, B.S.; Rosenwasser, L.J. Eosinophilic Bioactivities in Severe Asthma. *World Allergy Organ. J.* **2016**, *9*, 1–7. [CrossRef] [PubMed]

61. Berends, C.; Dijkhuizen, B.; Monchy, G.R.D.E.; Gerritsen, J.; Kauffman, H.F. Expression of CD35 (CRl) and CDllb (CR3) on Circulating Neutrophils and Eosinophils from Allergic Asthmatic Children. *Clin. Exp. Allergy* **1993**, *23*, 926–933. [CrossRef]

62. Dallaire, M.; Ferland, C.; Page, N.; Lavigne, S.; Davoine, F.; Laviolette, M. Endothelial Cells Modulate Eosinophil Surface Markers and Mediator Release. *Eur. Respir. J.* **2003**, *21*, 918–924. [CrossRef]

63. Lewis, S.M.; Treacher, D.F.; Edgeworth, J.; Mahalingam, G.; Brown, C.S.; Mare, T.A.; Stacey, M.; Beale, R.; Brown, K.A. Expression of CD11c and EMR2 on Neutrophils: Potential Diagnostic Biomarkers for Sepsis and Systemic Inflammation. *Clin. Exp. Immunol.* **2015**, *182*, 184–194. [CrossRef]

64. Ducker, P.; Skubitz, K. Subcellular Neutrophils Localization of CD66, CD67, and NCA in Human Neutrophils. *J. Leukoc. Biol.* **1992**, *52*, 11–16. [CrossRef]

65. Vidal, S.; Bellido-Casado, J.; Granel, C.; Crespo, A.; Plaza, V.; Juárez, C. Immunobiology Flow Cytometry Analysis of Leukocytes in Induced Sputum from Asthmatic Patients. *Immunobiology* **2012**, *217*, 692–697. [CrossRef]

66. Zissler, U.M.; Bieren, J.E.; Jakwerth, C.A.; Chaker, A.M.; Schmidt-Weber, C.B. Current and Future Biomarkers in Allergic Asthma. *Eur. J. Allergy Clin. Immunol.* **2016**, *71*, 475–494. [CrossRef] [PubMed]

67. Lakschevitz, F.S. Identification of CD Marker Expression and Neutrophil Surface Marker Changes in Health and Disease Using High-Throughput Screening Flow Cytometry. Master's Thesis, University of Toronto, Toronto, ON, Canada, 2016.

68. Kinhult, J.; Egesten, A.; Uddman, R.; Cardell, L.O. PACAP Enhances the Expression of CD11b, CD66b and CD63 in Human Neutrophils. *Peptides* **2002**, *23*, 1735–1739. [CrossRef]

69. Jin, C.; Shelburne, C.P.; Li, G.; Riebe, K.J.; Sempowski, G.D.; Foster, W.M.; Abraham, S.N. Particulate Allergens Potentiate Allergic Asthma in Mice through Sustained IgE-Mediated Mast Cell Activation. *J. Clin. Investig.* **2011**, *121*, 941–955. [CrossRef] [PubMed]

70. Mattsson, E.; Persson, T.; Andersson, P.; Rollof, J.; Egesten, A. Peptidoglycan Induces Mobilization of the Surface Marker for Activation Marker CD66b in Human Neutrophils but Not in Eosinophils. *Clin. Diagn. Lab. Immunol.* **2003**, *10*, 485–488. [CrossRef]

71. Banerjee, A.; Mondal, N.K.; Das, D.; Ray, M.R. Neutrophilic Inflammatory Response and Oxidative Stress in Premenopausal Women Chronically Exposed to Indoor Air Pollution from Biomass Burning. *Inflammation* **2012**, *35*, 671–683. [CrossRef]

72. Ishii, H.; Hayashi, S.; Hogg, J.C.; Fujii, T.; Goto, Y.; Sakamoto, N.; Mukae, H.; Vincent, R.; Van Eeden, S.F. Alveolar Macrophage-Epithelial Cell Interaction Following Exposure to Atmospheric Particles Induces the Release of Mediators Involved in Monocyte Mobilization and Recruitment. *Respir. Res.* **2005**, *6*, 1–12. [CrossRef] [PubMed]

73. Hiraiwa, K.; Van Eeden, S.F. Contribution of Lung Macrophages to the Inflammatory Responses Induced by Exposure to Air Pollutants. *Mediat. Inflamm.* **2013**, *2013*, 1–10. [CrossRef] [PubMed]

74. Pillay, J.; Kamp, V.M.; Pennings, M.; Oudijk, E.; Leenen, L.P.; Ulfman, L.H. Acute-Phase Concentrations of Soluble Fibrinogen Inhibit Neutrophil Adhesion under Flow Conditions in Vitro through Interactions with ICAM-1 and MAC-1 (CD11b/CD18). *J. Thromb. Haemost.* **2013**, *11*, 1172–1182. [CrossRef] [PubMed]

75. Medoff, B.D.; Seung, E.; Hong, S.; Thomas, Y.; Sandall, B.P.; Duffield, J.S.; Kuperman, A.; Erle, D.J.; Luster, A.D.; Medoff, B.D.; et al. CD11b+ Myeloid Cells Are the Key Mediators of Th2 Cell Homing into the Airway in Allergic Inflammation. *J. Immunol.* **2009**, *182*, 623–635. [CrossRef]

76. Guo, Y.; Hong, C.; Liu, Y.; Chen, H.; Huang, X.; Hong, M. Diagnostic Value of Fractional Exhaled Nitric Oxide for Asthma-Chronic Obstructive Pulmonary Disease Overlap Syndrome. *Medicine (Baltimore)* **2018**, *97*, 1–7. [CrossRef]

77. Kobayashi, S.; Hanagama, M.; Yamanda, S.; Ishida, M.; Yanai, M. Inflammatory Biomarkers in Asthma-COPD Overlap Syndrome. *Int. J. COPD* **2016**, *11*, 2117–2123. [CrossRef] [PubMed]

78. Alving, K.; Malinovschi, A. Basic Aspects of Exhaled Nitric Oxide. In *European Respiratory Monograph: Exhaled Biomarker*; Horvath, I., DeJongste, J.C., Eds.; European Respiratory Society: Sheffield, UK, 2010; pp. 1–31. [CrossRef]

79. Pelletier, M.; Maggi, L.; Micheletti, A.; Lazzeri, E.; Tamassia, N.; Costantini, C.; Cosmi, L.; Lunardi, C.; Annunziato, F.; Romagnani, S.; et al. Evidence for a Cross-Talk between Human Neutrophils and Th17 Cells. *Blood* **2010**, *115*, 335–343. [CrossRef]

80. Ray, A.; Kolls, J.K. Neutrophilic Inflammation in Asthma and Association with Disease Severity. *Trends Immunol.* **2017**, *38*, 942–954. [CrossRef] [PubMed]

81. Liu, C.; Flexeder, C.; Fuertes, E.; Cyrys, J.; Bauer, C.; Koletzko, S.; Hoffmann, B.; Von Berg, A.; Heinrich, J. Effects of Air Pollution on Exhaled Nitric Oxide in Children: Results from the GINIplus and LISAplus Studies. *Int. J. Hyg. Environ. Health* **2014**, *217*, 483–491. [CrossRef] [PubMed]

82. Liu, Z.; Hu, B.; Wang, L.; Wu, F.; Gao, W.; Wang, Y. Seasonal and Diurnal Variation in Particulate Matter (PM_{10} and $PM_{2.5}$) at an Urban Site of Beijing: Analyses from a 9-Year Study. *Environ. Sci. Pollut. Res.* **2015**, *22*, 627–642. [CrossRef] [PubMed]

83. Xu, H.; Guinot, B.; Sai, S.; Ho, H.; Li, Y.; Cao, J.; Shen, Z.; Niu, X.; Zhao, Z.; Liu, S. Evaluation on Exposures to Particulate Matter at a Junior Secondary School: A Comprehensive Study on Health Risks and Effective Inflammatory Responses in Northwestern China. *Environ. Geochem. Health* **2017**, 1–15. [CrossRef]

84. Weschler, C.J.; Wells, J.R.; Poppendieck, D.; Hubbard, H.; Pearce, T.A. Workgroup Report: Indoor Chemistry and Health. *Environ. Health Perspect.* **2006**, *114*, 442–446. [CrossRef]

85. Salthammer, T. Formaldehyde Sources, Formaldehyde Concentrations and Air Exchange Rates in European Housings. *Build. Environ.* **2019**, *150*, 219–232. [CrossRef]

86. Hua, A.K. Applied Chemometric Approach in Identification Sources of Air Quality Pattern in Selangor, Malaysia. *Sains Malays.* **2018**, *47*, 471–479. [CrossRef]

87. Askariyeh, M.H.; Venugopal, M.; Khreis, H.; Birt, A. Near-Road Traffic-Related Air Pollution: Resuspended PM2.5 from Highways and Arterials. *Int. J. Environ. Res. Public Health* **2020**, *17*, 2851. [CrossRef] [PubMed]

88. Othman, M.; Latif, M.T.; Matsumi, Y. The Exposure of Children to $PM_{2.5}$ and Dust in Indoor and Outdoor School Classrooms in Kuala Lumpur City Centre. *Ecotoxicol. Environ. Saf.* **2019**, *170*, 739–749. [CrossRef] [PubMed]

89. Rufo, J.C.; Madureira, J.; Fernandes, E.O.; Moreira, A.; Rufo, C.; Roberto, R. Volatile Organic Compounds in Asthma Diagnosis: A Systematic Review and Meta-Analysis. *Allergy* **2016**, *71*, 175–188. [CrossRef] [PubMed]

International Journal of
***Environmental Research
and Public Health***

rticle

dverse Birth Outcomes Due to Exposure to Household Air ollution from Unclean Cooking Fuel among Women of eproductive Age in Nigeria

ie Roberman [1], Theophilus I. Emeto [1,2] and Oyelola A. Adegboye [1,2,*]

[1] Public Health & Tropical Medicine, College of Public Health, Medical and Veterinary Sciences, James Cook University, Townsville, QLD 4811, Australia; jamie.roberman@my.jcu.edu.au (J.R.); theophilus.emeto@jcu.edu.au (T.I.E.)

[2] Australian Institute of Tropical Health and Medicine, James Cook University, Townsville, QLD 4811, Australia

* Correspondence: oyelola.adegboye@jcu.edu.au

Abstract: Exposure to household air pollution (HAP) from cooking with unclean fuels and indoor smoking has become a significant contributor to global mortality and morbidity, especially in low- and middle-income countries such as Nigeria. Growing evidence suggests that exposure to HAP disproportionately affects mothers and children and can increase risks of adverse birth outcomes. We aimed to quantify the association between HAP and adverse birth outcomes of stillbirth, preterm births, and low birth weight while controlling for geographic variability. This study is based on a cross-sectional survey of 127,545 birth records from 41,821 individual women collected as part of the 2018 Nigeria Demographic and Health Survey (NDHS) covering 2013–2018. We developed Bayesian structured additive regression models based on Bayesian splines for adverse birth outcomes. Our model includes the mother's level and household characteristics while correcting for spatial effects and multiple births per mother. Model parameters and inferences were based on a fully Bayesian approach via Markov Chain Monte Carlo (MCMC) simulations. We observe that unclean fuel is the primary source of cooking for 89.3% of the 41,821 surveyed women in the 2018 NDHS. Of all pregnancies, 14.9% resulted in at least one adverse birth outcome; 14.3% resulted in stillbirth, 7.3% resulted in an underweight birth, and 1% resulted in premature birth. We found that the risk of stillbirth is significantly higher for mothers using unclean cooking fuel. However, exposure to unclean fuel was not significantly associated with low birth weight and preterm birth. Mothers who attained at least primary education had reduced risk of stillbirth, while the risk of stillbirth increased with the increasing age of the mother. Mothers living in the Northern states had a significantly higher risk of adverse births outcomes in 2018. Our results show that decreasing national levels of adverse birth outcomes depends on working toward addressing the disparities between states.

Keywords: cooking fuel; household air pollution; preterm births; perinatal mortality; low birth weight; stillbirth; Nigeria

tion: Roberman, J.; Emeto, T.I.; gboye, O.A. Adverse Birth omes Due to Exposure to sehold Air Pollution from lean Cooking Fuel among Women eproductive Age in Nigeria. *Int. J. ron. Res. Public Health* **2021**, *18*, https://doi.org/10.3390/ h18020634

eived: 26 October 2020
epted: 8 January 2021
lished: 13 January 2021

lisher's Note: MDPI stays neu-
with regard to jurisdictional clai-
n published maps and institutio-
affiliations.

1. Introduction

Adverse birth outcomes (ABO) including low birth weight (LBW), preterm birth, and stillbirths represent an unmet public health need in Nigeria. The World Health Organization (WHO) classes infant birth weight of less than 2.5 kg as LBW [1]. LBW is a key risk factor for morbidity and mortality as well as being a predictor of survival, normal growth, and cognitive development in children [2,3]. In 2015, Nigeria had the third-highest number of neonatal deaths, and an estimated 313,700 stillborn deaths, the second-highest globally, and the highest number of maternal mortalities [4]. Hence, there is a need to explore the high incidence rate of these ABO in Nigeria compared with other countries.

Household air pollution (HAP) is a prominent global public health concern and is a leading environmental health risk globally [5]. In 2016, HAP was responsible for 7.7% of

J. Environ. Res. Public Health **2021**, *18*, 634. https://doi.org/10.3390/ijerph18020634   https://www.mdpi.com/journal/ijerph

the global mortality [6]. It is estimated that 3.8 million premature deaths occur each y
from illnesses attributable to HAP [7]. Of these deaths, 27% were due to pneumonia, 2
to ischaemic heart disease, 20% to chronic obstructive pulmonary disease (COPD), 18%
stroke, and 8% to lung cancer [7]. HAP has also been linked to pulmonary tuberculosis
eclampsia [9], and cataracts [10]. Regional variations in the particulate composition of H
has been reported within countries [11] and between countries [12], and may also depe
on locally accessible fuel type within regions [12,13]. Within Nigeria, geographic fact
also impact the rate of adverse birth outcomes as the North West region has a significa
higher risk for LBW [3]. Hence, it is important to highlight the effect of regional variat
on the prevalence of ABO due to HAP. Additionally, some studies have shown that pare
employment, education level, and other indicators of socio-economic status are associa
with LBW [14,15].

A primary source of HAP is the use of unclean fuel in cooking [7]. Common chro
respiratory symptoms, such as congestion and cough, are directly associated with
hours spent cooking food [16]. Using unclean fuel is especially common in developi
countries where biomass fuel sources are more affordable than healthier alternatives su
as electricity, natural gas, and liquefied petroleum gas [17]. Biomass fuel includes wo
crop residue, charcoal, coal, and dung [7]. The use of this type of fuel is most prevalent
Africa and Southwest Asia where over 60% of households cook with unclean fuels [18,
and there is evidence of geographical disparities in its use [20]. Nearly 70% of the Niger
population cook with unclean fuel [21].

Global evidence suggests an association of HAP with increased pregnancy compl
tions [22], preterm birth [23], and elevated risk of stillbirth [24,25]. It is hypothesized t
pollutants, including carbon monoxide, contained in HAP can be inhaled and absorb
into the mother's blood which could detrimentally affect the fetus [26,27]. HAP lik
potentiates ABO by promoting a pro-hypoxic phenotype including increased expression
Hofbauer cells, syncytial knots, and a compromised chorionic vascular density in utero [
Uterine hypoxia has also been associated with morbidity and mortality [29]. Both Hofba
cells [30] and syncytial knots [31,32] have been associated with ABOs such as pre-eclamp
and intrauterine growth retardation, while impaired chronic vascular density is linked
embryonic death [33]. In Nigeria, a limited randomized control trial demonstrated
increased risk of chronic hypoxia in pregnant women using unclean cooking fuel compa
to those using clean cooking fuel [28]. Another randomized trial showed an interventi
of ethanol as cooking fuel significantly increased mean birth weight and gestational age
delivery [34]. This may also mitigate the risk of cardiovascular disease such as hypert
sion by downregulating the expression of tumor necrosis factor alpha, interleukin 6, a
interleukin 8 in pregnant women [35].

Here, we explore the impact of HAP on adverse birth outcomes, such as stillbir
pregnancy duration, and LBW in Nigeria. We will evaluate the geographic distribution
the association to identify regions with more significant HAP effects on births. This stu
will add to the existing literature by analyzing the impact of HAP on ABO in Nigeria wh
accounting for geographic heterogeneity.

2. Materials and Methods

2.1. Data Source, Setting, and Variables

This study was based on the Nigerian Demographic and Health Survey (NDHS) pha
6 2013–2018 (hereafter, NDHS2018). NDHS is a cross-sectional nationally representati
survey that has been conducted every five years in Nigeria since 2003 [36]. The NDF
is funded by the United States Agency for International Development as part of t
worldwide Demographic and Health Surveys Program [36].

Nigeria is a West African country situated between latitudes 4° and 14° N and lon
tudes 2° and 15° E on the Gulf of Guinea, and has a total area of 923,768 km^2 (Figure
The NDHS2018 data collection occurred between 14 August and 29 December 2018, usi
a two-stage stratified cluster sampling framework. Nigeria's 36 states and capital territo

were separated into rural and urban areas and further grouped into six geopolitical zones, North-Central, North-East, North-West, South-East, South-West, and South-South, leading to a total of 74 sampling strata [37]. Further details of the NDHS sampling design can be found elsewhere [37]. Of the 40,666 occupied households selected for the sample, 40,427 were successfully interviewed, resulting in a response rate of 99% [37]. The resulting dataset consisted of 41,821 individual women (age 15–49) and 127,545 birth records.

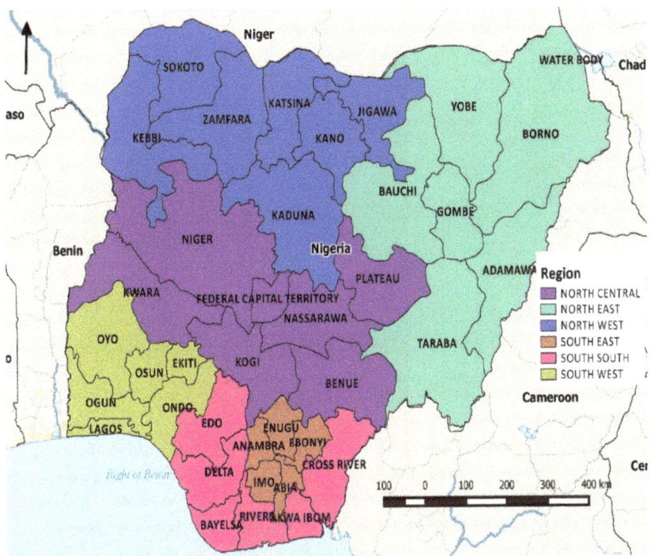

Figure 1. Map of Nigeria showing 36 states (plus federal capital territory) and the country's six geopolitical regions.

The outcome variables in this study were adverse pregnancy outcomes; birth status (stillbirth or alive, stillbirths were identified as pregnancies that did not result in the birth of a live child and included miscarriages), birth weight (children born under 2500 g were classified as LBW [1] or normal weight otherwise) and duration of pregnancy (term birth vs. preterm birth). Preterm births included those born before 37 weeks or 8.515 months of pregnancy [38]. The main exposure variables measuring HAP were cooking fuel type and mother's smoking habit. Respondents were asked about the primary source of fuel for household cooking. Responses included crops, animal dung, charcoal, kerosene, straw/shrubs/grass, coal, wood, liquefied petroleum gas (LPG), biogas, electricity, and natural gas. We classified LPG, biogas, electricity, and natural gas as "clean fuel" and the rest as "unclean" as per WHO guidelines [7].

Other variables explored include socio-economic, demographic, and geographical factors such as mother's age, educational attainment (no education, primary, secondary, higher), and wealth quintiles. As some mothers were identified with multiple birth records, we also controlled for the mother's effect on birth outcomes. This allowed us to account for the correlation between children with the same mother. We used the states to identify the effect of residence on adverse birth outcomes and the geopolitical regions as a fixed effect. See Table 1 for the descriptive summaries of the characteristics of the study participants.

Table 1. Descriptive summaries of the study participants.

Characteristics	n (%)			PR (95% CI) [b]	p-Value [b]
	Overall	Clean Cooking Fuel [a]	Unclean Cooking Fuel [a]		
		4410 (10.69%)	36,846 (89.31%)		
Demographic Variable (Mothers or Household) [c]	41,821				
Mother's Age, mean $\pm$ SD	35.9 $\pm$ 7.9	29.82 $\pm$ 9.04	29.1$\pm$ 9.8	NA	<0.0001
Mother's education					
No education	14,398 (34.43)	115 (0.28)	14,173 (34.35)	ref	<0.0001
Primary education	6383 (15.26)	252 (0.61)	6062 (14.69)	0.20 (0.16, 0.24)	
Secondary education	16,698 (39.93)	2189 (5.31)	14,219 (34.47)	0.05 (0.04, 0.06)	
Higher education	4342 (10.38)	1854 (4.49)	2392 (5.80)	0.01 (0.01, 0.01)	
Household wealth quintiles [c]					
Poorest	7747 (18.52)	5 (0.01)	7682 (18.62)	ref	<0.0001
Poorer	8346 (19.96)	11 (0.03)	8243 (19.98)	0.49 (0.15, 1.34)	
Middle	8859 (21.18)	91 (0.22)	8934 (20.93)	0.06 (0.02, 0.14)	
Richer	8840 (21.13)	656 (1.59)	8020 (19.44)	0.01 (0.00, 0.02)	
Richest	8029 (19.20)	3647 (8.84)	4267 (10.34)	0.00 (0.00, 0.00)	
Mother's exposure variable [c]					
Smoking Status					
Smoker	96 (0.23)	18 (0.04)	75 (0.18)	0.50 (0.30, 0.83)	0.0111
Non-smoker	41,725 (99.77)	4392 (10.65)	36,771 (89.13)	ref	
Adverse birth outcome [d]	127,545				
Birth status					
Alive	109,325 (85.701)	7626 (6.03)	100,669 (79.65)	ref	<0.0001
Stillbirth	18,220 (14.28)	503 (0.40)	17,598 (13.92)	2.40 (2.21, 2.62)	
Birth weight [e]					
Normal weight ($\geq$2500 g)	7166 (92.73)	1633 (21.60)	5373 (71.05)	ref	0.1250
Low birth weight (<2500 g)	562 (7.27)	113 (1.49)	442 (5.85)	1.17 (0.96, 1.43)	
Duration of pregnancy [f]					
Term birth	39,330 (99.00)	2856 (7.28)	35,972 (91.69)	ref	<0.0001
Premature	408 (1.00)	59 (0.15)	345 (0.88)	0.47 (0.36, 0.62)	

Characteristics	Overall	n (%)		PR (95% CI) [b]	p-Value [b]
		Clean Cooking Fuel [a]	Unclean Cooking Fuel [a]		
		4410 (10.69%)	36,846 (89.31%)		
Geographical variable [c]					
Region					
North-Central	7772 (18.60)	607 (1.47)	7074 (17.15)	7.21 (6.54, 7.97)	<0.0001
North-East	7639 (18.30)	86 (0.21)	7453 (18.07)	53.65 (43.37, 67.29)	
North-West	10,129 (24.20)	376 (0.91)	9703 (23.52)	15.98 (14.24, 17.97)	
South-East	5571 (13.30)	390 (0.95)	5004 (12.13)	ref	
South-South	5080 (12.10)	827 (2.00)	4181 (10.13)	3.13 (2.86, 3.43)	
South-West	5630 (13.50)	2124 (5.15)	3431 (8.37)	7.94 (7.08, 8.93)	

[a] Missing 565 cases. [b] Prevalence ratios (PR), comparing unclean vs. clean cooking fuel; p-value based on chi-square test. [c] At mother or household level, n = 41,821. [d] At child level, n = 127,545. [e] Missing birth weight records on 119,817 (93.9%) births. [f] Missing duration of pregnancy record on 87,807 (68.8%) pregnancies. Ref: Reference category.

2.2. Statistical Analysis

We used frequencies and percentages to describe categorical variables and me
and standard deviations for continuous variables. We tested the association betw
cooking fuel type and categorical variables (smoking status, education, wealth status) us
Pearson's chi-square test or Fisher's exact test and presented the estimated prevale
ratios (PR) with their 95% confidence interval (CI) [39,40]. We use a *t*-test to assess
difference in mother's age across adverse birth outcomes.

A Bayesian approach to generalized structured additive regression (STAR) [41–
was then used to estimate the association between adverse birth outcomes and HAP wh
adjusting for other predictors and region-specific terms. For a specific health outcome
child i, Y_i, the structured additive regression model with predictors is formulated as:

$$\Pr(Y_i | \text{predictors}) = h(\eta_i)$$

Using an appropriate link, $h(.)$, such as the probit link for ease of interpretation,

$$\eta = \gamma_0 + \gamma_1 \text{fuel} + \ldots + \gamma_k \text{COV} + f_1(\text{mother's cluster}) + f_2(\text{mother's age})$$
$$+ f_{unstr}(\text{STATE}) + f_{str}(\text{STATE})$$

where $\gamma_1 \ldots \gamma_k$ represents the fixed effect of the exposure cooking fuel variable and oth
covariates (COV) such as mother's age, mother's education, household wealth quin
geopolitical regions) modelled with Gaussian priors for continuous variables and diff
priors for categorical variables; function f_1 represents a random effect term for mothe
cluster identification; f_2 is the nonlinear effect of mother's age on health outcome model
with Penalize splines. Bayesian Penalized-spline priors was specified for the regressi
coefficients, f_1 for mother's random effect and f_2, for the nonlinear effects mother's a
with weakly informative inverse Gamma prior for the precision hyperpriors. f_{unstr} is
unstructured spatial effect with random effects term for the states in Nigeria, while
represents the structured spatial effect. We assumed i.i.d Gaussian random effects for
states in Nigeria in the unstructured spatial effect, $f_{unstr}(\text{STATE})$ and specified Mark
random field prior for structured spatial effect, $f_{str}(\text{STATE})$.

A unique model was fitted for each adverse birth outcome: birth status, birth weig
and pregnancy duration. All outcome variables were analyzed using STAR multivaria
logistic regression. Four models from simple to complex formulations were considered
select important predictors for each adverse birth outcome. Positive coefficients correspo
to an increased risk for the modelled outcome. Model 1 includes the exposure variab
types of cooking fuel, and spatial components (unstructured and structured). The mothe
random effect was added to Model 1 to form Model 2. An additional nonlinear effect of
mother's age effect was added to formulate Model 3. For the fourth model (Model 4),
included additional covariates, mother's education, and geopolitical regions. We exclud
household wealth quintiles from the multivariable Model 4, due to issues with collinear
with cooking fuel.

The Bayesian models were fitted via Markov Chain Monte Carlo (MCMC), carryi
out 25,000 iterations with a burn-in of 5000. Model selection was based on the small
deviance information criterion (DIC); we also report the posterior mean of deviance $(\overline{D})$ a
the effective number of parameters (pD). The posterior mean (p.mean) and 95% credib
intervals (CrI) were presented for the parameters included in the model [45,46].

All statistical analyses were implemented in R software version 3.6.2 [47], with packa
R2BayesX [41,42] and based on complete case data.

3. Results

3.1. Descriptive Summaries

Table 1 presents the descriptive summaries of the study population. Of the 41,8
women age 15–49 years surveyed (mean $\pm$ SD, 35.9 $\pm$ 7.9), 89.3% used unclean cooki
fuel as their primary source of cooking energy; however, only 0.2% of the responder

were smokers. A total of 127,545 birth records were included in the final analysis, out of which 14.3% (18,220) of all pregnancies resulted in stillbirth, 7.3% (562) were identified as LBW while 1.0% (408) of the children were born prematurely. There were significant associations between cooking fuel type and mother's age, mother's education, wealth quintiles, mother's smoking status, pregnancy duration, birth status, and geographical regions (Table 1). For example, compared to non-smoker, the prevalence ratio (PR) of smokers in households exposed to unclean cooking fuel was 0.50 (95% CI: 0.30, 0.83) compared to households using clean fuel. Similarly, compared to alive birth status, the prevalence of stillbirth status was higher (PR = 2.40, 95% CI: 2.21, 2.62) for household exposed to unclean cooking fuel compared to those in households using clean cooking fuel. Households with low birthweight were 17% more likely to use unclean cooking fuel, (PR = 1.17, 95% CI: 0.96, 1.43) however, this result was not statistically significant. In contrast, households with premature birth were 63% less likely to use unclean fuel (PR: 0.47, 95% CI: 0.36, 0.62). Figure 2 shows the rate of adverse birth outcomes per 100 pregnancies across the Nigerian states. Figure 3 shows the percentage of households primarily using unclean cooking fuel. Northern states appeared to have higher rates of ABO and higher dependence on unclean cooking fuel.

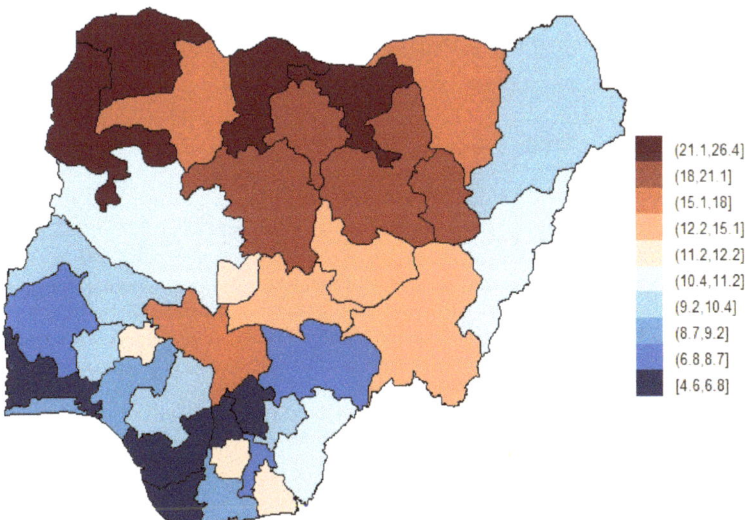

Figure 2. Geographical distribution of adverse birth outcomes. The prevalence is reported as the number of adverse birth outcomes per 100 pregnancies. Adverse birth outcomes include stillbirth, preterm or low birth weight (LBW).

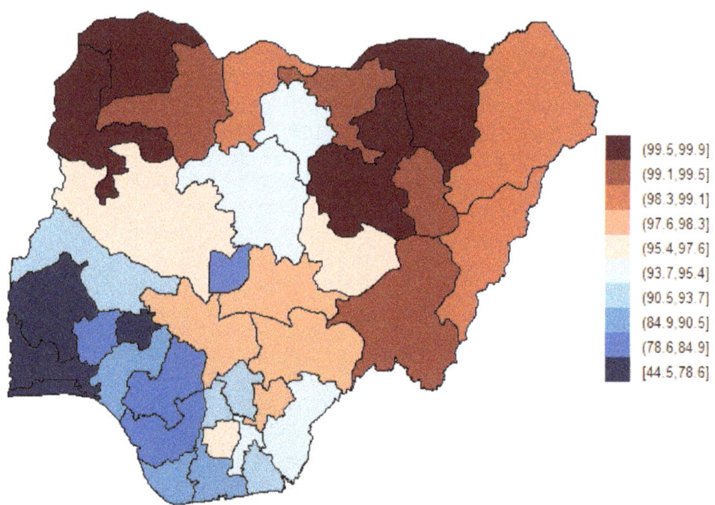

Figure 3. Geographical distribution of prevalence (%) use of unclean cooking fuel in Nigeria.

3.2. Association between HAP Exposure and Adverse Birth Outcomes

Table 1 and Table S1 summarize the relationship between the exposure variables and the pregnancy outcome variables of interest independently.

3.2.1. Prevalence of Stillbirth

Among pregnancies that resulted in stillbirth (17,598 + 503), 97.2% were from households with mothers who used unclean cooking fuel and 2.8% of the pregnancies were from households with mothers who used clean cooking fuel (Table 1 and Table S1). There was significant association between stillbirth and cooking fuel type ($p < 0.001$). Lower wealth quintiles, older age, and low education levels were associated with a larger proportion reported stillbirths ($p < 0.001$). On the other hand, smoking was not associated with birth outcomes (Table S1).

3.2.2. Birth Weight (Low Birth Weight vs. Normal Weight)

About 7.6% (442/5815) of mothers who used unclean fuel in this study had children born with LBW compared to 6.5% (113/1746) of births from mothers who use clean fuel. However, there was no significant association between the type of cooking fuel and birth weight. The bivariate analysis shows that only the mother's age and education level are independently significantly related to LBW (Table S1). Mothers who have not attained secondary or higher education level had an increased risk of an underweight birth. The smoking status of the mother was not significantly associated with any ABO.

3.2.3. Duration of Pregnancy (Preterm vs. Term Birth)

Preterm births accounted for 1.0% (408/39,738) of births in this study. Surprisingly, the proportion of preterm births in households where mothers used unclean fuel (0.95, 345/35,972) was significantly less than households where mothers used clean fuel (2.0%, 59/2856). Wealth status, education levels, and mother's age were significantly associated with preterm outcomes (Table S1).

3.3. Predictors of Adverse Birth Outcomes

Due to the small sample of mothers with smoking status, we focused only on cooking fuel as the main exposure variable. As shown in Table S2, the model diagnostic statistics for all models considered in this study for each birth outcome indicates that Model provides the best fit based on smaller DIC values for birth status and birth weight except the duration of pregnancy where Model 3 had the smallest DIC value. Table 2 presents the estimated posterior means and the 95% credible intervals (CrI) for Model 4 (which includes spatial components, mother's random effect term, the nonlinear effect of mother's age, and other covariates such as mother's education and geopolitical regions) for each ABO.

Results from Model 4 show that mothers using unclean cooking fuel are significantly at higher risk of stillbirth (0.14, 95% CrI: 0.08, 0.20) after adjusting for age, education, and region (Table 2). It also indicates that an increased level of education is protective against stillbirth, while an increase in mother's age significantly increases the risk of stillbirth (Figure 4A). Southern states tended to have a decreased risk of stillbirth and northern states had an increased risk (Figure 4B–D). However, the use of unclean cooking fuel by mothers was not significantly associated with any low birth weight and preterm birth. Women with higher education had an increased risk of preterm compared to no education but no association was found for low birth weight. Although there was a notable increase in the risk of low birth weight and preterm birth with an increase in mother's age, it was not statistically significant (Figures 5A and 6A). Spatial disparities were noted in Figure 5B–D and Figure 6B–D, generally suggesting increased risk in the northern parts of Nigeria.

Table 2. Parameter posterior mean and 95% credible interval (CrI) from multivariable analysis (Model 4) of adverse birth outcomes.

Characteristics	Birth Status: Stillbirth vs. Alive	Birth Weight: Low vs. Normal	Pregnancy Duration: Preterm vs. Term
	p Mean (95% CrI)	*p* Mean (95% CrI)	*p* Mean (95% CrI)
Cooking Fuel			
Clean (ref)			
Unclean	0.14 (0.08, 0.20)	−0.09 (−0.31, 0.10)	−0.01 (−0.33, 0.31)
Education			
No education (ref)			
Primary	−0.07 (−0.10, −0.03)	0.22 (−0.09, 0.49)	−0.04 (−0.33, 0.34)
Secondary	−0.23 (−0.26, −0.19)	0.37 (0.08, 0.68)	0.14 (−0.13, 0.42)
Higher	−0.39 (−0.45, −0.32)	0.36 (0.06, 0.65)	0.58 (0.24, 1.94)
Region			
North-Central (ref)			
North-East	0.262 (0.01, 0.44)	0.10 (−0.622, 0.88)	−0.01 (−0.64, 0.75)
North-West	0.36 (−0.02, 0.60)	−0.52 (−1.21, 0.19)	−0.42 −0.99, 0.09)
South-East	−0.16 (−0.44, 0.07)	0.23 (−0.57, 0.94)	−0.06 (−0.86, 0.70)
South-South	−0.15 (−0.35, 0.05)	0.20 (−0.44, 0.77)	−0.11 (−1.01, 0.57)
South-West	−0.10 (−0.34, 0.15)	−0.02 (−0.73, 0.67)	0.15 (−0.74, 0.77)

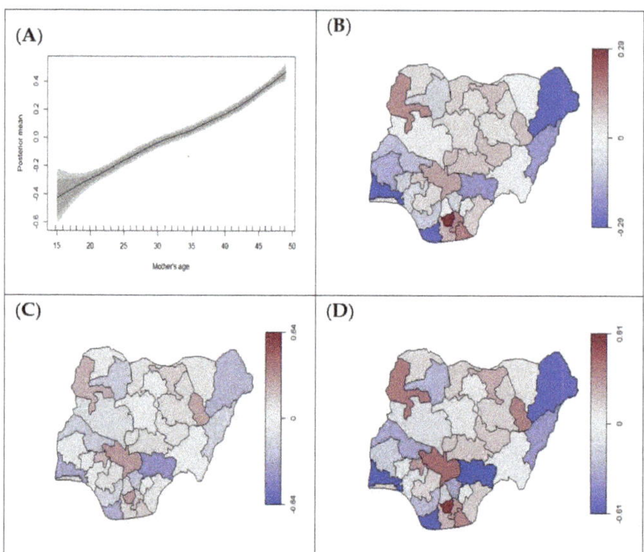

Figure 4. Estimated posterior means of (**A**) nonlinear effects of mother's age (black line) on status (stillbirth vs. alive) with 95% and 80% credible intervals, (**B**) structured, (**C**) unstructured (**D**) total spatial effects for the state. The scales in (**B–D**) represents the range of the posterior mean estimates of the spatial effect showing states with risk of stillbirth. The blue colour indicates low estimates, while the red colour signifies the states with high estimates.

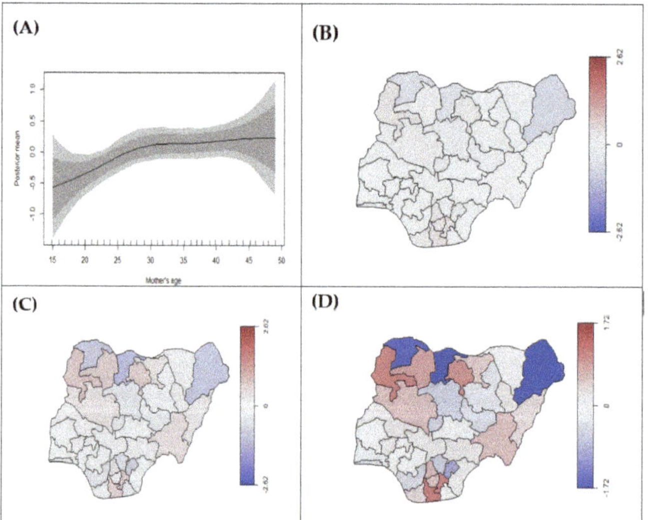

Figure 5. Estimated posterior means of (**A**) nonlinear effects of mother's age (black line) on birth weight (low birth weight vs. normal) with 95% and 80% credible intervals, (**B**) structured, (**C**) unstructured, (**D**) total-spatial effects for the state. The scales in (**B–D**) represents the range of the posterior mean estimates of the spatial effect showing states with risk of low birthweight. The blue colour indicates low estimates, while the red colour signifies the states with high estimates.

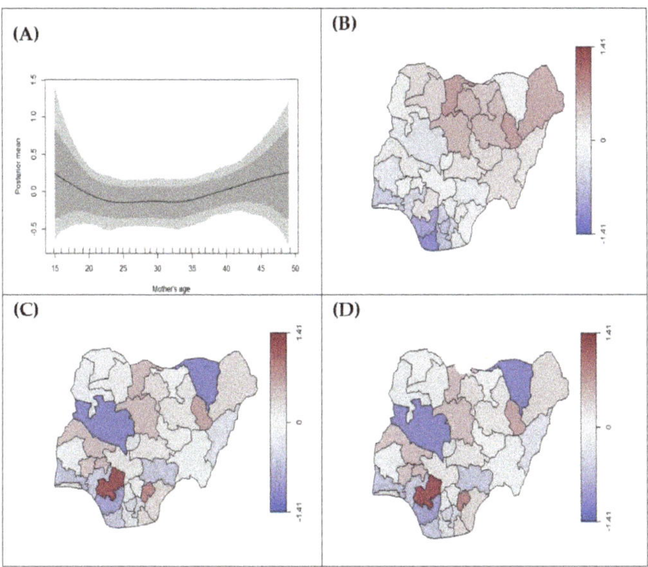

Figure 6. Estimated posterior means of (**A**) nonlinear effects of mother's age (black line) on pregnancy duration (preterm vs. term) with 95% and 80% credible intervals, (**B**) structured, (**C**) unstructured, (**D**) total spatial effects for the state. The scales in (**B–D**) represents the range of the posterior mean estimates of the spatial effect showing states with risk of preterm birth. The blue colour indicates low estimates, while the red colour signifies the states with high estimates.

4. Discussion

We have shown that 89.3% of the mothers included in this study primarily used unclean fuel for cooking. This suggests that a substantial proportion of the population is still dependent on unclean cooking fuel for cooking in Nigeria. This is a significant public health concern due to the associated risks associated with unclean cooking fuel [5–10,16]. In this study, we investigated the association between HAP and adverse birth outcomes, while accounting for geographical heterogeneity and mother's effect in our data. We were able to control for the mother's individual effect and analyze stillbirth, LBW, and preterm outcomes. In the bivariate analysis, unclean cooking fuel was associated with a higher risk of stillbirth and protective of preterm birth. This finding supports a previous study based in Argentina that concludes that lower socio-economic status and education levels are associated with higher rates of LBW and lower rates of preterm birth [15].

Our result indicates that HAP was associated with an increased risk of stillbirth but not for LBW and preterm birth. Overall, northern states had a higher rate of unclean cooking fuel use and a higher rate of adverse birth outcomes. Northern states were associated with a higher proportion of stillbirths and LBW than the southern states. Previous studies conducted in Nigeria showed a link between HAP and childhood acute respiratory tract infections, which was a leading cause of deaths in children under 5 years [48,49]. Our study extends these results to include the effects of HAP in Nigeria to adverse birth outcomes. This study contributes to previous research on HAP's association with pregnancy outcomes. Relevant literature offers contradictory findings on the significance of unclean cooking fuel when analyzing LBW. In a meta-analysis, it was concluded that indoor air pollution increased the risk of LBW [27]. A study in Bangladesh reported that indoor solid fuel use was significantly associated with LBW, but not with neonatal mortality or stillbirth when controlling for demographic variables [22]. Our study concluded that HAP did not significantly increase the risk of LBW. This may be due to the low prevalence of smoking in our sample, as smoking is correlated to increased risk for LBW [50]. This finding coincides with a Ghanaian cohort study that failed to support an association between HAP and L[ ] after adjusting for confounding variables [25].

Our study suggests that HAP is significantly associated with an increased risk[ ] stillbirth and is supported by a recent meta-analysis that concluded that HAP increa[ ] the risk of stillbirth in developing countries [27]. However, other studies did not fir[ ] significant relationship between HAP and stillbirth [22,25].In a cross-sectional study u[ ] the 2007 Ghana Maternal Health Survey, the authors concluded that unclean cooking[ ] could be on the causal pathway between lower socio-economic status and stillbirth[ ] It is hypothesized that smoke produced from unclean cooking fuels contains polluta[ ] including carbon monoxide, which can be inhaled and absorbed into the mother's bl[ ] and possibly cause detrimental effects on the fetus [26,27]. Our study suggests that th[ ] effects on the fetus can result in an increased risk for stillbirth.

Poverty has also been suggested to be a key factor in preventing access to clean[ ] but also compounds the burden associated with HAP by worsening access to adequ[ ] health care [51]. This may be a contributing factor to the regional disparity seen betw[ ] the mostly lower socio-economic demographic of the North compared to the mostly hig[ ] socio-economic demographic of the South. For example, a previous study conclu[ ] that the Northern region had the highest prevalence of underutilization of antenatal c[ ] services [52–54] and lowest immunization uptake [53,55,56]. When geographic variatio[ ] controlled for in the models, HAP continues to significantly increase the risk of stillb[ ] Addressing the use of unclean cooking fuel in Nigeria may lead to decreased rates[ ] stillbirth.

We acknowledge the following limitation in this study. The response rates for so[ ] adverse birth outcomes were relatively low. There was a 100% response rate for stillb[ ] outcomes, but there were low response rates for birth weight (6.1%) and pregnancy durat[ ] (31.2%). The smoking status of the mother was only recorded in a small proport[ ] (0.2%) of mothers who identified as smokers. Reporting pregnancy in duration in mon[ ] could introduce measurement bias in our analysis. Measuring pregnancy duratior[ ] weeks would result in a more accurate and sensitive model to variability in pregnar[ ] duration. We also could not distinguish when the pregnancy ended. There may be differ[ ] associations for perinatal mortality and miscarriages if analyzed separately. Amoun[ ] time spent around HAP could contribute to the magnitude of its effect on adverse bi[ ] outcomes, but we were unable to include this variable in this study. We could not cont[ ] for all possible confounding variables such as access to medical services, medical histc[ ] and BMI.

Despite these limitations, this study has several strengths. As the study was based[ ] a large representative dataset, the 2018 Nigeria DHS, we are confident that our sample is[ ] excellent representation of the Nigerian population. The use of STAR models for analy[ ] allows for flexible modelling of possible nonlinear effects of independent variables and[ ] geographical effects of the data [41–43,57].

5. Conclusions

This paper studied the association between HAP and adverse birth outcomes usi[ ] the 2018 NDHS. It highlighted the risk of adverse birth outcomes due to mothers usi[ ] unclean cooking fuel. In order to decrease the prevalence of adverse birth outcomes[ ] Nigeria, efforts should address the dependence on unclean cooking fuel. Disparities[ ] Nigerian states account for disproportionate risks of stillbirth and LBW, even when t[ ] effects of wealth and education are controlled for. This shows that decreasing natior[ ] levels of adverse birth outcomes depends on working toward addressing the disparit[ ] between states.

Further research should be performed to analyze the effects that our study could r[ ] control for, such as access to prenatal medical services, and the mother's medical histc[ ] This includes accounting for a combination of different fuels instead of studying only t[ ] primary cooking fuel. Analyzing dose-response using the combination of cooking fu[ ] would clarify the strength of the relationship between HAP and adverse birth outcomes. As our study is a cross-sectional study, we cannot analyze causal pathways between HAP, adverse birth outcomes, and other explanatory variables. Further research should be done to study these causal pathways and include a larger selection of adverse birth outcomes, as we limited our study to three outcomes.

Supplementary Materials: The following are available online at https://www.mdpi.com/1660-4601/18/2/634/s1, Table S1: Bivariate analysis of adverse birth outcomes and HAP exposure, mother's level, and household characteristics. Table S2: Criterion for model selection.

Author Contributions: Conceptualization, J.R., T.I.E., O.A.A.; methodology, J.R., T.I.E., O.A.A.; software, J.R., O.A.A.; validation, J.R., T.I.E., O.A.A.; formal analysis, J.R., O.A.A.; investigation, J.R., T.I.E., O.A.A.; resources, J.R., O.A.A.; data curation, J.R., O.A.A.; writing—original draft preparation, J.R., O.A.A.; writing—review and editing, J.R., T.I.E., O.A.A.; visualization, J.R., O.A.A.; supervision, T.I.E., O.A.A.; funding acquisition, O.A.A. All authors have read and agreed to the published version of the manuscript.

Funding: This research received no external funding.

Institutional Review Board Statement: This study did not require any formal ethics approval because it was a secondary analysis based on publicly available survey data sets from the DHS Program (www.dhsprogram.com).

Informed Consent Statement: Not applicable.

Data Availability Statement: The survey data sets used in this paper is publicly available from the DHS Program (www.dhsprogram.com).

Conflicts of Interest: The authors declare no conflict of interest.

References

World Health Organization. *WHA Global Nutrition Targets 2025: Low Birth Weight Policy Brief*; World Health Organization: Geneva, Switzerland, 2014.

Ballot, D.E.; Chirwa, T.F.; Cooper, P.A. Determinants of survival in very low birth weight neonates in a public sector hospital in Johannesburg. *BMC Pediatrics* **2010**, *10*, 30. [CrossRef] [PubMed]

Dahlui, M.; Azahar, N.; Oche, O.M.; Aziz, N.A. Risk factors for low birth weight in Nigeria: Evidence from the 2013 Nigeria Demographic and Health Survey. *Glob. Health Action* **2016**, *9*, 28822. [CrossRef] [PubMed]

Lawn, J.E.P.; Blencowe, H.M.; Waiswa, P.P.; Amouzou, A.P.; Mathers, C.P.; Hogan, D.P.; Flenady, V.P.; Frøen, J.F.P.; Qureshi, Z.U.B.M.; Calderwood, C.B.M.; et al. Stillbirths: Rates, risk factors, and acceleration towards 2030. *Lancetthe* **2016**, *387*, 587–603. [CrossRef]

World Health Organization. *Household Air Pollution—The World's Leading Environmental Health Risk*; World Health Organization: Geneva, Switzerland, 2017.

World Health Organization. *Global Health Observatory Data: Mortality from Household Air Pollution*; World Health Organization: Geneva, Switzerland, 2018.

World Health Organization. *Household Air Pollution and Health*; World Health Organization: Geneva, Switzerland, 2018.

Lakshmi, P.V.M.; Virdi, N.K.; Thakur, J.S.; Smith, K.R.; Bates, M.N.; Kumar, R. Biomass fuel and risk of tuberculosis: A case–control study from Northern India. *J. Epidemiol. Community Health* **2012**, *66*, 457–461. [CrossRef] [PubMed]

Agrawal, S.; Yamamoto, S. Effect of Indoor air pollution from biomass and solid fuel combustion on symptoms of preeclampsia/eclampsia in Indian women. *Indoor Air* **2015**, *25*, 341–352. [CrossRef] [PubMed]

Ravilla, T.D.; Gupta, S.; Ravindran, R.D.; Vashist, P.; Krishnan, T.; Maraini, G.; Chakravarthy, U.; Fletcher, A.E. Use of cooking fuels and cataract in a population-based study: The India eye disease study. *Environ. Health Perspect.* **2016**, *124*, 1857–1862. [CrossRef]

Dionisio, K.L.; Arku, R.E.; Hughes, A.F.; Vallarino, J.; Carmichael, H.; Spengler, J.D.; Agyei-Mensah, S.; Ezzati, M. Air Pollution in Accra Neighborhoods: Spatial, Socioeconomic, and Temporal Patterns. *Environ. Sci. Technol.* **2010**, *44*, 2270–2276. [CrossRef]

Rhodes, E.L.; Dreibelbis, R.; Klasen, E.; Naithani, N.; Baliddawa, J.; Menya, D.; Khatry, S.; Levy, S.; Tielsch, J.M.; Miranda, J.J. Behavioral attitudes and preferences in cooking practices with traditional open-fire stoves in Peru, Nepal, and Kenya: Implications for improved cookstove interventions. *Int. J. Environ. Res. Public Health* **2014**, *11*, 10310–10326. [CrossRef]

Wiedinmyer, C.; Dickinson, K.; Piedrahita, R.; Kanyomse, E.; Coffey, E.; Hannigan, M.; Alirigia, R.; Oduro, A. Rural–urban differences in cooking practices and exposures in Northern Ghana. *Environ. Res. Lett.* **2017**, *12*, 065009. [CrossRef]

14. Manyeh, A.K.; Kukula, V.; Odonkor, G.; Rosemond Akepene, E.; Adjei, A.; Solomon, N.-B.; Akpakli, D.E.; Gyapong Socioeconomic and demographic determinants of birth weight in southern rural Ghana: Evidence from Dodowa Health Demographic Surveillance System. *BMC Pregnancy Childbirth* **2016**, *16*. [CrossRef]
15. Urquia, M.L.; Urquia, M.L.; Frank, J.W.; Frank, J.W.; Alazraqui, M.; Alazraqui, M.; Guevel, C.; Guevel, C.; Spinelli, H.G.; Sp H.G. Contrasting socioeconomic gradients in small for gestational age and preterm birth in Argentina, 2003–2007. *Int. J. P Health* **2013**, *58*, 529–536. [CrossRef] [PubMed]
16. Juntarawijit, Y.; Juntarawijit, C. Cooking smoke exposure and respiratory symptoms among those responsible for house cooking: A study in Phitsanulok, Thailand. *Heliyon* **2019**, *5*, e01706. [CrossRef] [PubMed]
17. Apte, K.; Salvi, S. Household air pollution and its effects on health [version 1; peer review: 2 approved]. *F1000Research* **20** 2593. [CrossRef]
18. Bonjour, S.; Adair-Rohani, H.; Wolf, J.; Bruce, N.G.; Mehta, S.; Prüss-Ustün, A.; Lahiff, M.; Rehfuess, E.A.; Mishra, V.; Smith Solid fuel use for household cooking: Country and regional estimates for 1980–2010. *Environ. Health Perspect.* **2013**, *121*, 784 [CrossRef]
19. Matinga, M.N.; Annegarn, H.J.; Clancy, J.S. Healthcare provider views on the health effects of biomass fuel collection and u rural Eastern Cape, South Africa: An ethnographic study. *Soc. Sci. Med.* **2013**, *97*, 192–200. [CrossRef]
20. Ghimire, S.; Mishra, S.R.; Sharma, A.; Siweya, A.; Shrestha, N.; Adhikari, B. Geographic and socio-economic variation in mar of indoor air pollution in Nepal: Evidence from nationally-representative data. *BMC Public Health* **2019**, *19*, 195. [CrossRef]
21. Ezeh, O.K.I.; Ezeh, O.K.; Agho, K.E.M.; Agho, K.E.; Dibley, M.J.; Dibley, M.J.O.; Hall, J.J.; Hall, J.J.O.; Page, A.N.; Page, A.N.I. effect of solid fuel use on childhood mortality in Nigeria: Evidence from the 2013 cross-sectional household survey. *Env Health* **2014**, *13*, 113. [CrossRef]
22. Khan, M.N.; CZ, B.N.; Mofizul Islam, M.; Islam, M.R.; Rahman, M.M. Household air pollution from cooking and risk of adv health and birth outcomes in Bangladesh: A nationwide population-based study. *Environ. Health* **2017**, *16*, 57. [CrossRef]
23. Siddika, N.; Rantala, A.K.; Antikainen, H.; Balogun, H.; Amegah, A.K.; Ryti, N.R.I.; Kukkonen, J.; Sofiev, M.; Jaakkola, I Jaakkola, J.J.K. Short-term prenatal exposure to ambient air pollution and risk of preterm birth—A population-based cohort s in Finland. *Environ. Res.* **2020**, *184*, 109290. [CrossRef] [PubMed]
24. Amegah, A.K.; Nayha, S.; Jaakkola, J.J. Do biomass fuel use and consumption of unsafe water mediate educational inequaliti stillbirth risk? An analysis of the 2007 Ghana Maternal Health Survey. *BMJ Open* **2017**, *7*, e012348. [CrossRef] [PubMed]
25. Weber, E.; Adu-Bonsaffoh, K.; Vermeulen, R.; Klipstein-Grobusch, K.; Grobbee, D.E.; Browne, J.L.; Downward, G.S. Househ fuel use and adverse pregnancy outcomes in a Ghanaian cohort study. *Reprod Health* **2020**, *17*, 29. [CrossRef]
26. Bruce, N.; Perez-Padilla, R.; Albalak, R. Indoor air pollution in developing countries: A major environmental and public he challenge. *Bull. World Health Organ.* **2000**, *78*, 1078–1092. [CrossRef] [PubMed]
27. Pope, D.P.; Mishra, V.; Thompson, L.; Siddiqui, A.R.; Rehfuess, E.A.; Weber, M.; Bruce, N.G. Risk of low birth weight and stillb associated with indoor air pollution from solid fuel use in developing countries. *Epidemiol. Rev.* **2010**, *32*, 70–81. [Cross [PubMed]
28. Dutta, A.; Khramtsova, G.; Brito, K.; Alexander, D.; Mueller, A.; Chinthala, S.; Adu, D.; Ibigbami, T.; Olamijulo, J.; Odetu A.; et al. Household air pollution and chronic hypoxia in the placenta of pregnant Nigerian women: A randomized contro ethanol Cookstove intervention. *Sci. Total Environ.* **2018**, *619–620*, 212–220. [CrossRef] [PubMed]
29. Stanek, J. Hypoxic patterns of placental injury: A review. *Arch. Pathol. Lab. Med.* **2013**, *137*, 706–720. [CrossRef] [PubMed]
30. Grigoriadis, C.; Tympa, A.; Creatsa, M.; Bakas, P.; Liapis, A.; Kondi-Pafiti, A.; Creatsas, G. Hofbauer cells morphology and der in placentas from normal and pathological gestations. *Rev. Bras. Ginecol. Obs.* **2013**, *35*, 407–412. [CrossRef]
31. Tomas, S.; Prusac, I.K.; Roje, D.; Tadin, I. Trophoblast apoptosis in placentas from pregnancies complicated by preeclam Gynecol. Obstet. Investig. **2011**, *71*, 250–255. [CrossRef] [PubMed]
32. Heazell, A.; Moll, S.; Jones, C.; Baker, P.; Crocker, I. Formation of syncytial knots is increased by hyperoxia, hypoxia and reac oxygen species. *Placenta* **2007**, *28*, S33–S40. [CrossRef] [PubMed]
33. Meegdes, B.H.; Ingenhoes, R.; Peeters, L.L.; Exalto, N. Early pregnancy wastage: Relationship between chorionic vasculariza and embryonic development. *Fertil. Steril.* **1988**, *49*, 216–220. [CrossRef]
34. Alexander, D.A.; Northcross, A.; Karrison, T.; Morhasson-Bello, O.; Wilson, N.; Atalabi, O.M.; Dutta, A.; Adu, D.; Ibigbam Olamijulo, J.; et al. Pregnancy outcomes and ethanol cook stove intervention: A randomized-controlled trial in Ibadan, Nig Environ. Int. **2018**, *111*, 152–163. [CrossRef]
35. Olopade, C.O.; Frank, E.; Bartlett, E.; Alexander, D.; Dutta, A.; Ibigbami, T.; Adu, D.; Olamijulo, J.; Arinola, G.; Karrison, T.; e Effect of a clean stove intervention on inflammatory biomarkers in pregnant women in Ibadan, Nigeria: A randomized contro study. *Environ. Int.* **2017**, *98*, 181–190. [CrossRef]
36. The Demographic and Health Surveys (DHS) Program. Nigeria. (nd). Available online: www.dhsprogram.com/count (accessed on 11 January 2020).
37. National Population Commission (NPC) [NIGERIA] and ICF. *Nigeria Demographic and Health Survey 2018*; National Popula Commission (NPC) [NIGERIA] and ICF: Abuja, Nigeria; Rockville, MD, USA, 2019.
38. World Health Organization. New Global Estimates on Preterm Birth Published. Available online: https://www.who reproductivehealth/global-estimates-preterm-birth/en/ (accessed on 9 April 2020).

Coutinho, L.; Scazufca, M.; Menezes, P.R. Methods for estimating prevalence ratios in cross-sectional studies. *Rev. Saude Publica* **2008**, *42*, 992–998. [CrossRef] [PubMed]

Barros, A.J.; Hirakata, V.N. Alternatives for logistic regression in cross-sectional studies: An empirical comparison of models that directly estimate the prevalence ratio. *BMC Med. Res. Methodol.* **2003**, *3*, 21. [CrossRef] [PubMed]

Belitz, C.; Brezger, A.; Kneib, T.; Lang, S.; Umlauf, N. BayesX: Software for Bayesian Inference in Structured Additive Regression. Available online: http://www.bayesx.org (accessed on 10 October 2020).

Umlauf, N.; Adler, D.; Kneib, T.; Lang, S.; Zeileis, A. Structured additive regression models: An R interface to BayesX. *J. Stat. Softw.* **2015**, *63*, 1–46. [CrossRef]

Adegboye, O.A.; Gayawan, E.; Hanna, F. Spatial modelling of contribution of individual level risk factors for mortality from Middle East respiratory syndrome coronavirus in the Arabian Peninsula. *PLoS ONE* **2017**, *12*, e0181215. [CrossRef]

Khatab, K.; Adegboye, O.; Mohammed, T.I. Social and demographic factors associated with morbidities in young children in Egypt: A Bayesian geo-additive semi-parametric multinomial model. *PLoS ONE* **2016**, *11*, e0159173. [CrossRef]

Brooks, S.; Smith, J.; Vehtari, A.; Plummer, M.; Stone, M.; Robert, C.P.; Titterington, D.; Nelder, J.; Atkinson, A.; Dawid, A. Discussion on the paper by Spiegelhalter, Best, Carlin and van der Linde. *J. R. Stat. Soc. Ser. B Stat. Methodol.* **2002**, *64*, 616–639.

Spiegelhalter, D.J.; Best, N.G.; Carlin, B.P.; Van Der Linde, A. Bayesian measures of model complexity and fit. *J. R. Stat. Soc. Ser. B* **2002**, *64*, 583–639. [CrossRef]

R Core Team. *R: A Language and Environment for Statistical Computing*; R Foundation for Statistical Computing: Vienna, Austria, 2019.

Akinyemi, J.O.; Morakinyo, O.M. Household environment and symptoms of childhood acute respiratory tract infections in Nigeria, 2003–2013: A decade of progress and stagnation. *BMC Infect. Dis.* **2018**, *18*, 296. [CrossRef] [PubMed]

World Health Organization. *Global Health Observatory Data: Causes of Child Mortality*; World Health Organization: Geneva, Switzerland, 2019.

Ko, T.-J.; Tsai, L.-Y.; Chu, L.-C.; Yeh, S.-J.; Leung, C.; Chen, C.-Y.; Chou, H.-C.; Tsao, P.-N.; Chen, P.-C.; Hsieh, W.-S. Parental Smoking During Pregnancy and Its Association with Low Birth Weight, Small for Gestational Age, and Preterm Birth Offspring: A Birth Cohort Study. *Pediatrics Neonatol.* **2014**, *55*, 20–27. [CrossRef]

Lee, K.K.; Bing, R.; Kiang, J.; Bashir, S.; Spath, N.; Stelzle, D.; Mortimer, K.; Bularga, A.; Doudesis, D.; Joshi, S.S. Adverse health effects associated with household air pollution: A systematic review, meta-analysis, and burden estimation study. *Lancet Glob. Health* **2020**, *8*, e1427–e1434. [CrossRef]

Emmanuel Olorunleke, A.; Auta, A.; Khanal, V.; Olasunkanmi David, B.; Akuoko, C.P.; Kazeem, A.; Samson, J.T.; Zhao, Y. Prevalence and factors associated with underutilization of antenatal care services in Nigeria: A comparative study of rural and urban residences based on the 2013 Nigeria demographic and health survey. *PLoS ONE* **2018**, *13*, e0197324. [CrossRef]

Adedokun, S.T.; Uthman, O.A.; Adekanmbi, V.T.; Wiysonge, C.S. Incomplete childhood immunization in Nigeria: A multilevel analysis of individual and contextual factors. *BMC Public Health* **2017**, *17*, 236. [CrossRef]

Gayawan, E.; Omolofe, O.T. Analyzing Spatial Distribution of Antenatal Care Utilization in West Africa Using a Geo-additive Zero-Inflated Count Model. *Spat. Demogr.* **2016**, *4*, 245–262. [CrossRef]

Akwataghibe, N.N.; Ogunsola, E.A.; Broerse, J.E.; Popoola, O.A.; Agbo, A.I.; Dieleman, M.A. Exploring Factors Influencing Immunization Utilization in Nigeria—A Mixed Methods Study. *Front. Public Health* **2019**, *7*, 392. [CrossRef]

Adegboye, O.A.; Kotze, D.; Adegboye, O.A. Multi-year trend analysis of childhood immunization uptake and coverage in Nigeria. *J. Biosoc. Sci.* **2014**, *46*, 225–239. [CrossRef] [PubMed]

Brezger, A.; Lang, S. Generalized structured additive regression based on Bayesian P-splines. *Comput. Stat. Data Anal.* **2006**, *50*, 967–991. [CrossRef]

MDPI

St. Alban-Anlage 66

4052 Basel

Switzerland

Tel. +41 61 683 77 34

Fax +41 61 302 89 18

www.mdpi.com

International Journal of Environmental Research and Public Health Editorial Office

E-mail: ijerph@mdpi.com

www.mdpi.com/journal/ijerph